Praise for *Operatio[n...]*

"An eye-level view of mortal danger set [...] during World War II. . . . *Operation Pedesta[l...]* shepherding a frail cargo through all the fire and steel th[...] Mr. Hastings paints a portrait of naval combat with an artist's brush [...] by more than a half century of combat reportage. Compassionate toward men who braved bombs, torpedoes, fire, and a cruel sea, he showcases the Royal Navy—along with the merchant vessels it guarded—at its finest hour."
—*Wall Street Journal*

"Vividly chronicling the sinking of the aircraft carrier *Eagle*, Hastings initiates more than 250 pages of gripping fireworks and insights that continue well past August 15, when five battered merchantmen limped into Malta's harbor. . . . Another enthralling Hastings must-read." —*Kirkus Reviews* (starred review)

"Those who read Hastings' meticulously researched and clearly written account of Operation Pedestal will emerge with a greater appreciation of the dangers, continuous stress, and deprivations facing men who went to war at sea in WWII. . . . Hastings' treatment of this important but not often remarked on campaign belongs on the bookshelf of readers interested in World War II in Europe and of those looking for a reminder of what men can and are willing to do when the need is great and the cause is just." —*American Spectator*

"Military historian Hastings . . . delivers a sterling account of the August 1942 mission to bring food, oil, and other supplies to the besieged island of Malta. . . . Buoyed by prodigious research and vivid prose, this is a brilliant illumination of one of WWII's most dramatic episodes."
—*Publishers Weekly* (starred review)

"Hastings, author of bestsellers such as *All Hell Let Loose* and *The Secret War*, recalls an event that is often overlooked in World War II history: the four-day battle over a convoy relief mission that brought aviation fuel to besieged British forces in Malta. . . . Hastings should please his current fans and attract new devotees with this lucidly limned account, suitable for general readers and specialists alike." —Frederick J. Augustyn Jr., Library of Congress, Washington, DC

"[A] breathtakingly dramatic account. . . . Hastings details the violence and valor of that week with all the elegance for which he is famous. . . . In expertly recounting their courage and the horrors they faced, Max Hastings has helped ensure the well-deserved immortality of this band of heroes."
—*Objective Standard*

Operation Pedestal

The Fleet That Battled to Malta, 1942

Max Hastings

HARPER PERENNIAL

NEW YORK • LONDON • TORONTO • SYDNEY • NEW DELHI • AUCKLAND

For Nick Carter,
Britain's best chief of defence staff
thus far in this troubled century

'When this war is a misty memory in the minds of old men, they will still talk of the convoy for Malta which entered the Mediterranean in August 1942.'

Norman Smart, a war correspondent
accompanying the Pedestal fleet

Contents

Illustrations

Kenya (courtesy of Brian Crabb)

Eagle survivors transfer to a destroyer (courtesy of Brian Crabb)

Eagle crew are filmed for propaganda (Imperial War Museum/GM1422)

Lachlan Mackintosh (Imperial War Museum/A26912)

Kesselring with Rommel (Keystone/Getty Images)

Alberto da Zara with Wallis Spencer

Giacomo Metellini

German Ju88s in flight (Imperial War Museum/MH6115)

An Italian Sparrowhawk torpedo-bomber (Imperial War Museum/HU43447)

Ju87s in flight (Hulton Archive/Getty Images)

A depth-charge exploding (Imperial War Museum/A14362)

A rating making a signal (Imperial War Museum/A11181)

Cobalto moments before sinking (Imperial War Museum/HU53129)

Survivors on *Ithuriel* (Imperial War Museum/A11414)

The crew of *Wolverine* (Imperial War Museum/A12424)

Rearming a Hurricane (Imperial War Museum/A11164)

Harold Burrough on his bridge (courtesy of Brian Crabb)

Carriers in line with an Albacore in the foreground (Imperial War Museum/A11293)

Hugh Popham

Dickie Cork (Imperial War Museum/A11271)

Rodney Carver (Imperial War Museum/A11281)

Mike Crosley (courtesy of Joan Crosley)

Pilot smoking (Imperial War Museum/A11280)

Pilot with helmet and goggles (Imperial War Museum/A11277)

Martlet landing on *Indomitable* (courtesy of Brian Crabb)

An Italian Sparrowhawk plunging towards the sea (Imperial War Museum/FLM3795)

Dorset hit by air attacks (Imperial War Museum/A11173)

Fred and Minda Larsen (Larsen family collection)

Fred Larsen and Lonnie Dales (Larsen family collection)

Gerhart Suppiger

Fred Riley

David MacFarlane (Imperial War Museum/GM1699)

Dudley Mason (Imperial War Museum/HU43092)

Richard Wren

Ledbury (Imperial War Museum/A30687)

Harold Burrough on his bridge (Imperial War Museum/UKY425)

Alf Russell (courtesy of Brian Crabb)

Eddie Baines and Milford Haven (Imperial War Museum/HU70699)

Ohio is hit (Imperial War Museum/HU47560)

Bramham stands by *Deucalion* (Imperial War Museum/A11188)

Ohio completes her last voyage (Imperial War Museum/GM1480)

Melbourne Star returns to Grand Harbour (Imperial War Museum/GM1429)

Survivors disembark from *Ledbury* (Imperial War Museum/GM1483)

Introduction

On 10 August 1942, the largest fleet the Royal Navy had committed to action since Jutland in 1916 entered the Mediterranean to fight a four-day battle that became an epic of courage, determination and sacrifice. The objective of Operation Pedestal was to pass through to beleaguered Malta fourteen merchant vessels. Their ordeal, together with that of the fifty-odd ships of their protective naval force, deserves to be much better known to posterity than it is. Neglect stems chiefly from the fact that at the heart of Pedestal was a convoy. The word conjures up images of lumbering merchantmen, escorted by a handful of destroyers and corvettes. Yet this action engaged on the British side two battleships, four aircraft-carriers, seven cruisers and thirty-two destroyers, together with a hundred naval and RAF aircraft, eight submarines, two minesweepers and a bevy of smaller craft, almost all the survivors of which came home with gun barrels worn out, ammunition almost exhausted, men absolutely so. A separate book could be written about the experiences of every ship's company through those August days. No comparable British naval force would be sent into action again, save for bombardment support of invasions, and the Pacific Fleet in the dying days of the war. Meanwhile, against Pedestal Germany and

Italy deployed more than six hundred aircraft, twenty-one submarines and two score torpedo-boats. The best of the Italian battlefleet put to sea.

Among perhaps twenty thousand men – I have not attempted an exact count – who passed the Straits of Gibraltar on 10 August under the command of Vice-Admiral Neville Syfret on the flag bridge of the 34,000-ton battleship *Nelson* were almost a thousand members of the British Merchant Navy and US Mercantile Marine. Those who cherish the memory of what such seamen contributed to allied victory in the Second World War sometimes grieve that their contribution has been garlanded with fewer laurels than it deserves. Here I have done my best to do justice to the achievements of the civilian seafarers. Most performed superbly; a minority who failed to do so showed themselves no worse than a like proportion of landsmen faced with similar challenges on battlefields ashore.

I believe the Royal Navy to have been Britain's outstanding fighting service of the Second World War, just as the US Navy was America's. Although it suffered many losses and setbacks, its overall record of achievement was remarkable. Its warships conducted evacuations, and later invasions; contributed as much as did the RAF's Fighter Command to render impossible Hitler's Operation Sealion, a landing in Britain; imposed blockades upon the enemy; held open sea lanes first to sustain national survival; afterwards, to convey and support the armies that achieved victory. The Pedestal fleet sailed at the midpoint of the global struggle: almost three years after it began, three more before it concluded. Churchill's famous 'end of the beginning' speech still lay more than three months in the future.

The Royal Navy's official historian characterizes this period as 'the high-water mark of the flood tide of Axis success'. The outcome of the greatest conflict in history still hung in the balance, before the full strengths of the United States and Soviet Union had been brought to bear. German and Italian troops and aircraft controlled most of the Mediterranean littoral. Rommel's tanks stood at the gates of Egypt. The Japanese threatened India, Australasia and the North American continent. Axis forces were on the Don. The pivotal clash at Stalingrad had barely begun. A British intelligence officer wrote of the Eastern Front: 'The Germans have most things in their favour ... They possess a better fighting machine.' Much of the innovative technology of land, sea and air warfare, which between 1943 and 1945 would contribute so much towards turning the tide of war decisively in favour of the allies, had yet to be introduced into service. Mass – of soldiers, tanks, warships, aircraft – such as would also be created in full measure, had still to come to the allied forces.

In July 1942 the Royal Navy suffered the most discreditable defeat of its war: under a supposed threat from German capital ships, escorts abandoned PQ17 to Murmansk, after receiving an order from Admiral Sir Dudley Pound, the ailing First Sea Lord: 'convoy is to scatter'. Twenty-four out of thirty-six merchantmen, most of them American, were sunk by U-boats or the Luftwaffe, and only eleven reached Russian ports. Pound should have been sacked for this gross misjudgement, but the blow such action would have inflicted upon confidence in the direction of the war effort seemed too great to be borne at a low point in Britain's fortunes.

Count Galeazzo Ciano, Mussolini's son-in-law and foreign minister, wrote in his diary on 4 August: 'The prospects [in the Mediterranean theatre] are good because English reinforcements are slower than had been foreseen while our reinforcements, especially the Germans, are arriving regularly. Operations in Russia are developing well. [Italy's Comando

Supremo] thinks that Russia will probably be detached from the Allied camp, after which Great Britain and America will be obliged to come to terms.'

It was against this background that the saga of Pedestal unfolded. Beyond the Straits of Gibraltar the island of Malta, less than sixty miles south of Sicily, was the only surviving British bastion in the central Mediterranean. It had been subjected to blockade and bombardment so relentless that the population and garrison were half starved; its usefulness as a naval and air base was almost extinguished. It was argued by some unsentimental allied officers that the island might best be surrendered to the enemy, who could then accept the burden of feeding its people. If the Russians pulled through, and when the full weight of American industrial might was committed, the issue of who held this Mediterranean pimple would become unimportant.

In 1940 Churchill had abandoned the Channel Islands to the Nazis as indefensible, and nobody argued; other, more strategically significant imperial outposts had been relinquished since. The British were obliged to supply their armies in the Middle and Far East by dispatching ships around the African continent, vastly lengthening voyages. With or without Malta, in 1942 the Mediterranean was too perilous to be used by allied convoys to Egypt: in the three months between June and September, 262 merchant vessels had to be employed carrying military cargoes from Britain on almost interminable passages via the Cape of Good Hope. What price the island, when every operation to feed its inhabitants, supply fuel and ammunition to its defenders required prodigious commitments and losses?

In the Second World War, however, as in all conflicts, huge moral issues were at stake, beyond the territorial and strategic ones. Among Winston Churchill's foremost qualities as Britain's warlord was his understanding of the importance of sustaining an appearance of momentum in the war effort, even when substance was lacking. Strip the global struggle to its essentials: between the June 1940 Nazi triumph in France and the allied landings in Normandy four years later, the British and American armies achieved little of importance against the Germans. The North African and Italian land campaigns were indispensable preliminaries to what came later for the Western allies, but remained marginal by comparison with the titanic clashes in the east, between the hosts of Hitler and Stalin.

For Churchill, 1942 was almost as perilous as 1940, and politically much bleaker. The British people were weary, especially of the defeats that seemed to be all that their bellicose prime minister could contrive. Most recent were the humiliating surrenders at Singapore and Tobruk, where imperial forces were brought low by smaller numbers of Japanese and German enemies. The post-war official history of the Mediterranean and Middle East campaigns subtitles that period, culminating in September 1942, 'British Fortunes Reach Their Lowest Ebb'. Churchill observed bitterly to John Kennedy, the army's director of military operations: 'If Rommel's army were all Germans [instead of mostly Italians] they would beat us.'

The strategic-bomber assault on Germany was trumpeted by air-power enthusiasts, and by Churchill himself, when he had little else. In 1942, however, the RAF and the USAAF were still gathering strength; they could do nothing to blunt German battlefield prowess. The navy's struggles to hold open global convoy routes continued despite murderous losses. Hitler's grasp on the European continent was unshaken, indeed for two more years remained almost unshakeable. Senior allied officers still anticipated Russian defeat. Ernest Bevin, the doughty Minister of Labour, observed fiercely at a June War Cabinet meeting: 'We must have a victory! What the British public wants is a victory!'

Amid such grim realities, which provoked bitter words in both high places and low, Churchill and his warlords were confronted by tidings that, unless supplies of food, fuel and ammunition could be shipped to Malta before the leaves fell at home, its surrender was inevitable. Since 1815, the island had been viewed as one of the jewels in Britain's crown of empire: way station to the East; polo ground for generations of naval officers and drunk tank for their sailors; a sunbaked fortress that showcased the Fleet's dominance of the Mediterranean. Malta's loss to the Axis, coming after so many other humiliations, would be a crushing blow to national spirit, heedless of its disputable importance to the outcome of the war. Across the Atlantic, America's President Franklin Roosevelt loyally echoed Churchill's rhetoric, calling the island 'one bright flame in the darkness'.

Thus, in the summer of 1942, the decision was taken that Malta must be sent sustenance at almost any cost. Planning began, to dispatch to the Mediterranean the most heavily escorted British convoy of the Second World War. It would be composed of a single US-built tanker, named *Ohio*, together with thirteen fast modern cargo-liners, two of them American, carrying mixed loads of vital commodities, all the ships being fitted with anti-aircraft weapons. A task force was concentrated in Scottish anchorages, to be joined by more warships in the course of its passage to Gibraltar. The cast was thus assembled for one of the bloodiest air–sea battles of the war in the West. In the wake of the previous month's Arctic disaster, almost every officer of the Royal Navy was imbued with a determination to erase its shameful memory. Destroyer captain Roger Hill said:

'There was a strong touch of desperation and bloody-minded-ness following PQ17,' in which his own ship had taken part.

This, then, is the story that I shall unfold in these pages, my first full-length narrative about the war at sea, after writing many books about the deeds of British and American armies and air forces. The saga of Pedestal offers a focal point around which to explore the wartime ethos of the Royal Navy. Although every modern conflict has been minuscule in scale alongside that of 1939–45, it may be of some slight value to this narrative that in 1966 I submerged on exercise aboard a British submarine of wartime vintage. I have been a spectator of carrier operations, bounced across the sea at forty knots, clinging to a Brave-class torpedo-boat.

In 1982, as a correspondent I witnessed the tumult of the Falklands War, clad in helmet and anti-flash gear under air attack, such as caused me to resemble the men who sailed to Malta, in appearance if not in fortitude. Readers should forgive the shameless snobbery of an officer who confided to me before the shooting started in the South Atlantic: 'You are about to be privileged to witness, for the last time in history, the Royal Navy going to war with its ships commanded by gentlemen.'

I have seen aircraft shot down; watched ships burn, explode, founder; lost sleep to the clanging echoes of anti-frogman 'scaring charges' exploding in the darkness of an anchorage. Anachronistically, in 1982 many ships' gunners were firing Bofors and Bren guns akin to those their forebears had manned on Pedestal. I have gazed in awe at the spectacle of ships surmounting stormy seas; even more so at their crews, doing fine and brave things amid the shock of battle. The captains who fought in the Mediterranean would have recognized as blood-brethren those who fought in the South Atlantic forty years on.

As for Malta, I first landed on that sunny island in 1963, shortly before it became independent from Britain, and I have renewed the happy acquaintance several times since. Only the tiresome circumstance of CV-19 prevented me from making another planned visit in 2020.

A note of caution: any history or biography that announces itself as definitive should be consigned to the wastebin, because such accolades are never justified. This narrative relies significantly on personal accounts, often contradictory and sometimes demonstrably wrong. Modern oral historians and TV documentary-makers treat with reverence the tales of very old men and women, but the memories of such witnesses often play them false. Reading the testimony given by Pedestal survivors to twentieth- and early twenty-first-century authors, a sceptic, such as every decent historian should be, guesses that some men embroidered their own roles, indeed fabricated anecdotes. At this distance of time, a writer can only use instinct and experience to guess whose words deserve belief.

The official documents and reports on Pedestal omit or skate over several contentious issues and episodes. It was much easier for an admiral to lament, for instance, the navy's lack of good carrier fighters than to criticize the behaviour of named officers. Doubts persist about exact timings of some events. Thus this narrative offers a plausible version of one of the most significant naval actions of the Second World War, which readers may find as thrilling and moving as I do. Not for a moment, however, can it be considered definitive. Only the angels can aspire to achieve that.

MAX HASTINGS

Chilton Foliat, West Berkshire and Datai, Langkawi, Malaysia
October 2020

The Pedestal Fleet

Most Atlantic convoys in the Second World War were composed of thirty to sixty merchant vessels, protected by perhaps six or eight naval escorts. The force engaged in Pedestal was so vastly larger, the events in which it participated of such magnitude, that it seems mistaken in the narrative that follows to characterize Vice-Admiral Syfret's command as anything save a fleet. All groups of warships committed to operations were given single-letter code designations, not least for convenience in signals. The Pedestal fleet was thus dubbed Force F, until it divided at the entrance to the Sicilian Narrows; Syfret's capital ships were thereafter known as Force Z, while Rear-Admiral Burrough's cruiser squadron and escorts were designated as Force X. The fuelling ships, fleet oilers and escorts, were branded Force R. Thereafter, on the last stage of the passage to Malta, Pedestal become an authentic convoy operation, albeit one like none other.

Force F (codename for the entire fleet), which became **Force Z** after leaving the convoy on the evening of 13 August 1942

Acting Vice-Admiral Neville Syfret

Battleships: *Nelson* (33,300 tons, commissioned 1927), *Rodney* (33,730 tons, commissioned 1927)

Aircraft-carriers: *Eagle* (21,850 tons, commissioned 1924). Carried twenty Hurricanes

Victorious (23,207 tons, commissioned 1941), flagship of **Rear-Admiral Lumley Lyster**, commanding Pedestal's air operations; carried five Hurricanes, sixteen Fairey Fulmars, twelve Fairey Albacores

Indomitable (23,000 tons, commissioned 1940), flagship of **Rear-Admiral Denis Boyd**. *Indomitable* joined the fleet from Freetown, carried twenty-two Hurricanes, nine Grumman Martlets, nineteen Fairey Albacores

+ the carrier *Furious* (19,513 tons, commissioned 1917) accompanied the fleet through the first two days of its Mediterranean passage to execute Operation Bellows, a fly-off of thirty-seven Spitfires to Malta

+ the carrier *Argus* (14,450 tons, commissioned 1918) accompanied the fleet to Gibraltar carrying six Hurricanes, with escorts *Abbeydale*, *Burdock*, *Armeria*

Anti-aircraft cruisers: *Charybdis* (5,600 tons, commissioned 1941), *Phoebe* (5,600 tons, commissioned 1940)

Light cruiser: *Sirius* (5,600 tons, commissioned 1942)

19th Destroyer Flotilla: *Laforey, Lightning, Lookout, Quentin, Tartar, Eskimo, Somali, Wishart, Zetland, Ithuriel, Antelope, Vansittart*

Additional escorts: *Keppel, Westcott, Venomous, Malcolm, Wolverine, Amazon, Wrestler, Vidette*

Ocean tug: *Jaunty*

Force R

Royal Fleet Auxiliary oilers *Brown Ranger* and *Dingledale* accompanied the fleet with tug *Salvonia*, escorted by corvettes *Jonquil, Spiraea, Geranium, Coltsfoot*

Force X (escort for convoy through to Malta from 1850 on 13 August)

10th Cruiser Squadron (commanded by **Rear-Admiral Harold Burrough**): *Nigeria* (8,530 tons, commissioned 1940), *Kenya* (8,530 tons, commissioned 1940), *Manchester* (9,400 tons, commissioned 1938), *Cairo* (4,190 tons, commissioned 1919)

6th Destroyer Flotilla: *Ashanti, Intrepid, Icarus, Foresight, Fury,* Derwent, Bramham, Bicester, Ledbury, Pathfinder, Penn, Wilton*

---

* The destroyers *Intrepid, Icarus, Foresight* and *Fury* were specially fitted with high-speed minesweeping equipment, to lead the convoy through the notoriously dangerous Axis minefield off the Skerki Banks, between Sicily and Tunis.

The merchant vessels: *Deucalion* (7,516 tons), *Empire Hope*
(12,688 tons), *Wairangi* (12,436 tons), *Clan Ferguson*
(7,347 tons), *Brisbane Star* (11,076 tons), *Waimarama*
(12,843 tons), *Melbourne Star* (12,086 tons), *Almeria Lykes*
(US) (7,773 tons), *Santa Elisa* (US) (6,085 tons), *Dorset*
(10,624 tons), *Glenorchy* (8,982 tons), *Port Chalmers* (8,535
tons), *Rochester Castle* (7,795 tons), *Ohio* (9,265 tons)

Malta minesweeper force: *Speedy, Hebe, Hythe, Rye* and
motor launches *ML121* and *ML135*

HM submarines deployed: north of Sicily *Safari* and
Unbroken; between Malta and Tunisia *United, P222,
Uproar, Ultimatum, Unruffled, Utmost; Una* transported
from Malta a party of the Special Boat Section to attack
Sicilian airfields on the night of 11/12 August.

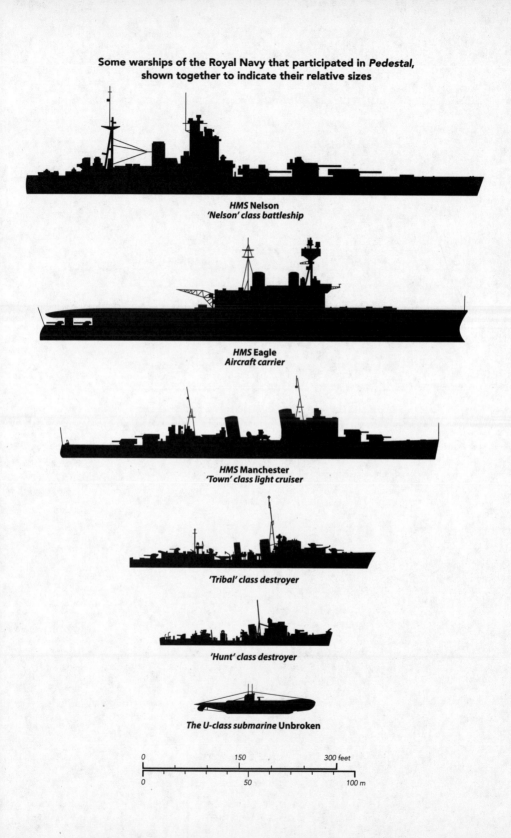

Some warships of the Royal Navy that participated in *Pedestal*, shown together to indicate their relative sizes

HMS Nelson
'Nelson' class battleship

HMS Eagle
Aircraft carrier

HMS Manchester
'Town' class light cruiser

'Tribal' class destroyer

'Hunt' class destroyer

The U-class submarine Unbroken

0   150   300 feet

0   50   100 m

Glossary

Abwehr – German intelligence organization directed by Admiral Wilhelm Canaris.

Asdic – Royal Navy submarine-detection technology, a form of what is today known as sonar, of which a primitive version entered service in 1918. A transmitter/receiver mounted beneath an escort's hull emitted sound waves through the water which returned a double echo to the warship on encountering a substantial object, a submarine being the most conspicuous possibility. Asdic's effectiveness and directional precision varied immensely according to local sea conditions – water density, salinity and suchlike – together with an operator's skill in distinguishing between a U-boat, a school of dolphins and natural underwater features.

Capital ship – Term used to characterize battleships and aircraft-carriers, a navy's largest units.

Carley floats – Primitive rafts of various sizes, constructed of copper tubing encased in cork or kapok, the largest of which could accommodate twenty-five survivors in acute discomfort. Warships carried few launches or lifeboats, and were seldom able to hoist out and lower them before a stricken ship sank. Thus Carleys, named for their 1903

American inventor Thomas Carley, were fitted to the upperworks of all British and American vessels. In emergencies, they could be almost instantly unlashed and lowered, and saved many lives. They carried water and emergency rations for their occupants. Nets were attached, which hung down in the water, enabling additional survivors to cling on for as long as their strength permitted.

C.S.10 – Signal call-sign of Harold Burrough, as flag officer 10th Cruiser Squadron.

E-boat – German high-speed torpedo-boats, somewhat larger (108 feet) than their British **MTB** and **MGB** (71.6 feet) – motor torpedo-boat and motor gunboat – or USN **PT boat** (77 feet) – patrol torpedo-boat – counterparts. The version deployed in the Mediterranean displaced 100 tons and carried crews of twenty-four. Such vessels were capable of almost forty knots in calm sea conditions.

FAA – Fleet Air Arm, known within the Royal Navy as 'the Branch'.

FAMs and **PACs** (fast aerial mines, and parachute and cable rockets) – Exotic defensive weapons fitted to Pedestal's merchant vessels. FAMs and PACs could be launched into the sky from ships' upperworks, in hopes of deterring enemy pilots, though there is no record that they inflicted damage on attacking aircraft.

H.E. (hydrophone effect) – The noise of a ship's propeller, audible to submarine hydrophone-operators. H.E. often provided a submarine captain with the first indication of an approaching vessel.

HO (Hostilities Only) – The designation for ratings who had volunteered for service in the Royal Navy for the duration of the war.

Knot – Unit of measurement by which ships assess their speed in nautical miles, each 1.15 land miles. Thus 20 knots represents just over 23mph.

MAS boat (*motoscafo armato silurante*) – Smaller and slightly faster Italian equivalent of the British MTB, wooden-hulled and displacing 25 tons, very effectively deployed against Pedestal. British officers dubbed all such 'mosquito craft' E-boats.

MS boat (*motosiluranto*) – Similar to the MAS but larger, displacing 63 tons, two of which the Italians used in the Sicilian Narrows clashes.

NLO (naval liaison officer) – Carried on the bridge of every Pedestal merchant vessel, along with a small Royal Navy signalling party.

Paravane – A First World War invention, in appearance resembling an undersized torpedo, that could be lowered into the sea on a winch and cable, to stream at an angle off the bows of a ship, in waters mined by the enemy. The device might sever a mine's anchoring cable, or explode it, as occurred several times during Pedestal.

Pipe – Naval term for a bosun's whistle deployed over the broadcast system, on large warships sometimes replaced by a Royal Marine bugle call, which preceded an order, for instance a call to action stations.

PO and **CPO** (Petty Officer and Chief Petty Officer) – Non-commissioned naval ranks.

RDF (radio direction-finding) – Wartime British cover name for radar.

RNR (Royal Naval Reserve) – An RNR officer or 'Rocky', a retired or superannuated Regular, provided almost all the naval liaison officers for service on Pedestal's merchant-ships. Some Merchant Navy officers had also trained as pre-war reservists, then joined or commanded escorts.

RNVR (Royal Naval Volunteer Reserve) – Officers commissioned from civilian life, solely for the duration of hostilities, who seldom qualified for commands larger than those of corvettes, or attained ranks above lieutenant-commander, the equivalent of an army major.

Scuttle – Naval term for a porthole or window in a ship's side, confusingly also used as a verb to describe the deliberate self-sinking of a damaged vessel, or one threatened with capture.

Ultra – Intelligence material derived from enemy signal traffic decrypted at Bletchley Park.

Watches – Ships' crews have for centuries been organized into 'watches', to keep their respective stations manned around the clock. Normal routine in the Second World War required each officer and man to serve four hours on, eight hours off, which sounds undemanding until it is remembered that dawn and dusk action stations, together with emergencies imposed by each enemy attack, were superimposed on this schedule. Every man of every crew was summoned to duty at such times, with off-watch stokers, for instance, mustered at readiness as members of damage-control parties. During Pedestal, most of the participants thought themselves fortunate if they achieved three or four hours' sleep a night, often prostrate on deck beside their action stations.

Wren – Vernacular for a member of the Women's Royal Naval Service, widely regarded socially as the smartest of Britain's wartime women's services.

Wherever possible in this narrative, in the interests of accessibility I omit men's ranks and numerical designations of aircraft. I cannot, alas, escape some of the Royal Navy's confusing terminology – for instance, designating the executive officer of a major warship its Commander, though he does not command and is subordinate to the captain. Ships are always characterized as female, and thus referred to as 'she', while small craft including torpedo-boats and submarines are addressed as 'it'.

Some civilian readers may chafe at timings given by the 24-hour operational clock, as in 1100 for eleven o'clock, but this offers the least confusing chronology and was used by all the participants in Pedestal.

1

'It Would Be a Disaster of the First Magnitude'

1 MALTA

The Second World War evolved in some ways that none of the belligerents wished or anticipated, even those who started it. Hitler, preoccupied with his Slavonic and Anglo-Saxon foes, had no desire to fight a campaign around the Inland Sea. While he relished the direction of continental and race wars, in the words of Douglas Porch 'the Mediterranean bored and exasperated him'. Mussolini's flatulent ambitions were responsible for the launching of Italian campaigns in Libya, Albania and Greece, from which his forces had to be rescued by the Wehrmacht. Yet Hitler's folly was that, having felt obliged to save his fellow dictator from humiliation, he declined to support the consequent operations in the Mediterranean theatre sufficiently strongly to secure a victory there, though this was probably within his reach.

Germany's admirals had been attracted since 1940 to a belief that here was a region in which, alongside the paper-powerful Italian Navy, they could achieve a big success. The Kriegsmarine's staff asserted at a conference with Hitler that the British fleet must be 'driven out of the whole Mediterranean … destroyed'. To that end, in late 1941 more than twenty

U-boats, a quarter of Admiral Karl Dönitz's operational strength, were shifted from the North Atlantic to the Mediterranean, and still more boats to the West African coast, to attack shipping destined for the Middle East. This change of priorities delighted the British Admiralty, preoccupied with protecting convoys from North America. In December, the primacy of the Atlantic was once again recognized in Berlin. But the Kriegsmarine's faith persisted, in the possibility of a Mediterranean triumph.

Hitler's admirals may have been right. It is possible to imagine a scenario whereby in May 1941 German paratroops seized Malta in place of Crete, the latter being strategically virtually worthless to both sides. With air superiority over almost the entire Mediterranean and a mere two or three extra divisions for Rommel, Axis forces could almost certainly have driven the British out of Egypt, crushed their last active field army and inflicted a humiliation that might have brought down Churchill's government.

Instead, however, Germany invaded Russia while Malta remained in British hands, albeit precariously. A combination of intelligence about sailings provided by Ultra decrypts; Gibraltar-, Malta- and Egypt-based surface warships, submarines and aircraft; the Axis shortage of oil; together with attrition of Italy's cargo shipping, kept Gen. Erwin Rommel's forces in North Africa on short commons. The war cost Italy three-quarters of its merchant fleet, prompting an ailing Mussolini to say 'my illness has a name: convoys'.

The little campaign fought by the British Eighth Army against the Afrika Korps and Italian troops sufficed to sustain an illusion of momentum in the British war effort, though until November 1942 never much more than twenty divisions were engaged on the two sides, compared with the four hundred Axis and Soviet formations locked in a death grapple in Russia. Though there were moments when the British feared that they faced absolute defeat in North Africa, Rommel was never quite strong enough to impose this at the end of an over-stretched supply line. As long as the 'Desert Fox' was kept out of Alexandria and Cairo, the Royal Navy sustained warships and especially submarines in the Mediterranean, and the RAF denied mastery of the desert sky to the Axis, the struggle in the theatre served allied interests better than those of Hitler or Mussolini. This seldom seemed the case, however, to those charged with defending allied bastions by land, sea and air.

The British sentimentalize relations between themselves and the subject peoples of their empire, often with little justifica-tion. Nonetheless the affection and loyalty displayed towards the 'Mother Country' by the three hundred thousand inhabit-ants of Malta seemed sincere, indeed enthusiastic. Their feel-ings were strengthened by disdain for their northerly neighbours, though pre-war Malta depended upon Italy for 70 per cent of its food, fertilizer and animal fodder. As for the colonial masters, they cherished this rocky pimple as one of the few places in the Mediterranean and Middle East where they were liked. A London-born resident described the Maltese people as 'kindly and smiling and good-mannered, tolerant of the strangeness of the British ("we have got used to you, you have been with us a long time now")'. For Malta, as a commu-nity conscious of its smallness and isolation, even in the last years of empire the instinct was to cling to nurse for fear of something worse.

On 20 December 1940, twenty thousand Maltese lined Grand Harbour to cheer the arrival of Admiral Sir Andrew Cunningham, commander-in-chief of the Mediterranean

Fleet, aboard the battleship *Warspite*. By contrast, in Egypt, Syria, Lebanon, Iraq, Iran, Palestine and other countries subject to wartime British thraldom, troops often had to be deployed to maintain control in the face of local hostility. In North Africa, following the 1942 allied invasion of Morocco and Algeria, some Arabs volunteered to fight alongside the Germans.

The novelist and traveller Paul Theroux has written, probably only half in jest, that Malta glories in war, because only then does this island eighteen miles by eight show its full worth. Before 1939, however, of Britain's three services the navy alone considered it defensible if Mussolini declared war. And even the admirals preferred to fight from their French ally's bases at Toulon, Tunis, Corsica, rather than from an anchorage within twenty minutes' flight time of Italian airfields. Malta's 1941–2 RAF chief, Air Vice-Marshal Hugh Pughe Lloyd, wrote: 'I could not imagine fighting without France as our friend. When we entered the war with her, I thought we should see it through together.'

After the French ports were lost in 1940, however, followed by those of Greece, Malta became the only haven between Gibraltar and Alexandria to which the Royal Navy retained access. 'To Italian seamen', wrote Mussolini's Admiral Aldo Cocchia, 'it seemed an act of historical injustice that this hostile island should exist in the middle of *Mare Nostrum* and on the direct route between Sicily and North Africa.' Hugh Lloyd, 'Huff-Puff' as he was nicknamed by his aircrew, saw matters the other way around: 'Malta was a lonely place. There was not a friend at any point of our compass – there was hostility everywhere.'

Until the winter of 1941, the perils and privations that war imposed upon the islanders seemed bearable. Mussolini's bombing inflicted only intermittent pain. Britain closed Malta's two Italian schools, forbade use of the Italian language in the law courts. In the first nine months of 1941, twenty-four British ships delivered 146,000 tons of cargo. The Maltese even contrived to admire the stoicism of their British governor, Major-Gen. William Dobbie, a Plymouth Brother who once retired to his balcony during a conference on air-raid precautions and fell to his knees, before returning to report, 'God has spoken. So be it.' Dobbie was not without wit, however. A much decorated veteran of both the Boer War and the Great War, on 11 November 1918 he chanced to have been duty staff officer at Field-Marshal Sir Douglas Haig's headquarters on the Western Front and thus signed the ceasefire order signalled to all units of the British Army. When asked later what he had done during the struggle, Dobbie replied, 'I stopped the bloody thing!'

But if Malta's predicament seemed endurable for most of 1941, when confronting the Italians alone, it became infinitely less so thereafter as the Germans infused their accustomed dynamism into the Mediterranean campaign. Hitler's regional commander, Albert Kesselring, wrote: 'The war ... was not being taken by the Italians with the seriousness demanded by their responsibility to the soldiers at the front ... Peacetime working conditions prevailed even during the most critical periods ... Their submarines' exhibition diving and the stunt flying of their airmen were no fit preparation for the real thing ... One had only to watch a simple guard change to see that the Italian soldier had no enthusiasm for his profession.'

Kesselring, a Bavarian schoolmaster's son born in 1885, was a much abler campaign director than Rommel, though lacking the latter's charisma. His skills as a soldier – he had been a First World War gunner officer on the Western Front – exceeded his grasp of air strategy. He had transferred to the Luftwaffe only in 1933, and bore a considerable share of blame for its failure in the Battle of Britain. He was nonetheless a brilliant organizer, of iron will. An ardent, if non-ideological Nazi, he remained loyal to Hitler to the end, partly because of an obsessive loathing of communists. His personal life remains an enigma. His memoirs make no mention of Pauline, the wife he acquired in a curious 1910 arranged marriage. When they remained childless, he adopted a son.

He was commanding the Luftwaffe's Luftflotte 2 in Russia when abruptly summoned to Berlin in November 1941, to receive orders for a new posting as Commander-in-chief South. He was briefed by Hitler personally, in the presence of Göring. Malta, said the Führer, must be neutralized, to secure the passage of Axis supply convoys across the Mediterranean. Kesselring responded that the obvious answer was to occupy the island, only to be silenced by Hitler, who said that no forces were available to accomplish this: air power must instead suffice, to render the island untenable by British naval and air forces. Kesselring later testified that he did not know that Germany's navy chief, Grand Admiral Erich Raeder, shared his own belief that occupation was the only solution to the Malta problem: the Führer took care not to tell him.

The field-marshal, as Kesselring now became, commanded Hitler's confidence because he possessed the quality esteemed above all others at his conference table: optimism. He never despaired. On arrival in Rome on 28 November, however, he immediately encountered the difficulties of coalition command. Count Ugo Cavallero, the Italian chief of staff, declined to surrender his air resources to German control. The Regia Aeronautica doggedly insisted upon communicating through its own wireless wavelengths, even when its aircraft were airborne alongside those of the Luftwaffe. Initially, Kesselring could provide only limited air support for southbound Axis convoys to North Africa, which suffered severely from British attacks. He raged at obvious allied foreknowledge of shipping movements, which he attributed to Italian traitors: 'though we could not prove it, we suspected that the times of our convoy sailings were betrayed ... We know now that the treachery of Admiral [Franco] Maugeri [chief of Italian naval intelligence] was responsible for the sinking of many ships and the loss of many lives.' The German C-in-C never guessed the truth, about the triumphs of allied codebreaking at Bletchley Park.

It was a contradiction in Kesselring that, despite the many war crimes against Italian civilians which he presided over, he was among the few Germans who liked Mussolini's country and people, and handled Italian commanders with charm and skill. One night in January 1942, the airman piloted through darkness to a conference a visibly nervous Cavallero in his tiny Storch, because no other plane was available. Kesselring wrote: 'the *abracci* – embraces – and *baci* – kisses – that followed our safe landing are no fancy of my imagination'.

From the day that he assumed command until August 1942, he continued to lobby for the seizure of Malta by amphibious landings spearheaded by *Fallschirmjäger* – paratroops. There were meetings in Berlin at which he argued the case passion-ately, influenced by a dislike of Rommel, whose prestige he considered to outweigh his real ability. Kesselring was frus-trated by the fact that despite his title as C-in-C he lacked direct authority over the 'Desert Fox', who supposedly reported to the Italian Comando Supremo. At one point Hitler was moved to soothe his southern warlord with a disingenuous assurance about the capture of Malta: 'Keep your shirt on, Field-Marshal Kesselring. I'm going to do it!'

On 31 December 1941, Kesselring issued a 'directive for the battle against Malta'. He later claimed that he would never have thought of launching an air blitz had he not assumed that his bombers would be blasting open the way for a subsequent invasion. His order asserted that neutralization of the island was 'the indispensable precondition ... for establishing secure lines of communication between Italy and North Africa'. Two bomber and two fighter *Geschwader* were committed. The German and Italian navies were ordered to cordon off the island, blocking its sea supply routes with mines and submarines.

Kesselring unleashed 250 bombers and 160 fighters, flying from three airfields in Sicily, all within a hundred miles of Malta. The island's people afterwards called the last months of 1941 and first of 1942 the 'Black Winter'. The Italians bombed them by day and the Germans by night, devastating great swathes of the island and destroying 15,500 houses. Over a thousand people were killed, 4,500 injured. On the three airfields, anti-blast pens for the RAF's planes were built from petrol tins filled with the rubble of Maltese homes.

Far more bombs rained down on the island than the Luftwaffe had unleashed on London: Valletta escaped the fires that erupted during Britain's blitz only because its buildings were almost entirely constructed of rock. After the war, Kesselring asserted sanctimoniously that his orders for Malta included a directive that the town was to be spared, writing: 'it is to the credit of the Luftwaffe that it restricted the battle to purely military targets, a fact that has been acknowledged by the British'. This was a travesty. The Nazi warlord's claim was set at naught by the damage or destruction visited upon seventy-eight churches, thirteen hospitals, twenty-one schools and the opera house that was Valletta's pride, together with five hundred civilian deaths in April 1942 alone. 'The destruction is inconceivable,' wrote army chief Gen. Sir Alan Brooke during a visit, 'and reminds one of Ypres, Arras, Lens at their worst during the last war.'

The plight of the Maltese was pitiable, beneath this sustained onslaught. Late in March 1942, an air-raid shelter received a direct hit from a 500-pound bomb, which killed 122 people. Anti-aircraft fire was no more effective on the island than anywhere else in the world, while obsolescent Hurricanes and Beaufighters were no match for the Germans' Junkers, Heinkels and Messerschmitts. By 1 April the RAF had lost only twenty fighters in the air but 126 on the ground. Grand Harbour became a lagoon of stagnant oil from sunken ships, amid which bobbed debris and decomposing corpses. The air bombardment rendered Malta unusable for British air and naval attacks on the Axis supply line from Italy to North Africa: as a result, in February and March, 90 per cent of supplies shipped south across the Mediterranean reached Rommel safely. A Maltese wrote that his island had been reduced to 'the great kingdom of terror'.

In February 2,099 German and 791 Italian sorties were flown against the island. During Kesselring's 'big April', the assault intensified dramatically: the island was sometimes attacked by two hundred aircraft within twenty-four hours – twice by three hundred. The RAF's sixty operational fighters were by the middle of the month reduced to six. On the 20th, forty-six Spitfires flown from the US carrier *Wasp*, generously loaned by President Franklin Roosevelt, landed successfully on the island, but bombing promptly destroyed almost all of them on the ground – a disaster that was scarcely an endorsement for Hugh Lloyd's direction of the air defences. The picture brightened for the British only on 28 April, when many of Kesselring's squadrons were sent north to support Hitler's summer offensive in Russia. Moreover, on 9 May a second consignment of sixty Spitfires flown from *Wasp* arrived, and survived. But routine attrition, in the harsh operating conditions prevailing on the island, caused its RAF strength to be depleted by an average of seventeen aircraft a week, almost heedless of the level of enemy activity.

On 15 April, King George VI paid homage to Maltese staunchness by awarding the entire island the medal for courage that bore his name. Yet graffiti appeared on ruined walls in Valletta: '*Hobz, mux George Cross*' – bread, not medals. Far more even than bombing, starvation threatened to break the spirit and resistance of Malta, its garrison and people alike. The island lies some eight hundred miles north-west of Alexandria – over nine hundred miles from Gibraltar – fifty hours' fast steaming, at least a hundred hours for ships zigzagging and fighting.

The passage from the west was menaced by Axis air bases on the islands of Sardinia, Sicily, Pantelleria; also on the North African mainland and the toe of Italy. In the eastern Mediterranean, the enemy held Greece and Crete, and varying portions of the Libyan littoral. British transport aircraft could deliver small quantities of urgent medical stores and mail. Big Thames-class submarines ferried eighty tons apiece of torpedoes, ammunition and special lubricants. The fast minelayers *Welshman* and *Manxman* each carried 350 tons, making five runs between them, at speeds of almost forty knots. But none of these expedients could feed three hundred thousand people. Only large merchant-ships could do that, and it was impossible for them to sail to Malta save in convoys protected by powerful escorts. Between February and August 1942, eighty-five British cargo vessels sailed for the island, of which twenty-four were sunk and eleven forced to turn back. One-third of all food and war materiel dispatched from Egypt was lost, and 43 per cent of those shipped from Britain.

In the months preceding Operation Pedestal, a February attempt to run three merchantmen to the island from Alexandria failed miserably, all being lost to air attack. In March, another such sally was made from the east, four ships setting forth escorted by Rear-Admiral Sir Philip Vian's cruiser squadron and a bevy of destroyers. Two cargoes reached Grand Harbour intact, and some oil was salvaged from a third before she sank inshore. The fourth vessel foundered after being bombed just twenty miles from safety. Worse followed: the two arriving ships fell victim to air attack before they could be unloaded. To achieve this unhappy outcome, the navy lost one destroyer sunk, another fourteen warships damaged.

The offload calamity exacerbated tensions between the three services, and intensified criticism of Malta's leadership. The clever, sharp-tongued Air Chief Marshal Sir Arthur Tedder, RAF supremo in the Mediterranean, thought the naval staff on the island 'deplorably weak'. Hugh Lloyd sent an outspoken signal to his chief, highlighting failure to organize the harbour, to handle the incoming ships with the necessary urgency: 'Too many old and worn-out men, some far too bomb-shy. Civilian stevedores did practically no work. They must be conscripted and worked under guard. Suggest impartial investigation to wake this place up. People want a leader and someone with energy. Excuses are given for failure to take off cargo, but I accept none.'

Dobbie, disgusted in turn by what he saw as the airman's disloyalty, demanded that Lloyd should be sacked. Tedder refused.

A local woman wrote, 'Only those who have suffered hunger can understand what this really means. Mothers went from door to door with young children begging for food ... Rations for two weeks lasted for only four days. We were allowed four ounces of bread per day for women and children and six ounces for men.' Pasta ran out in April. In the absence of Italian manure, potatoes could no longer be grown. The Maltese diet had been reduced to bread, a little olive oil and tomato paste, some preserved fish and rice, home-brewed wine known as 'screech'. The island's fruit trees were stripped bare.

Every inhabitant, after almost interminable queueing, qualified for a daily free meal from so-called Victory Kitchens, 'if you can call a piece of goat floating in muggy gravy a meal!' as a local woman wrote. 'The bravest thing I ever did was to eat that goat.' She described how working members of a family returned home of an evening for their suppers: 'this in itself was an ordeal because while eating one had to suffer the eyes of the young children sitting around ... They never asked for food, they just stood looking at you with pleading eyes.' Men scoured the streets for cigarette ends. Office letters were typed on toilet paper. Shops closed. People wore shoes that lacked soles. A sack of flour, rightfully costing two shillings, commanded £40 on the black market. Only a bumper crop of tomatoes averted absolute starvation.

Few people, either military or civilian, discovered much private glory in garrisoning a 'beleaguered fortress', as Churchill and the British press characterized the island. Local churchman Canon Nicholls lost forty pounds in weight and wrote in his diary, 'Oh, what a glorious relief it will be to quit Malta. Life is terribly monotonous.' The spirit of Nicholls and his wife was already sorely tried by the loss of a son, Anthony, killed serving with the Royal Navy. The canon deplored the looting and pilferage from ships: 'it has assumed the character of a national scandal'. The governor issued an order of the day urging soldiers to abandon their own thieving practices, and to stop Maltese pillage of stores.

RAF pilots, on whom the island's survival depended, ate little better than Maltese civilians. They subsisted on a diet of stew and pilchards in tomato sauce, infested with blowflies and sandflies from the moment the tins were opened. At Luqa, aircrew professed gratitude for supplementary fare provided by rats, killed by a local farmer and sold to the messes. Among civilians tuberculosis, pellagra and dysentery were widespread, together with a milder form of the latter, 'Malta Dog', which plagued soldiers and aircrew. Moreover, the threat of enemy invasion hung constantly over the island. RAF chief Hugh Lloyd wrote: 'as each dawn broke, I expected the Axis assault'.

On 17 March 1942 Germany's Lt. Gen. Enno von Rintelen presented Mussolini's Comando Supremo with detailed plans for an occupation of Malta. Yet Italian commanders remained hesitant. By this time, Italy's leader – 'Il Duce' – had considerably soured on both the war and his Teutonic ally, but had no notion of how to escape from either. He observed bitterly to Ciano: 'Among the graveyards will some day be built the most important of all, one in which to bury German promises.' Hitler, in his turn, had become resigned to the fact that his ally was a burden upon the Nazi war effort, rather than an asset. It would have better served his global purposes if he had persuaded his fellow dictator to remain neutral, instead of becoming a belligerent.

Early in April Mussolini approved the Malta invasion plan, but Hitler and Gen. Alfred Jodl, Wehrmacht chief of operations, remained sceptical. On 27 March Jodl wrote to Kesselring: 'One can hardly tell the Italians they may as well drop their preparations to take Malta because they won't get it anyway.' The field-marshal was instructed to inform the

Comando Supremo that, if an invasion was launched, Germany could contribute only one or two paratroop regiments and some E-boats; most of the aircraft would have to be Italian. With the utmost reluctance, in April Kesselring told Rommel that he recognized the impossibility of simultaneously invading Malta – Operation Herkules – and supporting Operation Theseus, the latest Axis offensive in Libya. The commander-in-chief accepted that, granted the Führer's preference for North Africa, this must come first, as it did.

Yet even if the Axis declined to storm the island, like a besieged fortress of old it must be surrendered to the enemy if its occupants could no longer be fed. On 7 May 1942, Dobbie was removed from his post. The prime minister dispatched a personal telegram to Gen. Viscount Gort, governor of Gibraltar, ordering him immediately to assume the role of governor, C-in-C and supreme commander on Malta: 'every effort must be made to prolong the resistance of the fortress to the utmost limit (stop) We recognise you are taking over a most anxious and dangerous situation at a late stage (stop) We are sure that you are the man to save the fortress and we shall strive hard to sustain you.'

Gort had commanded the British Expeditionary Force in France through its 1940 retreat to Dunkirk without exhibiting much military genius, but instead some staunchness in the face of adversity. This latter was the quality now deemed necessary for the governorship of Malta. A taciturn Grenadier who wore a 1918 Victoria Cross, Gort bicycled through the blitzed streets of Valletta to perform his duties. These were largely ceremonial and morale-boosting, for he possessed neither means nor opportunity to influence Malta's fate by generalship. The hapless governor was haunted by fear that he had been obliged to accept a poisoned chalice; that, just as Gen. Arthur

Percival had incurred the odium of surrendering Singapore a few months earlier, Gort, in his turn, would be obliged to present Axis generals with the keys to Malta when the island's store cupboards were at last empty.

Tedder replaced the exhausted Hugh Lloyd as air commander, sending in his place Keith Park, the New Zealander who had commanded 11 Fighter Group with distinction throughout the Battle of Britain. Park reported to Cairo that Gort repeatedly voiced fears that a German prisoner-of-war camp – 'the cooler' – appeared his likeliest destination, a fate the general was desperate to escape. The spiky Tedder later commented in his memoirs: 'Such an attitude became doubly deplorable at a time when morale among troops and populace could hardly have been higher.' Those words reflected an excessively sunny view of the island's mood, but he added: 'Personal intrigues of the kind frequently found under the Dobbie regime continued under Gort ... The defensive, defeatist atmosphere was more pronounced than ever.' Malta's garrison, he thought, seemed psychologically prepared to 'go into the bag'. RAF aircrew fantasized about flying away south to Africa, as far as their fuel tanks would take them, once enemy occupation became inevitable.

The struggle around the island assumed the character of a private war, in which picaresque episodes took place unlike anything anywhere else. One day in July, a Luqa-based RAF Beaufort was shot down off western Greece while attacking an Axis convoy. An Italian floatplane pilot retrieved its four-man crew from their dinghy, plied them with brandy and cigarettes, then landed to overnight in Greece. Next day the same floatplane ferried the fliers towards Italy. In the air, the Beaufort crew decided that they must now seize their only opportunity to avoid a PoW camp.

New Zealander Jim Wilkinson smashed his fist into the face of the wireless-operator, then tossed the man's pistol to South African pilot Ted Strever. Bill Dunsmore and Arthur Brown overpowered the rest of the crew, whom they tied up with their own belts. The only casualty was an Italian corporal acting as armed guard, who was violently sick. The hijackers then induced the pilot to set a course for Malta. Offshore, they narrowly escaped being shot down by Spitfires, but made a safe descent after running out of fuel; the captured seaplane was towed in by a launch. The Beaufort crew sought to lavish good things on their Italian prisoners, whose kindly treatment they felt conscience-stricken for repaying so roughly. The enemy airmen, however, seemed not unhappy to find themselves safe, albeit hungry, in British hands. Here was a droll moment in a grim story.

In June, a new convoy-relief operation was launched, code-named Harpoon, involving seven supply ships that sailed from Alexandria simultaneously with six from Gibraltar, to divide the enemy's fire. Churchill messaged Roosevelt on the 7th: 'If we cannot get this convoy through ... the doom of the fortress cannot long be deferred. The surrender of Malta and over thirty thousand of our best troops would be a major disaster in itself. Its consequences will be the free passage of whatever troops Hitler chooses to send to Africa, with further results only too plainly foreseeable ... I had hoped that we might by now have regained some of the landing grounds in Cyrenaica, but the battle sways to and fro as yet without decision.'

After losing most of Harpoon's eastbound vessels en route from Gibraltar, including the American tanker *Kentucky*, yet another avoidable disaster took place as the survivors approached Grand Harbour. There was confusion about the location of the mineswept channel through the defences. Three destroyers, a minesweeper and a merchantman struck British mines. Just two cargoes out of six arrived intact, while two destroyers were lost, several other warships damaged.

The westbound convoy from Alexandria, escorted by seven cruisers and twenty-six destroyers, suffered attacks by aircraft, E-boats, submarines and the Italian fleet. Forty Stuka dive-bombers hurled themselves against a single merchantman, which sank. Mediterranean C-in-C Admiral Sir Henry Harwood eventually ordered the survivors to turn back, having lost two merchantmen sunk and two damaged, a cruiser and three destroyers sunk, two more cruisers damaged.

In all these operations, the poorly armed ships of the Royal Navy and Merchant Navy endured an intensity of battle, and especially of air attack, which strained their crews to the limits. Gunfire from a heaving, floating platform seldom offered an effective response to aircraft, and almost none against dive-bombers descending on a near-vertical. Airborne fighters, and fighters alone, could protect fleets and convoys. In the first seven months of 1942, just five merchantmen reached Malta, none of them a tanker, and only two cargoes were successfully offloaded. Tensions ran high between Britain's three services about responsibility for the navy's fearful losses. The RAF's Tedder was serenaded as he left the Mohamed Ali Club in Cairo by drunken sailors and Marines chanting: 'Roll out the *Nelson*, the *Rodney* and the *Hood*, since the whole [fucking] air force is no [fucking] good!'

In those months, however, Russia dominated Hitler's warmaking. By the end of April, much of Kesselring's Mediterranean air force had been redeployed north, to support the Wehrmacht's summer offensive. German bomber strength in the theatre shrank by half, fighter numbers by more than one-third. The field-marshal professed to believe that his campaign against

Malta had fulfilled its objectives: that the island was neutralized as a British base. Kesselring wrote: 'Thanks to [the blitz's] success, our ascendancy at sea and in the air in the supply lanes between Italy and North Africa was assured.'

Following Rommel's June capture of Tobruk, the commander-in-chief renewed his pleas to Berlin to exploit dominance of the North Africa coastline by taking Malta through a *coup de main*: 'It would have been easy to capture the island after the bomber assault. That this did not happen was a grave mistake … which came home to roost later.' To cope with Axis supply problems 'we just had to have Malta, yet the withdrawal of the forces [to Russia] destined for the invasion [of the island] made this impossible'. II Fliegerkorps dwindled from 434 aircraft in April to 216 on 20 May. Kesselring fumed that this force 'proved too weak to neutralize the island fortress or to deprive it of supplies'. Again, he pressed for outright seizure of the island, and again he was overruled at a key conference with Rommel and the Italian chiefs at Sidi Barrani on 26 June. Cavallero and his colleagues accepted the desert commander's assurance that he could get to Cairo in ten days, with or without Malta.

On 21 July 1942 the German naval war staff signalled to Rome: 'Operation Herkules [the Malta invasion] to remain in abeyance until the termination of Theseus [Rommel's dash for Cairo]'. Professor Sir Michael Howard, in his *Grand Strategy* volume of the British official war history, describes Berlin's decision to seek a decisive outcome in Egypt, while leaving Malta in British hands, as 'one of the most disastrous Hitler ever made'. If this exaggerates, it reflects the body of scholarly opinion which holds that seizure of the island could have transformed the strategic picture. It remains a focus of dispute between historians whether Axis troops could have achieved this. Reinhard Stumpf, one of the authors of Germany's authoritative Potsdam war history, writes: 'The answer, in my opinion, is no.' This writer disagrees. The British soldiers garrisoning Malta were no elite, as a small force of German paratroopers discovered in October 1943 when they easily overcame many of these same men, by then transferred east to defend Leros. Kesselring's forces could almost certainly have secured Malta in the spring or early summer of 1942, easing Rommel's logistical nightmare.

Kesselring ordered a new summer air offensive against the island, of which he says nothing in his memoirs because it failed, and on 23 July it was broken off. British fighter reinforcements to the island drastically reduced the effectiveness of Axis raids. The field marshal's post-war account of the decision-making is self-serving. There is no doubt of the sincerity of his enthusiasm for capturing Malta. But with his usual desire to please his Führer and, indeed, Italy's Duce, he failed frankly to confront them with the strategic consequences of weakening Axis air strength in the Mediterranean even as Tedder and Park were increasing their own.

Yet for Britain the threat persisted, that Malta would be lost through starvation. On 18 July, the prime minister signalled Gort: 'I should be very glad to hear how you are getting on. It is a great comfort to me to feel that you are in full control of this vital island fortress. You may be sure we shall do everything to help you.' The governor replied at length three days later, observing that 'morale remains good, but the people have a great deal to put up with ... I have formed a great admiration for the cheerfulness of the Maltese people and their stoical determination to withstand everything that is humanly possible to endure, sooner than surrender to the despised Italians.'

Nonetheless, Gort emphasized the shortage of aviation fuel and questioned whether it was wise to expend substantial quantities conducting ineffectual bomber sorties: 'As I see it, the question is either that Malta must make a supreme effort to help Middle East and be ready to exhaust herself in the process, or else she must lie low, being content with defensive fighting only, until her stocks are replenished ... I am doubtful whether the people of Malta will be prepared again to sustain such a heavy bombardment from the air as they suffered last spring.'

Gort's administration had been instructed by the chiefs of staff to issue and deplete food stocks at a rate that would enable the Maltese people and garrison to support existence until a terminal 'target date' of late September. This might be managed, said the governor; nonetheless 'to be told you will not starve, but to be conscious that your stomach has an aching void, is apt to leave the average person discontented'.

Twenty thousand tons of food a month, including a hundred tons of flour a day, were necessary to feed Malta. Only a small fraction of this amount was reaching the island, which thus became dependent upon its shrinking reserves. The garrison's daily ration shrank to one and a half slices of bread with jam for breakfast; bully beef and one slice of bread for lunch; the same with two slices of bread for supper. Gort himself, once nicknamed 'Fat Boy' by his fellow officers, could no more be thus characterized. A naval officer lunching at the Governor's Palace found himself served a vegetable omelette. Malta's offensive value for attacking Rommel's supply line was drastically reduced. British forces sank nineteen Axis merchant vessels of significant size during July and August, but during the same period sixty enemy cargoes reached North Africa unscathed, together with twenty-five cargo-carrying submarines and fifty-three barges. Moreover, the outcome of the entire war seemed still to hang in the balance, with Russia's survival as uncertain as ever.

Yet the phrase in Gort's letter to Downing Street which made the deepest impression was that in which the governor told Churchill: 'you can rest assured that everyone in Malta is determined to stand up to privation so as to endure *as long as possible* [author's emphasis]'. This did not set an exact term on the limit of Malta's resistance, but made it plain that, when food resources were exhausted, as by the end of September they would be, the island must be surrendered to the enemy. The civilian population could not be permitted to starve to death or resort to cannibalism, as Stalin a few months earlier had mandated as the fate of hundreds of thousands of the citizens of besieged Leningrad.

The price of fighting under the banner of Western liberal democracy was that the war must be conducted on civilized terms. No man, woman or child could be expected to exceed the bounds of reasonable behaviour in sustaining the allied cause. Herein lay the justification for many of the British military surrenders made over the previous three years, and also the explanation for the prime minister's extreme displeasure towards the commanders who presided over them. Himself a hero, the unwillingness of some others to play the hero was a source of bewilderment as well as anger to him. This was the way it was, however. If the civilian population of Malta confronted the prospect of death by inanition – as by autumn they must, failing relief – then Gort would surrender the island to the Axis, whatever contrary imprecations might reach him from London. And who can say that Gort's assessment of the island's predicament, and of his own duty, was mistaken or ignoble?

Axis intelligence was often imperfect, but it was plain to Jodl in Berlin, to Kesselring and to his Italian counterparts in Rome that the British would strain every sinew to relieve Malta. Some of the *Geschwader* sent to Russia were permitted to

return to the Mediterranean theatre. The inertia that had overtaken the British Army in Egypt made it possible to divert
aircraft that would otherwise have been required to support
Rommel to address the Royal Navy. Kesselring's air commanders could call upon more than six hundred German and Italian
bombers, fighters, dive-bombers and torpedo planes to contest
a British lunge across the Mediterranean.

Submarines, E-boats and the Italian Fleet were summoned
to readiness. Despite the wrangles between Berlin and Rome,
and the tensions among Italian and German regional
commanders, the fundamental was plain: it was a vital objective of the Axis, for reasons of global prestige as well as regional
supremacy, to frustrate a new British attempt to relieve Malta.
If a fleet was coming, its ships must at all costs be denied
passage to the lonely rock at the heart of the Mediterranean,
where its half-starved people cherished uncertain hopes,
increasingly urgent fears.

2 CHURCHILL'S COMMITMENT

At all times and seasons, Winston Churchill was an advocate
of boldness, at sea as also on land and in the air. One day in
May 1941, when the navy was suffering terribly in the struggle
for Crete, his private secretary commiserated on the loss of
two cruisers. He received the robust prime ministerial
response: 'what do you think we build the ships for?' Churchill
went on to deplore the Admiralty's insistence upon treating its
fighting fleets as if they were too precious to hazard.

In May 1942, he asserted that the loss of Malta would be 'a
disaster of the first magnitude to the British Empire'. The question of whether the island was, indeed, a vital allied strategic
interest or instead a mere imperial sacred cow has divided

historians since 1945. Correlli Barnett argues that 'the island served less as a British strategic asset than as a hostage to the enemy'. For Britain's leaders it had become, he says, 'a matter of prestige and pride, a symbol of heroic resistance ... Like Verdun and Stalingrad the "George Cross island" triggered powerful emotions which dictated that it must be held no matter what the cost in resources, risk, losses, and distortions of the balance of strategy as a whole.' By contrast, the post-war British official historian wrote: 'on the success or failure of [operations to resupply the island] would hang the fate of Malta and hence in all probability the Nile Valley'. Douglas Porch argues: 'The Mediterranean was not the *decisive* theatre of the war, [but] it was the *pivotal* theatre, a requirement for allied success,' with Malta a vital piece in the British campaign against Rommel's supply line.

That spring of 1942, the British chiefs of staff went so far as to urge that Eighth Army should fast-forward an offensive in Libya, not with much hope of crushing Rommel's Afrika Korps, but instead to relieve pressure on the beleaguered island. In March, they signalled Gen. Sir Claude Auchinleck, British C-in-C in Cairo: 'the dominant factor in the Mediterranean and Middle East ... is Malta'. Thus, they said, it was a matter of the utmost urgency for Eighth Army to seize Cyrenaican airfields lost in the previous year's retreat, to allow British fighters to regain a radius of operations over the south central Mediterranean. On 10 May Churchill dispatched a sonorous signal to Auchinleck in the name of the War Cabinet and chiefs of staff: 'we are determined that Malta shall not be allowed to fall without a battle being fought by your whole Army for its retention ... Its possession would give the enemy a clear and sure bridge to Africa with all the consequences flowing from that.'

Auchinleck deployed his army to fight a defensive action against an Axis assault that he knew was coming. Though Rommel's forces were numerically weaker, not only did the British suffer defeat in the Battle of Gazala, which began on 26 May; they were thrown back further than at any time in the desert war since 1940, to El Alamein, gateway to Egypt, with the loss of Tobruk, Churchill's proclaimed 'fortress', though it was nothing of the kind. Far from fulfilling British hopes that the battle would relieve pressure on Malta, its outcome dramatically increased this. Hitler rewarded the victorious Rommel with a field-marshal's baton.

The prime minister returned from a June visit to President Roosevelt in Washington to find parliament and the British people disconsolate after the procession of disasters that had befallen their cause since New Year. Chris Gould, a twenty-five-year-old Londoner serving as an able seaman on the destroyer *Lightning*, soon to sail for Malta, wrote to his fiancée Vera: 'I was on watch for Churchill's speech but he gives me the pip anyway, so I didn't miss much. I don't need him to give me moral support and if a vote was cast in the services he would not be there anyway. The navy has a special dislike for him as we do all his dirty work.'

This was by no means untrue. Admiral Sir Andrew Cunningham was Britain's most admired fighting sailor, but his impatience with the prime minister was well known: 'what a drag on the wheel of war this man is!' When Churchill proposed that Lord Louis Mountbatten, whose most notable achievement as chief of Combined Operations was his sponsorship of the disastrous Dieppe raid, should become a member of the chiefs of staffs' committee, First Sea Lord Sir Dudley Pound remonstrated. There was, he wrote, 'a very widespread belief, not only in the Services but also in the country, that you do override the opinion of your professional advisers'. This was 'doing a great deal of harm in undermining confidence in the Service leaders'. Both Pound and Cunningham smarted over Churchill's insistence on appointing as Mediterranean commander-in-chief, in April 1942, Sir Henry Harwood, a notoriously stupid officer who had secured the prime minister's confidence by a single headline action, the December 1939 Battle of the River Plate.

Even before PQ17, the navy lamented the requirement to dispatch convoys to Russia, because these put at hazard half the Home Fleet, to assist people with whom the admirals had no sympathy, without hope of securing a scintilla of Soviet gratitude. In the matter of Malta, also, it is doubtful that their Lordships of the Admiralty would have been willing to risk so many ships had not the prime minister insisted.

To Churchill's conviction of the strategic importance of the island was added, in the summer of 1942, a political impera- tive. There was pressure, even within the government, for him to relinquish his role as minister of defence, a post that he held alongside the premiership. Confidence in his direction of the war had sunk to its lowest ebb. With the United States commit- ted to the allied cause, ultimate victory seemed assured. But what would a pauper's share in such an outcome avail a humil- iated British nation, dependent on the Russians and Americans to win battles that its own armed forces apparently could not? In such a climate of public despondency, Churchill's authority would be even further weakened if he suffered another disaster such as must be the loss of Malta, after he had so often proclaimed the island's importance. On 10 June the prime minister dispatched a stern minute to the Admiralty, demand- ing that 'Lord Gort must be able to tell the Maltese that the Navy would never abandon Malta.'

The genesis of Pedestal was a chiefs of staffs' meeting on the night of 15 June, attended by the prime minister, following the failure of Operation Harpoon. Churchill said 'the Naval Staff should now tackle the problem of running in a large convoy to Malta from the West during the July dark period'. Pound responded that his officers were already studying this question, for which it would be necessary to recall a battleship destined for the Indian Ocean, and to begin loading the merchant vessels as soon as possible. Sir Charles Portal, for the RAF, said that it might be possible to divert some bombers to fly supplies into Malta, but they could move no more than thirty tons a day, 'a drop in the ocean', and could do nothing to alleviate the critical fuel shortage.

Churchill said that the American naval triumph at Midway a few days earlier had reduced the Japanese strategic threat in the Pacific, and made it less urgent to reinforce Admiral Sir James Somerville's Eastern Fleet. The meeting concluded with an instruction that the Admiralty should advance proposals for 'the largest possible convoy to be run into Malta from the west'. It proved too much of a stretch to prepare a big convoy for a July sailing. Instead August was appointed – perilously close to the September 'target date' at which Malta's reserves of foodstuffs would be exhausted.

The War Cabinet formally endorsed the commitment to sail a big convoy, composed of the fastest merchant-ships that could be found, which must include a tanker. For the cause of saving Malta, the Americans had already leased them *Kentucky*, which lay on the bottom of the Mediterranean. They now replaced her with a sister ship: the 9,625-ton *Ohio*, less than two years old, the strongest, best-constructed such vessel in the world. She was sailed across the Atlantic by an American crew, then on 9 July transferred to British registry.

As for escorts, no mere bevy of destroyers and corvettes could suffice. Italian battleships and big-gun cruisers could sortie against Pedestal. Thus their British counterparts must be ready to confront such a threat. Naval fighters were also essential, so the strongest possible carrier group must be assembled. British cruisers might have to fight an Italian squadron alone, when the capital ships retreated. And such ships also possessed command-and-control facilities that destroyers lacked.

The battleships *Nelson* and *Rodney*, which had been earmarked to reinforce Somerville in the east, now instead received warning orders for the Mediterranean. For the worst of reasons, it became possible to gather a much larger cruiser and destroyer force than had been expected: following the July disaster to Arctic convoy PQ17, further sailings to Russia were postponed until September. Acting Vice-Admiral Neville Syfret, commander of the legendary Force H, which had operated out of Gibraltar and in the Indian Ocean, was ordered to take command of the fleet being assembled. He flew home from Africa on 13 July to conduct planning at Norfolk House, St James's Square. Syfret's standing with the prime minister and the Admiralty was high, following his successful command two months earlier of the amphibious descent on Vichy French Madagascar. He also had experience of running a 1941 convoy to Malta, under Somerville's command – 'Uncle Jim' thought well of him.

Syfret was joined by Rear-Admiral Harold Burrough, flag officer of the 10th Cruiser Squadron, and by carrier specialist Rear-Admiral Lumley Lyster, stage manager of the Fleet Air Arm's 1940 triumph against Italian capital ships at Taranto. Burrough was Syfret's senior in the Navy List, but agreed to serve under his flag for Pedestal: as commander of the close escort through to Malta, Burrough would anyway bear the weightiest responsibility. Days of grave, intense debate took place across chart tables and before wall plans, with the admirals surrounded by a throng of staff officers of every possible specialization – gunnery, air operations, anti-submarine tactics, refuelling at sea.

Often when a fleet sailed, commanders were uncertain whether it would have to fight. This time, there was no doubt: the enemy would deploy against Syfret every weapon in his armoury, except possibly the Italian battleships. Pound, the First Sea Lord, was a sick man who must at this time have been harrowed by consciousness of personal responsibility for the PQ17 disaster. Now, he nursed private gloom about Operation Pedestal, saying later to one of its senior officers, 'when we sent you out to Malta, we wrote you off'.

On 1 August the prime minister briefed the War Cabinet about the operation, mentioning some dilemmas as yet unresolved: 'The Admiralty had to decide whether the heavy escort should accompany the convoy right through to Malta at the risk of our two sixteen-inch [gun] battleships being heavily attacked by air ... If ill befell our heavy ships in the narrow waters approaching Malta, the whole balance of naval power would be affected. On the other hand, if it was decided not to risk the heavy ships during the final stage and the convoy suffered severe losses from attack by enemy surface ships, some searching questions would be asked. In view of the grave issues involved, he asked the War Cabinet to support any decision which might be taken.'

The Admiralty's draft plan for Pedestal was created by Captains Jocelyn Bethell and Edward Pleydell-Bouverie. Four options were initially considered, designated A, B, C, D. The first three required a request to the Americans for the use of their carrier *Ranger*, with five US Navy escorts. Then came

Gort's message, reporting that Malta could hold out until September, and thus that a convoy might be postponed until August. This made possible more training and preparation, together with the assembly of an entirely British carrier group, avoiding the embarrassments and difficulties inseparable from all requests to Washington for resources, especially major naval units. Following PQ17, with which American ships had sailed for the first time under British orders, the US Navy's respect for its allied counterpart had fallen to a low point.

The staff worked upon three assumptions, which were accepted by the operational commanders. First, the fleet would time its approach to the Sicilian Narrows, most perilous sector of the passage because closest to enemy air bases, between 10 and 16 August, in the dark of the moon. Nothing could be done to hide Syfret's force in daylight, but it might profit from eight hours of immunity from air attack. Second, the capital ships were deemed too precious and vulnerable to be risked in the Narrows. As the prime minister had warned the War Cabinet, this meant that the convoy would have to endure at least one full day at the mercy of Kesselring's aircraft, without benefit of either carrier fighter cover or Malta-based RAF support.

Finally, a route was appointed which hugged the North African coast around Cap Bon and neutral Vichy French Tunisia. This was significantly longer than the straight course Harold Burrough had steered a year earlier, when he led a convoy on a successful dash for Malta. Burrough's instinct was now to do the same again, but the planners highlighted the dramatic increase in Axis air strength since September 1941. They thus successfully urged adoption of a course that kept the convoy an additional fifty miles from Kesselring's bases.

The prime minister had made a personal decision to place on the table of the Mediterranean naval war a huge stake, a battlefleet such as recent experience showed to be immensely vulnerable to air and submarine attack. Less than two months earlier, in the Battle of Midway which proved a turning point of the Pacific war, five aircraft-carriers – four Japanese, one American – had been destroyed by aircraft. More than a few influential figures in Britain's corridors of power, some of them wearing naval uniform, privately believed that Pedestal would risk too much for too little; that succouring Malta was not worth the possible loss of yet more precious capital ships, above all carriers.

Among the doubters was Harwood, the Mediterranean C-in-C. When Rommel's Afrika Korps closed to within sixty miles of Alexandria, the admiral ordered an evacuation of the Fleet's shore base there, on 2 July shifting his headquarters to Haifa in so precipitate a fashion that the White Ensign still flew over the abandoned camp at Sidi Bishr when local Arabs burst through the fences to loot it. Preparations were made to sabotage Egyptian harbour facilities. Harwood justified himself by recalling the misfortunes caused by failure five months earlier to evacuate the Singapore naval base until it was too late, but it became plain that his haste invited ignominy.

Whatever the balance of forces in the Mediterranean, however, there was one matter that Winston Churchill understood better than his service chiefs: war is a clash of wills. Even though for two years Britain had been powerless to challenge Germany's mastery of the continent, and would remain so for two more, it was vital that it should be seen to fight somewhere. If the Royal Navy sought to remain merely 'a fleet in being' – a phrase often used in discussing the First World War balance between the British and Germans – it would show itself no better than Mussolini's navy. Since 1940, Churchill had insisted that coastal convoys continue to pass down the

Channel and up the North Sea, at painful cost in shipping losses, not because commerce made this necessary, but because he deemed it vital to assert Britain's freedom to navigate its neighbouring seas.

In the late summer of 1942, only in the Mediterranean theatre could the army and navy engage the forces of the European dictators. If this was to be Britain's sole significant campaign, save for the bomber offensive and the struggle to hold open the Atlantic passage, then it must be fought to the bitter end. Regardless of whether or not loss of Malta would be a critical blow to the allied cause, it must be a calamity for the prestige and self-respect of the British Empire. Thus the prime minister seems right, in the eyes of posterity – and his trial before a jury composed of future generations was a prospect never far from his mind between 1940 and 1945 – to have committed the Royal Navy to an operation that precipitated one of the fiercest sea battles of the war.

2

Men and Ships

1 THE FLEET

Vice-Admiral Neville Syfret may have welcomed the distinction of being granted such a critical command as that of the fleet for Pedestal. He would have been less than human, however, had he not reflected upon the reception he might face on his return, should the operation prove a disaster. Almost three decades earlier, Churchill exaggerated when he claimed that the Grand Fleet's C-in-C Admiral Sir John Jellicoe was 'the only man who could lose the war in an afternoon'. Nonetheless, in August 1942 a few bad days in the Mediterranean could grievously damage Britain's precarious standing, especially in the eyes of Americans.

The Admiralty's Strategic Report declared bleakly: 'We are now approaching a crisis in the war and we must face the facts. We have lost control of sea communications over very wide areas … Our whole power to wage an offensive by sea, on land and in the air, both at home and abroad, is being undermined by our weakness at sea both in ships and in the air … Our merchant ship losses are immense; the tanker situation is grave, and the supplies reaching this country are much less than we require … The war has shown that ships alone cannot control sea communications in an area which can be effectively covered by enemy air forces.'

Syfret would need iron nerve to withstand the buffeting the enemy would inflict upon his ships, in a theatre where the Axis deployed such strength. Like most senior commanders of the Second World War, the admiral was a Victorian, born in Cape Town in 1889, scion of an old-established South African banking house, almost a teenager when the old Queen died. His early career followed an accustomed pattern for officers of his generation. After Britannia Naval College, he became a gunnery specialist, serving in light cruisers during the First World War, and scarcely injured his popularity by playing cricket for the navy. Married with two children, in the early months of the Second World War he was fortunate enough to impress the First Lord, when serving as naval secretary at the Admiralty during Churchill's tenure.

Syfret had held admiral's rank for over a year, gaining experience of fighting convoys through both the Arctic and the Mediterranean. His personality seems to have made little mark upon the leaders of the Grand Alliance, who recorded no anecdotage about him. A senior officer wrote in a confidential report shortly before the war that he found Syfret 'very quiet and perhaps rather shy'. The commander of *Nelson* grumbled about the self-indulgent quantity of baggage his forty-strong staff brought aboard when the battleship became flagship for Pedestal 'more wine, food and mess traps than any admiral we've ever had'.

Be that as it may, Andrew Cunningham called Syfret 'a tower of strength, a man of great ability and of quick and sound decision ... His great knowledge and charm of manner made him a delightful comrade.' Sir John Tovey, commanding the Home Fleet, wrote in March 1942 that he found him 'a clear thinker,

quick in decision, handles ships well, a most delightful personality, inspires confidence and gets the best out of his subordinates'. Pedestal's admiral was an efficient, experienced,
rock-steady commander, though not an ebullient or cerebral
one, well-suited to guiding his force through the ordeal by fire
that it must face.

Beyond two battleships, Syfret would lead into the
Mediterranean three fleet-carriers, a concentration of naval air
power such as Britain had never before assembled, essential to
keep airborne a fighter force capable of meeting massed air
attacks. The Royal Navy's 'flattops' were significantly smaller
than those of the US Navy, reflecting their proprietor's lesser
resources: *Indomitable*'s flight-deck was 754 feet long to
Yorktown's 824 feet, and after the latter's sinking at Midway her
successors were built even bigger. British ships then, and for
the rest of the war, carried fewer aircraft: *Indomitable*'s complement was forty-five against *Yorktown*'s ninety; they also
marshalled and flew them off significantly more slowly. *Eagle*
was an old vessel, destined to become a Chilean battleship
when laid down in 1913. Belatedly converted to a British flight
platform, she had already done stalwart Second World War
service.

Victorious and *Indomitable* were modern ships, boasting
armoured flight-decks that shrank their aircraft capacity but
importantly increased their survivability under air attack. A
few months earlier '*Indom*' had the good fortune to run
aground in the West Indies just before sailing to join *Prince of
Wales* and *Repulse* in the Far East, where she would almost
certainly have shared the fate of those huge warships, sunk by
Japanese torpedo-bombers on 10 December. As it was, after
service in the Madagascar campaign the 23,000-ton carrier
was now available to Syfret.

Yet there was much less cause for satisfaction about the aircraft these ships embarked – Albacore biplanes and Fulmar, Hurricane and Martlet fighters. On 2 May 1942 Somerville, the Eastern Fleet's C-in-C, responded acidly to an Admiralty inquiry prompted by the prime minister about why his carriers had attempted so few offensive operations: 'the unattractive policy you refer to is forced upon me by the very unattractive aircraft with which my carriers are equipped ... I ... [would have thought] by now it would be appreciated that our Fleet Air Arm, suffering as it does from arrested development for so many years, would not be able to compete on all-round terms with a [Japanese] FAA which has devoted itself to producing aircraft fit for sailors to fly in.'

Some hundreds of the Hawker Hurricanes which had contributed so much to winning the Battle of Britain were modified for naval use, but they struggled to catch German and even some Italian planes. Moreover, the carriers' flight-deck lifts were designed for folding-wing aircraft, and it was difficult on *Indomitable*, impossible on *Victorious*, to strike fixed-wing Hurricanes below, to the hangar deck. Thus *Victorious* carried only five Hurricanes, together with sixteen Fulmars and twelve Albacores, these last being embarked to attempt torpedo strikes in the event of a sortie by the Italian fleet. Pedestal's air contingent was the strongest the Royal Navy could provide in August 1942, but was sadly inadequate for the challenge it must face in the Mediterranean, fighting shore-based enemy aircraft, most of which flew higher and faster.

There was a late change of plan, to attach two more carriers to Syfret's command: the elderly *Argus* would accompany the fleet for exercises on the Atlantic passage south, carrying a reserve of aircraft. More important, to redress Malta's shortage of fighters, the Admiralty decided to tack on to Pedestal a

subordinate operation, codenamed Bellows, to fly more
Spitfires to the island as soon as the fleet reached a point in the
Mediterranean within the fighters' extreme reach of six
hundred miles. The old *Furious* was tasked to fulfil this ferry
role. Syfret, with more than enough responsibilities already,
responded irritably to taking on Bellows. He would thus lead
into the Mediterranean all but two of the Royal Navy's entire
strength of fleet carriers – four had already been sunk since
1939. He complained of the 'general unsettling effect on all,
which last-minute changes always cause'.

The other major warships which joined the fleet were a
mixture of so-called anti-aircraft cruisers and six-inch-gunned
light cruisers of over eight thousand tons, each crewed by
seven or eight hundred men. These relatively big and expensive warships were often a disappointment to all their owners
in the Second World War. Their guns seldom fulfilled the role
for which admiralties built them – to fight similar enemy
warships. And they proved as vulnerable to submarines and
aircraft as were their bigger sisters, the battleships – twenty-eight Royal Navy cruisers were sunk between 1939 and
1945, all but one by bomb, mine or torpedo.

Providing anti-submarine protection, together with sheep-
dog services for the merchant-ships, were more than thirty
smaller vessels, destroyers. It is a reflection of the social ethos
of the Royal Navy of that era that one of its most numerous
classes of warship, strongly represented in Syfret's fleet, took
names from foxhunts – thus the 1,050-ton *Ledbury*, *Bramham*,
Bicester and suchlike. Another class, the bigger and faster
1,800-ton Tribals, crewed by around two hundred men, were
mostly named for peoples against whom British imperial
forces had fought bloody wars, for instance *Ashanti*, *Zulu*,
Somali.

By 1942 the Admiralty was painfully conscious how inadequately its warships were protected against air attack. Beyond their main gun and torpedo armament, all carried four- or six-barrelled pom-poms, which the Americans called 'Chicago pianos', automatic weapons firing low-velocity two-pound shells. The Swiss-designed 20mm Oerlikon cannon, fitted with a sixty-round drum magazine, was manned by a gunner attached to a shoulder-harness – these latter were lavishly distributed around warships and merchant vessels. Both were chuck-and-chance-it weapons, to borrow a fisherman's phrase, which might hit an enemy aircraft only if the latter obligingly flew into their line of fire. The 40mm Swedish-designed Bofors was a much better gun, fitted to a few warships and now mounted on all Pedestal's merchant-ships, manned by mostly army gunners. None of them was radar-controlled, but against massed attackers a blind barrage, filling as much sky as possible with shells, offered a better defence than attempts at sniping a chosen bomber.

Rear-Admiral Denis Boyd, the senior officer responsible for carriers in the Mediterranean, had himself endured many air attacks as captain of *Illustrious*. Boyd, who sailed as an extra flag officer and anti-aircraft expert with Pedestal, wrote in February 1942: 'War experience creates grave doubts whether existing battleships and large cruisers are of suitable design to fulfil their functions. As much as we found in the last war that capital ships must have a destroyer screen to protect them from submarines, so in this war we have found that the capital ships must have fighter protection when within range of enemy bombers.

'Whatever the progress of anti-aircraft gunnery may be, it cannot alone counter the air menace, and can only increase the cost to the enemy or force upon him the need for a heavier scale of attack.'

Boyd asked fancifully whether any means could be devised
of enabling fighters to fly off battleships and cruisers. He might
more rationally have concluded that the Royal Navy desperately needed carriers rather than battleships, but was stuck
with an embarrassment of the latter. Under enemy air attack,
the ships of Syfret's fleet would pour forth noisy, spectacular
displays of shells and tracer automatic fire; but these would
bring down enemy aircraft only at the odd lucky moment. The
guns boosted the morale of ships' crews, and deterred cautious
enemy pilots. Ships' speed and manoeuvrability, however,
provided their best defences against air attack: destroyers were
sometimes dubbed 'greyhounds of the ocean'; cruisers were
fast but less nimble. Merchantmen, battleships and carriers
were neither.

As for the men who would sail these ships into battle, in
September 1941 the opinion survey organization Mass
Observation invited its correspondents, members of the British
public, to comment upon the Royal Navy. Most declared that
they did not know a lot about it: in contrast to the show-offs of
the RAF, the 'Silent Service' prided itself upon not enlighten-
ing them. The great Cunningham wrote grumpily in his
memoirs: 'publicity is anathema to most naval officers and I
was no exception'. People nonetheless considered Britain's
seafaring arm 'anonymous and efficient'. One respondent said,
'the navy is doing a tremendous job with its usual silence and
courage'. Another said sailors did 'a lot of dull, dangerous work
bravely and well', which was not far wrong. The monotony of
sea service is hard to overstate, with months of comfortless
routine interspersed with only occasional spasms of action and
the possibility of choking, watery death.

The war brought together in His Majesty's ships a mingling

of humanity. Like members of the RAF, all were volunteers: sailors made a choice, to spend the war amid the dank, hammock-huddled mess decks of a warship, affecting anachronistic uniforms with bellbottom trousers and HMS pillbox hats held in place by coy little straps, rather than in a muddy slit trench. Following the destroyer *Lightning's* commissioning, a senior petty officer mustered the fifteen members of one mess and asked how many had been to sea before. Only three hands went up, and this was typical. Vincent Shackleton was a former clerical assistant in a Bradford wool mill. On enlistment, he wrote, 'we came together, clerks, refuse collectors, bank managers, mechanics, teachers, fitters and others representing a whole gamut of occupations.

'Within a month we had been sorted into some category or trade. We learned a new language, came to accept the discipline without question, were conditioned to carry out orders without questioning their validity, and accepted an unvarying routine. Some of the lads had never made a bed in their lives. Others were incapable of sewing on a button. They came from the far north of Scotland, from Dawlish in Devon, from the slums of Liverpool, and fifteen of them came from "the smoke" [London] with every variety of cockney accent. We had to learn to communicate, to get on together, to eat and sleep as a community, to accept the principles of fair shares for all, and preservation of a high standard of cleanliness and appearance which was gently, but very firmly, forced upon us. There were some rebels, but they were few and far between.'

For every new rating turned ashore for incompetence or indiscipline, five were lost to seasickness. Some embryo gunners proved literally gun-shy. Conspicuous characteristics of the British of that era, especially remarked by Americans, were their terrible teeth – often a complete lack of them.

Destroyer captain Peter Gretton described how he found it necessary to relieve several toothless telephonists, whose lisps 'made them unintelligible even to their fellow Merseysiders'.

The culture of the Royal Navy was formed by the experience of living at sea, in enforced intimacy, for months and even years on end. A warship was far more densely crewed than a merchant vessel, because of the requirement for men to serve her weapons and fighting equipment around the clock: a cruiser might carry eight hundred, while a cargo-liner of almost twice her tonnage would have fewer than a hundred. A warship's existence resembled that of a boys' boarding school in its noisy, odorous, often naked masculinity, which embraced officers and other ranks alike.

Many 'Hostilities Only' ratings – the civilians enlisted only for the duration of war, who dominated every warship since the navy's eightfold increase in strength – professed themselves shocked by the lack of brains they perceived among long-service men. HO signalman Harold Osborne described his messmates as 'a poor crowd … who I'm afraid turned me against naval life … They seemed solid from the neck up and all you could get out of them was the blaming of Hostilities Only men for everything.'

When *Nelson*'s paymaster organized a spelling contest to amuse off-duty sailors, the ship's Commander deplored his choice of words as much too difficult for most of the crew. At the concerts held on bigger vessels, men were roused to hysterical mirth by touchingly simple jests. One skit on *Nelson* involved a sailor asking the captain for compassionate leave because his wife was impregnable, to be corrected by the Master at Arms, who interjects, 'He means stagnant, sir!'

Their favoured games were children's games. John Somers, aboard a cruiser, wrote: 'What excitement you can have playing uckers!' – naval ludo – 'We have an enlarged version of this: a four-foot-square board on deck, dice three inches square shaken out of a bucket, the players all dressed in fancy-dress made of signal bunting. Usually one mess deck plays another, Seamen v. Stokers or Marines v. Torpedo men or the petty officers' mess will put out a team. All helps to relieve the tedium.' Many of the men who sailed with Pedestal were of an age, and disposition, to find farts irresistibly funny.

During brief interludes ashore, many officers and ratings contracted hasty marriages, prompted by their consciousness of looming remoteness and plausible extinction. A pilot on *Eagle* described his motives for joining hands with a Wren that summer of 1942: 'We both felt that life hung on a thread and the thread might part before we had tasted life's goodness. Life was for living while it was still possible. Our thoughts went no deeper. Our eyes saw no further. The long future was a hazy vision, nothing discernible. We lived only for the present.' The young man was too coy to mention sex, but this of course was a prominent motive for wartime unions, in that era before reliable birth-control methods, when many 'good' girls were unwilling to go to bed with a man without a ring on their finger. The senior pilot of one of *Indomitable*'s squadrons married the daughter of a leading local family in Jamaica during the ship's enforced sojourn there in the winter of 1941. The couple had known each other just three weekends, and the carrier sailed back to the war immediately following their brief honeymoon. When the pilot was killed, a comrade reflected cynically, 'what a super wedding for a marriage which was to last just two days, for all practical purposes'. He mused about the unknown fate of the bride.

While both the army and RAF had their intellectuals, some in the highest ranks, the navy boasted few. One such was the brilliant Admiral Sir William Fisher, who commanded the Mediterranean Fleet in 1932 and might have been First Sea Lord, a step-change improvement on wartime incumbent Sir Dudley Pound, had not ill-health obliged him to retire: Fisher died in 1937, aged sixty-two, unknown outside the service. His successors, the Royal Navy's fleet commanders of the Second World War, prided themselves upon being doers rather than thinkers.

Likewise at a humbler level, in the mess decks, sailor Richard Walker noted: 'The chief drawback of warfare is not so much the long periods of inaction, not even the infliction of discipline, but intellectual isolation. You can't go to a concert or look at a picture. If you want a book your choice is limited … You can't think, because it's time for "hands fall in" or you must go on watch. You can't talk sensibly, because the main subjects are either sport or naval personalities. Your sense of values falters and loses its balance and you do not know what is worthwhile and what is not.'

Service at sea, almost perpetually at hazard from the elements or the enemy, was a lonely, monastic, exhausting business. Women were a remote presence in hundreds of thousands of young hearts, as frequently represented in their dreams by a beloved mother or sister as by a wife or girlfriend. An officer wrote long afterwards, 'whenever I hear Vera Lynn singing her wartime songs I am carried back to the heaving messdeck of a destroyer, and see again the tired men, the sea-sick, hushed and listening, some with trembling lips and moist eyes'.

Donald Goodbrand, a telegraphist, noted that relationships aboard seemed overwhelmingly based on association in a given function, rather than upon real intimacy: 'People didn't talk much about their private lives, the world of their wives, sweethearts, parents, children or domestic circumstances ... The magical quality of the here and now took over, and thrust all other elements into the background, even photographs of loved ones. Nobody's business but one's own.' A long-serving engineer officer who had come up 'through the hawsehole' – commissioned from the lower deck – wrote: 'we lived rather narrowly and independently, as might the members of some religious order, deprived of many normal and human relationships'.

And so to leadership. It was a tribal service: the father of John Manners, gunnery officer on *Eskimo*, was acting as a convoy commodore after years of regular naval duty and retirement. Manners' three brothers were also at sea, while his sister was a Wren, and such dynasties were common. There was a significant 1939 argument between Britain's three services when an attempt was made to standardize the criteria for awarding decorations for bravery. The RAF and army agreed that awards should be made for single acts. The navy dissented, saying that such a policy would promote exhibition-ism and recklessness by warship captains. Admirals said they deplored medal-seeking officers who showed off, gratuitously risking their ships. It was for this alleged failing that the navy so deeply disliked Churchill's favourite silk-smooth seadog, Captain and later Admiral of the Fleet Lord Louis Mountbatten.

The navy's overall wartime performance was outstanding, but by no means all its officers were supermen. Peter Gretton wrote about escort captains in the early war years: 'The good regulars were desperately few ... [There were] many retired officers and many incompetents. There were a few young lieu-tenants in their first commands.' Reserve officer Alec Hughes wrote bitterly about the arrogance of some regulars: 'Right through the Navy one finds the same thing – the permanents always have an eye to the future – even if that may be to the detriment of the present execution of the war. Most of the Reserves feel as I do, I think.'

Some captains on Pedestal were warmly respected by their men, such as *Kenya*'s Alf Russell. *Indomitable*'s Tom Troubridge was a descendant of one of Nelson's 'band of brothers': rotund, irrepressibly cheerful, a surprisingly agile deck-hockey player. He knew the enemy intimately, having served as a pre-war naval attaché in Berlin: Admiral Raeder personally presented Troubridge with a piece of silver plate at his farewell dinner. More exotically, his stepmother was the sculptor Una Vincenzo, who abandoned his father, Admiral Sir Ernest Troubridge, to form one of the most famous lesbian partnerships of the century with Radclyffe Hall. A carrier pilot characterized Troubridge as 'a heroic figure as he stood beaming on the bridge, a tiny pair of binoculars balanced on his massive chest'. Humphrey Jacomb, the battleship *Nelson*'s captain, was likewise admired as a ship-handler, popular below decks because he chatted to humble stokers and seamen, though forever pacing his bridge to assuage the pain of arthritis that never gave him peace.

Lt. Cmdr. James Swain of the destroyer *Penn* once faced a strike, of seamen aggrieved by alleged injustices relating to leave. Naval police and special investigators sought the instigators of the mutiny – for such it was – of which details are unimportant; what was significant was the fact of the ignition of such anger in *Penn*'s mess decks, a refusal to accept the captain's arbitration. On another warship, rating Charlie Thomas wrote bitterly: 'I have no faith in a Navy which, at a time when the fate of our Empire hangs in the balance, prefers to give commissions to civilians whose only qualifications are social position and approved education.' Likewise Charles

Hutchinson: 'What I've seen of this war so far has opened my eyes … I'm convinced that most of the men who are supposed to be educated and act as leaders are no better than any working man in England – just appearance, that's all – and when it calls for anything else it's lacking.'

Peter Gretton, commanding *Wolverine*, was one of many officers who suffered pangs of guilt about the chasm between the dinners served to himself and his officers in the wardroom, and the coarse fare ladled out in seamen's and stokers' messes: 'we had cabins, food produced for us, a wine bar, and someone to make one's bed and wash one's underclothes … It was a very different life, though of course we had the responsibility, which perhaps is the worst strain of all'. Seaman Chris Gould wrote sourly to his adored fiancée Vera in south London, from the destroyer *Lightning*, bound for Pedestal: 'Our food has been lousy for the last week or so. The bread is off before it's made – well, nearly everything is off. The stokers' mess deck has bugs in it and our stupid first lieutenant just laughed when they told him.' Her sister ship *Laforey* had recently required fumigation, to address a plague of cockroaches.

Gould, writing to Vera again: 'we went to sea on Sunday night. I was very miserable and longed for you so much, as the conditions aboard disgusted me so much more than usual.' Others were happier. Petty officer Reg Coaker, who had been serving on a battleship, was now transferred to another Pedestal ship, the little Hunt-class *Bramham*, which he much preferred: 'I didn't want any more of this bugle-blowing and tannoy-sounding stuff.'

Those were days in which the Royal Navy still prided itself upon being officered, in some measure at least, by gentlemen. The warships escorting Pedestal boasted in their wardrooms some figures with such names as the Marquis of Milford

Haven, first lieutenant of *Bramham*; David Maitland-Makgill-Crichton, *Ithuriel*'s captain; *Amazon*'s Lord Teynham, and other impeccably blue-blooded figures. Peter Gretton was an ardent foxhunter. It would be hard to overstate the strain of leadership responsibility through month upon month at sea. One of Roger Hill's sub-lieutenants on *Ledbury* was an excellent officer, but still visibly stressed after surviving a cruiser's sinking off Crete. Hill was apprehensive about how this man would respond to another round of heavy bombing, such as they all knew must come. Some officers and indeed ratings cracked. The previous commanding officer of an air squadron embarked for Pedestal had to be sent home after losing his nerve, then shot himself on the Gosport ferry. A chief of staff to the flag officer of Force H hanged himself one night in his cabin.

A sailor who broke under the strain of action aboard a warship could not run far, but it was not unknown for men to quit their posts under fire, then scuttle below decks. In action, John Manners once found a gunlayer standing mute, passive, paralysed. Though acquitted of cowardice by a court-martial, the man was removed from his ship. Another such episode, on a cruiser during an air attack, caused her captain to display special severity at the subsequent disciplinary hearing because the man was deemed to have betrayed his class. He told the defaulter: 'the whole ship's company know your background, and they are intently watching this case. The very fact that you come from a public school makes it impossible for me to show leniency.' The man was sentenced to thirty days in the cells. Another rating was discharged from the navy before he was twenty, having been traumatized by experiencing his destroyer's sinking: he was far from alone in discovering such a breaking point.

Men's destinies in action were entirely determined by their captains' decisions: even the most senior subordinate officers were almost as much prisoners as were galley slaves of old. If a ship's commanding officer decided to be brave his sailors, unlike soldiers who might flee, had no choice save to accompany him to their common fate. It is unknown how the crew of the little destroyer *Glowworm* felt on 8 April 1940 when their captain decided to ram the huge German cruiser *Admiral Hipper*, winning himself a posthumous VC; but it seems reasonable to guess that, on a free vote, not all would have endorsed his decision. Contrarily if a captain flinched, the ship's company were inescapably complicit in his disgrace. Senior officers, especially aboard the smaller vessels, never forgot that the eyes of every man were upon them; that the example from the top was critical, to make a ship an effective fighting unit.

Able seaman Fred Jewett on *Ashanti* professed absolute confidence in his thirty-eight-year-old captain, Richard Onslow, sometimes nicknamed 'the Horse' because he affected a straw panama hat, and in the crew of 'that silly little ship … But it was a wonderful ship.' As Onslow watched bombs fall around them, 'he was so cool and calm and collected, you'd think he was going for a walk down the park'. War correspondent Anthony Kimmins likewise spoke of Onslow in his chair on the bridge, feet up, 'looking to all the world as if he was happily driving a tractor, harvesting'. The 'harvester' emerged from the war with four DSOs.

And somehow, out of this conglomeration of the willing and unwilling, the brave and not so brave, the proficient and the fumblers, the wartime Royal Navy contrived something that was often wonderful. J. P. W. Mallalieu, himself a wartime seaman, wrote in his best-selling wartime novel *Very Ordinary*

Seaman that sailors are 'dirty, unromantic and damp'. He was wrong about the middle bit. There was indeed a romance, a profound sense of the glorious history of their service, which suffused a host of those who fought its ships, from admirals to junior ratings. One of the latter, about to sail for the Mediterranean aboard a light cruiser, spent his last day ashore in Plymouth. Charlie Thomas then wrote in his diary: 'Whenever I visit … the town, I think of the thousands of sailors from Drake's time until now, who have trodden the same stones. It is impossible not to be stirred by the thought that along these narrow lanes the men went to join the ships which fought the Spanish Armada, with Blake against Van Tromp, with Rodney at the "Saints", with Howe on the "Glorious First of June", at Trafalgar, and not so many years ago at Jutland. I wonder if, in years to come, other sailors will think of us?'

Yes, they would.

2 THE CONVOY

The fourteen merchant vessels that constituted the purpose of Pedestal were chosen because they were fast, strongly constructed and available in British ports. There were to be no tramps or sluggards, such as restricted the pace of many Atlantic convoys to eight knots. Most, like *Rochester Castle* and the Blue Star Line's two ships, were much grander than mere freighters: they took pride in their aesthetic beauty as cargo-liners, albeit drab-painted in wartime grey instead of their accustomed liveries in primary colours. Each had a personality of her own, strongly influenced by the character of her Master – and, in some degree, of her owners.

Among the Pedestal ships *Deucalion* and *Glenorchy*, both owned by Alfred Holt of Liverpool, were constructed to especially high technical standards. In 1939 British owners controlled a third of the world's tonnage. If there was class distinction in the Royal Navy, among shipping companies and their employees there was an established hierarchy of respectability and importance. The Hogarth Line's men were known as 'Hungry Hughies'; Turnbull & Scott was nicknamed 'Terror and Starvation', Ellerman & Bucknall 'Eggs and Bacon'; P&O 'Penury and Ostentation'.

Merchant Navy officers on smart ships owned by rich companies enjoyed comfortable quarters and a relatively privileged lifestyle afloat, waited upon by servants. Humble seamen, stokers and greasers saw little of this, however. When they visited vessels of the US Mercantile Marine, they were awed by the good things available to their American counterparts but denied to themselves. Shipowners of all nationalities shared an unenviable reputation for cupidity, exploitation and monopolistic practices. A 1935 Board of Trade report found that the fatal accident rate per thousand men employed at sea was ten times that in the cotton industry, almost twice that of the railways, nearly three times that of coal mines. It seems no exaggeration to assert that, while maritime trade was an important success story for Britain, shipowners as a class brought shame upon capitalism, and did not become less cynical or greedy with the coming of war. This is not their story, however. It is, instead, that of those who served them, in response to the call of the sea rather than for relatively meagre cash rewards.

Britain in 1942 had around 120,000 sailors afloat in its merchant service, with another 34,000 between jobs or otherwise temporarily ashore. Some 45,000 hailed from south-east Asia or China, the latter often especially highly regarded, and there were a further seven thousand of other nationalities. A man who found himself temporarily jobless, and thus listed as available on the roster of the Seamen's Pool, had a right to decline the first two ships he was offered, but was obliged to accept the third. An experienced hand might earn £12 a month, slightly more than half the pay of an RAF sergeant pilot, or of an American seaman, and double that of a naval rating. A chief officer – the Master's second-in-command – could expect £30; cadets received only fifteen shillings.

Yet, despite its critical contribution to the prosperity of the British Empire, the Merchant Navy commanded little respect ashore. On leave, many men hid their MN badges, instead told girls that they were factory workers. Patrick Fyrth, a P&O cadet, described an angry mother expelling him from her house: 'Get out! My daughter's a good girl! She won't go with sailors.' In peacetime, most of those who went to sea did so because they lacked education to aspire to anything better – few of the service's officers would have been characterized by their naval counterparts as 'gentlemen'. In war, some men facing conscription concluded that the Merchant Navy offered a better life than did the likely alternative, infantry service. They may have regretted their choice, after experiencing the food served in most messes: lumps of iron-hard meat and floury potatoes, rice into which tins of condensed milk had been emptied, tinned carrots, swedes and tomatoes.

Seventeen out of twenty British wartime convoys never encountered air or submarine attack. Patrick Fyrth wrote: 'with the weather fine and the enemy out of range convoy work was quite pleasant. You kept distance from the ship ahead by calling down to the engine-room "up two revs" or "down two revs". The best way … was to watch the swirl of the wake ahead, which showed an instant increase or decrease in speed; at night, you could see the flame of a match three miles away.'

Those lines may suggest that war service in the Merchant Navy was a soft option. Yet seamen were required to ply the ocean year in and year out. The passage of time, as well as distance, caused the statistics measuring their prospects of survival to worsen relentlessly. More than thirty thousand men perished, roughly one in four of those who sailed. This made their roles far more hazardous than those of British and American soldiers – and of the Royal Navy or US Navy.

Much depended, of course, on what ships and which convoys men joined. Every transport company in the twenty-first-century world assures its staff and passengers that 'your safety is our first concern'. It is a fundamental reality of wars that this priority is abolished. Every man of the belligerent nations who put to sea between 1939 and 1945 did so in the knowledge that enemies would do their utmost to see that they drowned, burned or were blown up in the course of their passages. Yet the crews now ordered to sail with Pedestal were especially conspicuous losers in the lottery of destinies. Malta vied with Murmansk as the toughest destination for warships and merchant vessels alike. Whatever glory was to be won would be conferred upon the men of the Royal Navy, the warriors, rather than on themselves.

Once ordered to sail, however, most accepted their fate without much fuss. Steward Jim Parry fell into conversation with a docker, whom he told that he was about to join *Empire Hope*. 'Blimey,' the man said, 'you don't want to go on that. You know where that's going?' No, said Parry, 'and don't care either'. The docker said that he and his gang had spent the day loading cased aviation spirit for Malta. Parry and his mates shrugged 'nothing to do with us – we're just crew'. The bemused docker turned away, muttering, 'Oh, well, please yourself, mate.'

Assistant steward Ray Morton, aged eighteen, was dispatched to *Ohio* on Clydeside by the Newcastle Seamen's Pool, with a pay increase to £14 a month and £4 'War Risk' supplement. Having experienced three years of rationing, and after sailing in a succession of 'rust-buckets' since he started seafaring life as a cabin boy, he was awed by the comparative luxury of the American-built vessel: 'two-berth cabins and food we had only dreamed about. She had been provisioned in the USA and had grapefruit in the cool rooms, a dozen varieties of cereal for breakfast, followed by bacon and eggs! A whole variety of meat and fish and ice-creams!' Another eighteen-year-old, seaman Allan Shaw, enthused about the coffee machines bubbling around the clock: 'It all seemed too good to be true.'

The initial plan for the convoy nominated eleven merchant vessels, four of them veterans of earlier Malta runs – *Deucalion*, *Melbourne Star*, *Clan Ferguson* and *Port Chalmers*. Immense pains were taken with the preparation of their cargoes. Each ship was loaded with a mix of coal, aviation spirit, cooking kerosene, flour and other food, medical supplies, weapons and ammunition, so that some fraction of each vital commodity would reach the island, aboard whichever ships got through. In fields beside rail sidings near Didcot in Berkshire, the apportioned tonnages were assembled beneath tarpaulins in vast dumps, each one labelled for a specified vessel, then taken by goods trains to their several respective ports.

The Admiralty's certainty that some must be lost was reflected in the size of the convoy. Valletta was a 'lighterage' port: almost every ship reaching Grand Harbour was obliged to discharge its burden by derrick into lighters for transfer to the shore, a slow and cumbersome business even when fast-track procedures were established in readiness, as was the case with the Malta authorities' planning for Operation Ceres, the unloading of Pedestal vessels, even before they left Gourock. If all the designated ships, or even most, had got through to Malta, it would have been hard to give them effective anti-aircraft protection through the many days and nights that would have been needed to empty them.

The convoy and its offload were organized on the harsh, realistic assumption that most of the participating ships would repose on the bottom of the Mediterranean before lighterage became an issue. The sole exception to the mixed-cargo principle was the 12,843-ton Shaw Savill Line motor ship *Waimarama*, built in 1938, which had just completed an Atlantic run from New York. Now loaded with eleven thousand tons of military stores and fuel, all explosive, she became explicitly what other vessels were in somewhat lesser degree, a floating bomb. The courage of her crew, in agreeing to sail bearing such a burden, merits special respect. Quixotically, to *Melbourne Star*'s cargo were added a hundred carrier pigeons.

After the battle was over, when those involved were struggling to wash the blood from their hands, there were recriminations, especially among merchant seamen, about the poor security that attended the mustering of the convoy – cargoes on various docksides plainly labelled with their Mediterranean destination. A consignment of tubes 'For the Senior Naval Officer Malta' was loaded aboard *Deucalion*, with only the thinnest and most transparent overlay of paint. A naval officer wrote reproachfully later, 'dockers are not so foolish that they cannot deduce the destination from the nature of the cargo'.

Yet it is most unlikely that poor dock security or enemy agents alerted Axis commanders to Pedestal. Instead, a combination of common sense and decrypts of British naval signal traffic provided the information. The enemy knew that Malta was starving, that Churchill would go to any lengths to ship supplies to the island. The British sometimes like to delude themselves that only their own intelligence service was clever enough to break wartime enemy signal traffic, at Bletchley Park.

In truth, however, both the Italians and Germans read significant British and especially convoy codes and ciphers. These were almost certainly the sources of their foreknowledge of Pedestal. By the end of July Kesselring, together with the senior German naval officer in the Mediterranean, fifty-one-year-old former submariner Rear-Admiral Eberhard Weichold, and his Italian counterparts, were sure that a British convoy would soon make a dash for Malta. An Axis intelligence report stated: 'A large-scale Allied operation [is] about to break into the Mediterranean. Large merchant ships and fleet units [are] being fetched from far and wide in preparation.'

Ohio was the most important vessel committed. A personal letter of welcome was sent to Captain Sverre Petersen, the tanker's Master, on behalf of Lord Leathers, minister of war transport, following the ship's arrival on Clydeside, having borne across the Atlantic the first cargo of oil to be carried to Britain in an American bottom: 'This special United States assistance in the rebuilding of United Kingdom oil stocks is greatly appreciated and valued … I trust your stay on this side may be a pleasurable one.' *Ohio*'s strength derived from her electrically-welded hull. In trials, she had attained a speed of over nineteen knots. Twenty-three transverse bulkheads and thirty-three separate cargo tanks gave her the potential to survive damage that would have sent most other ships straight to the bottom. Since the sinking of her sisters *Kentucky* and *Oklahoma*, she was the toughest tanker afloat.

When President Roosevelt personally authorized the ship's lease to Britain for the Malta run, both Churchill's ministers and her owners the Texas oil company sought to retain the American crew, to handle her sophisticated systems. However, the US Navy's cantankerous chief, Admiral Ernest King, a committed Anglophobe, ruled otherwise: if the British wanted the tanker to run the gauntlet across the Mediterranean, their own seamen must man her, he said. It is not known why he made this ruling for *Ohio* while leaving Americans crewing two other ships.

Petersen, the tanker's Master, was replaced by forty-year-old veteran seaman Dudley Mason of the Eagle Oil and Shipping Company. Mason was the son of a chauffeur and housemaid, brought up alongside the Surrey family with which both his parents were 'in service'. He left school at fifteen, first went to sea as an apprentice in 1920. He spent many years as a mate, but had completed only one passage as an acting Master before spending three months at home in Surrey, awaiting a new posting. During that period he was divorced from his wife Elsie after eighteen years of marriage. In those days such a domestic upheaval was relatively unusual and thus traumatic. It can hardly have failed to open emotional wounds in this lean, undemonstrative Merchant Navy officer, and also perhaps explains his long sojourn ashore, before Pedestal.

Another, much more experienced captain was originally nominated for *Ohio*, but dropped out for reasons unknown – most likely unwillingness to command a supremely vulnerable vessel on such a perilous voyage – to be replaced at short notice by Mason. The new Master was obliged to accept a scratch crew, assembled from the Seamen's Pool, most from Scotland and the north of England. He was given some choice of officers, however, and chose well. The ship's American chief engineer stayed aboard the ship for several days, to brief his British replacement, forty-year-old Jim Wyld. The latter gazed in wonder at the new tanker's state-of-the-art engine-room, then declared it 'the most beautiful place he had ever seen': Wyld knew the new captain, having served with him on other ships. So had twenty-six-year-old Edinburgh man Doug Gray, who became chief officer. By 27 June, Mason's fifty-three-man complement was complete. There could be little team spirit among such a group, but they were the best willing to make the voyage.

Ohio's revolutionary welded construction made the ship strong, but further modifications were now introduced to increase the ship's survivability under attack: her two turbines were set on rubber mountings, to cushion the shock of near-misses, and the steam lines were supported by wooden props and springs. An emergency generator was installed, such as other Pedestal ships would also have profited from: loss of power, especially to keep fire hoses functioning, would prove a critical vulnerability on damaged vessels. All this work required a month in a Clydeside dockyard, and was completed only when the tanker was moved to an oiling wharf on 24 July, just before German intelligence alerted Admiral Weichold in Rome that a big allied convoy would soon enter the Mediterranean. Four days later, *Ohio* took on her cargo: 1,705 tons of diesel; 1,894 tons of kerosene – paraffin in British parlance; 8,695 tons of fuel oils, mostly aviation spirit; 1,300 tons of bunker fuel; 15 tons of lubricating oils.

Every merchant vessel in the convoy was fitted with an unprecedented battery of armament to supplement the ineffectual six-inch gun already mounted on each stern. Typically, a vessel's defences were reinforced by a 40mm Bofors with a thousand rounds of ammunition; six 20mm Oerlikons with three thousand rounds; two .50-calibre Brownings and two Marlin machine-guns, each with six thousand rounds. They were also issued with snowflake illuminant rockets, together with depth-charges to be used as scuttling charges, if it proved necessary to abandon the ship, minesweeping paravanes with cables and cutting devices for severing mine wires.

They received PACs, ingenious progeny of the prime minister's scientific adviser Lord Cherwell, which launched wires and parachutes into the sky to entangle unwary dive-bombers, though there is no evidence that they ever achieved such a feat. Armour plating was fitted around ships' bridges. Most crews were issued with gas masks, which nobody ever used, and helmets that almost every man wore in action. Each seaman who joined Pedestal, together with the detachments of army gunners, was required to sign their respective Ship's Articles – essentially, her disciplinary code. Some soldiers came aboard as mere passengers, destined to join the Malta garrison, but they were quickly disabused of any illusion that they might enjoy a Mediterranean cruise: all were earmarked to support guns' crews, refill ammunition drums or otherwise join the defence of their ships.

Three more vessels were added to the initial convoy plan at the last minute: *Wairangi*, *Empire Hope* and a second American-crewed ship, *Santa Elisa*. All were chosen because, besides chancing to be available, they were capable of sixteen knots. If such a powerful covering fleet was to be deployed, it made sense to match this with the largest possible cargo-carrying contingent. Among those least excited by their participation were the crews of the two American vessels. The Master of *Almeria Lykes*, fifty-nine-year-old Willie Henderson, born in London but long resident in Houston, told his newly appointed British naval liaison officer, Hugh Marshall, that he and his midshipman had best mess separately, because many of the Americans aboard had no time for 'Limeys'.

Gerhart Suppiger from Illinois was a twenty-three-year-old US Navy ensign, whose diary reveals his dismay at finding himself on the eastern side of the Atlantic. He wrote: 'this Europe is certainly a hellish place and no place to be in wartime. The British show very much class distinction. I certainly look forward to returning to the United States!' Before that happy day could come, however, Suppiger was obliged to assume command of a fourteen-strong American naval armed guard which, with some British Army gunners and US Army wireless-operators, joined *Santa Elisa*. The callow, gangling young officer meant well, but found his new shipmates as unsympathetic as they found him.

The ship's fifty-six-strong crew, captained by thirty-three-year-old Brooklyn policeman's son Theodore 'Tommy' Thomson, for whom this was his first command, displayed no more enthusiasm for Suppiger and his men than he did for them when he wrote: 'this outfit is a lousy bunch, drunk, greedy, jealous, hypocritical and untrustworthy – I sure will be glad to get away from them!... I think they are a bunch of misfits and loafers ... The talk about the bravery of the merchant marine is a lot of bunk. They are in it to get out of the Army and for the tremendous pay it offers.'

Suppiger noted sulkily that *Santa Elisa*'s assistant engineer earned $600 a month, while he himself, as a junior naval officer, received a mere $125. When the touchy ensign visited the US Navy Liaison Office ashore to complain about having been issued the wrong ammunition for the ship's Oerlikon guns, he was shocked to find its staff drinking hard, in female company. Suppiger's subsequent Pedestal narrative reflects his immaturity and sense of grievance. He was thus far probably correct, however, in supposing that *Santa Elisa*'s crew lacked unity, were suspicious of all things British and cherished little desire to become heroes or interest in the fate of Malta.

There was an exception, however. Fred Larsen was born in Newark, New Jersey twenty-seven years earlier to a Norwegian father and Irish mother. When he was not quite four, almost his entire close family was wiped out in the flu pandemic that swept the world in 1918–19. Thereafter, he grew up in Norway, with an uncle whom he learned to love. As a customs inspector, this man spent much time aboard ships, often with his nephew in tow. When Fred was fourteen Uncle John, too, died, leaving the teenager to fend for himself. Three years later, he became a deckhand, then attended Norway's Mates' and Masters' Maritime College. In 1939 he married his long-time love Minda, and left her behind to have their baby while he went back to sea.

In April 1940, when the Germans invaded Norway, he was a quartermaster aboard a United States Lines ship. He began working his way through a State Department paperchase, to secure the documentation to enable Minda and their baby son Jan to join him in America. Love for his wife and hatred for the Germans became the ruling passions of this tough, taciturn, driven young man's life. In May 1941, he joined *Santa Elisa* as Third Officer. Grace Lines had just commissioned the brand-new 8,380-ton freighter. Larsen's pay at last enabled him to send Minda PanAm clipper tickets from Lisbon to New York. Three months later, however, Germany declared war on the US. Fred lost all contact with his family in Norway.

In January 1942, *Santa Elisa* survived a serious fire on board, prompted by a collision with another vessel off Atlantic City. Fred Larsen, wearing oxygen-breathing apparatus, played a notable part in quelling flames in the hold, along with Tommy Thompson, then the ship's mate. The vessel was obliged to spend three months drydocked for repair. In May, now with Thompson as captain, *Santa Elisa* sailed for the British Isles. Unbeknown to the Norwegian-American, even as his ship docked, Minda and their son were beginning a trek across Europe in which, astonishingly, the Nazi authorities acquiesced. This ended with her arrival in neutral Lisbon on 11 June, a time when half the peoples of occupied Europe would have sold their souls to reach such sanctuary. Months, and a sea battle, would intervene before Fred Larsen learned of her escape. Meanwhile Gerhart Suppiger noted of the mate: 'he has a fanatical hatred for the Germans'.

Santa Elisa's crew complained bitterly as they loaded coal, filthiest and thus least popular of cargoes, at a pier in Newport, Wales, during early June. They protested even more vigorously when they were suddenly ordered to discharge it. Then, in the last days of June and first of July, *Santa Elisa* took aboard 6,200 tons of aircraft parts, bombs and mines at the bottom of her holds; sacked flour and grain above, along with coal and 1,300 tons of aviation fuel in notoriously leaky five-gallon tins, ninety thousand of them.

Meanwhile aboard *Ohio* there was a sublimely petty disciplinary squabble on 31 July, after a messman named Byrne was seen tossing plates into the sea. The ship's log noted that the accused crewman 'freely admitted that he intentionally threw those aluminium dishes over the ship's side because he did not wish to wash them. He stated that he was willing to pay for them,' a view he changed somewhat when he heard that his pay would be docked a guinea apiece. The incident was significant only in exploding the myth that crews for Pedestal were an elite. For William Isard, a thirty-year-old British Army gunner drafted to *Santa Elisa*, this was his sixth convoy. He could not, however, be described as a distinguished addition to the crew:

the Surrey man was in constant disciplinary trouble ashore, mostly for drunkenness.

Every merchant vessel was assigned a naval liaison officer, most often a 'Rocky' – reservist – together with naval or army signallers equipped with wireless, Aldis lamp and flag paraphernalia. Reservist Commander Arthur Venables was appointed convoy commodore: he joined *Port Chalmers*. Before sailing, every merchant vessel's armament and fire-fighting gear were minutely checked by naval officers.

Although most crews assumed that they were bound for Malta, some speculated wildly about other possibilities, Murmansk being the least popular. They were told that any man who wished to go ashore, rather than join a notably dangerous operation, might do so. Since, however, for security reasons such a choice required acceptance of six weeks' confinement in an army camp, incommunicado and unpaid, nobody made it. US Navy ensign Suppiger became censorious about the amount of drinking aboard *Santa Elisa*, his dismay surely influenced by the fact that two mates, having staggered back on board after a night's revelry ashore, tossed over the side a seabag containing Suppiger's laundry. On the Fourth of July, some crew had to be bailed out of a local jail, following a punch-up with British sailors. The ensign also deplored the captain's decision to allow the purser to bring on to *Santa Elisa* cases of whisky, which he later sold on board – American merchant vessels customarily followed US Navy practice by sailing dry.

Tensions aboard *Santa Elisa* rose higher when it was found that leaking petrol tins were causing a build-up of gas in the holds. Fred Larsen rigged up an electrically powered ventilator that ran twice a day to clear the fumes. Sacks of coal were heaped atop the hatches, to provide some insulation against bomb penetration. Crewmen practised stripping and firing the Bofors and Oerlikons, and tested machine-guns in a field ashore. They also practised aircraft recognition, though the crews of every ship in the convoy, Royal Navy and merchant vessels alike, were to prove lethally ill-disciplined in this respect. They conducted lifeboat drills which were especially important for the fourteen additional British gunners drafted to the ship, men of the army and Royal Marines, some of whom had never before been to sea.

In the last days of July, all the merchant vessels committed to Pedestal made a rendezvous with their close escort in the Clyde estuary. Many captains and even crews of the destroyers that joined the Malta-bound fleet had been hapless participants in PQ17 two months earlier. Cdr. 'Jackie' Broome of the destroyer *Keppel* observed wryly that it was a case of 'from polar bears to pineapples'. Roger Hill, commanding *Ledbury*, wrote: 'This was wonderful news – a chance to redeem ourselves and to go to the Mediterranean with warm weather, sunshine and a warm sea if you were sunk.' This last factor was important: whereas aircrew and seamen who survived crashes or sinkings in the Arctic or even Atlantic faced a quick death in the water, those cast adrift in the summer inland sea often lived until rescued.

Aboard *Nigeria* on the afternoon of Sunday 2 August, Rear-Admiral Harold Burrough briefed the merchant vessel captains and their naval liaison officers – each ship had been joined by her own NLO and signalling staff. This conference took place in the ship's empty seaplane hangar: all the cruisers' usual Walruses save one had been disembarked, because experience showed that the elderly planes and their high-octane fuel posed more of a fire risk than their usefulness justified. The hangar was now instead occupied by two big tables, one bearing a chart of the Mediterranean, the other furnished with crude wooden ship models.

That Sunday, there was no danger that Burrough would have an inattentive audience. Some fifty men, many middle-aged, all of them matching the formality of the naval officers in full uniform by wearing their best civilian suits, ties, shoes polished as brightly as Burrough's own, together with hats – this was still an age when every respectable man covered his head in public places. There was old Richard Wren of *Rochester Castle*, with his moustache and little goatee beard; the Americans Willie Henderson and 'Tommy' Thompson. Forty-seven year-old New Zealander David MacFarlane, 'Captain Mac' of *Melbourne Star*, had received an OBE a few months earlier for his service as a convoy commodore on the Malta run, winning the warm applause of Admiral Syfret. War correspondent Anthony Kimmins remarked of *Ohio*'s Dudley Mason that in his trilby 'he looked more like a successful draper listening to the football results' than Master of a tanker.

The gay war correspondent Godfrey Winn, who had been a horrified witness of PQ17, was asked afterwards by a friend how he had found life with the Royal Navy. Winn responded theatrically: 'admirals are heavenly, but captains make me shy'. Winn might thus have felt somewhat awed by the assembly aboard *Nigeria*. Burrough himself, who would play at least as important a role in what followed as did Syfret, had no illusions about the hazards, but welcomed the transfer to the Mediterranean for the same reason as Roger Hill, 'frankly because it was warmer'.

A bluff, burly, rugby enthusiast, Burrough was born in 1888, one of eleven children of a Herefordshire parson. He later remembered his childhood as a Victorian rural idyll with Christian overtones: his sister played the organ at their father's church, in which he himself was a chorister. He was reared to revere the navy, in which several forebears had served. Harold's elder siblings attended Marlborough, Shrewsbury or Bruton public schools, but when money ran out to pay fees for the younger children, entry into the Royal Navy through the cadet ship *Britannia* held no terrors for him.

Despite chronic seasickness, which he never entirely overcame, Burrough prospered in the service, enjoying adventures ashore as a spectator of several revolutions, notably Mexico's in 1914. That year also, he married a slight but notably forceful Canadian girl named Nellie Outkit, whom he had met years earlier as a midshipman in Halifax, Nova Scotia: the couple raised five children. Like Syfret, he became a gunnery specialist, and in 1916 served as gunnery officer of the light cruiser *Southampton* at Jutland. He was a witness to the terrifying spectacle of three of Vice-Admiral Beatty's battlecruisers blowing up in succession, 'a very terrible sight', with the loss of almost all their crews. The decks of his own cruiser were ravaged by German gunfire which inflicted eighty casualties, nearly half of them fatal, 'almost all my guns' crews'. To the end of his days, he could recite from memory the names of those men. Thus, in August 1942, he was no stranger to the carnage of battle, to sinking ships, to maimed and drowning sailors. No intellectual he, Burrough read nothing beyond *The Times* newspaper, especially its sports pages, and never fulfilled a promise to address the novels of Anthony Trollope when he quit the sea.

That Sunday afternoon in *Nigeria*'s echoing hangar, the admiral began: 'Gentlemen, it is our great privilege to be chosen to go to Malta.' Tommy Thompson of *Santa Elisa* said later: 'for a moment no one said a word. We knew Malta was at the end of its endurance, and this was the last, desperate attempt to get through.' Burrough continued: 'Don't bother about the convoy number – WS21S. That's just to confuse the Hun in case he hears about it' – Winston's Specials was the designation customarily given to convoys around the Cape – 'You're going to Malta all right, by way of Gibraltar. The operation is to be called "Pedestal".' This was soon shortened by the young and irreverent to 'Ped'.

Burrough's briefing lasted two hours, and addressed chiefly procedures and alternative formations for a variety of scenarios at sea – for instance, maintaining four or five columns at the outset, reforming into two on approaching the minefields of the Sicilian Narrows. The ships would respond to a submarine alarm – signalled by two thunderous blasts on *Nigeria*'s siren, repeated throughout the fleet – by making emergency turns. The seventy-odd ships, of which some would join at Gibraltar, having sailed around the Cape from the Indian Ocean, would zigzag day and night, while seeking to sustain a speed of sixteen knots – more than eighteen miles an hour, in landsman's language.

On the Atlantic passage to Gibraltar, all ships would intensively exercise station-keeping, aircraft operations and gunnery, a programme labelled Operation Berserk. Pedestal was scheduled to cross the Mediterranean, and especially to brave the Sicilian Narrows, between 10 and 16 August, when the nights were moonless. Burrough concluded that they expected to approach the Skerki Banks, most dangerous point of the passage, on 12 August: 'You know what the Twelfth is. That's when grouse-shooting starts. We should find plenty of birds in the Mediterranean!'

Burrough asserted later that he was 'most impressed with the cheerful and determined manner in which the Masters went out to make [Pedestal] a success'. It was true that the captains said all the right things to the admiral as they descended the gangway of the big cruiser, attended by the courtesies of Marine salutes and bosuns' pipes. Privately, however, many were sensibly apprehensive about what they were being asked to do. Thompson of *Santa Elisa* noted: 'The Admiral might have said it was our great privilege to commit suicide. But we all nodded our heads, accepted our orders, and said "Thank you, sir!"'

He and the crews of the two American vessels should be forgiven their wariness about the operation. It is a hard thing to be called upon to risk one's life among British comrades whom one has no basis in experience for trusting, in a cause that no words of Franklin Roosevelt or Winston Churchill had yet persuaded many Americans was authentically their own as well as that of the British Empire. Each captain was handed a sealed envelope, not to be opened until his ship approached the Mediterranean, containing a message from A. V. Alexander, First Lord of the Admiralty. This expressed his gratitude that they had undertaken the task and emphasized the dire predicament of Malta: 'her courage is worthy of yours … We wish you all God speed and good luck.'

Even as Burrough was briefing captains, elsewhere on *Nigeria* his senior signals officer was addressing merchant-ship wireless-operators and yeomen, both civilian and naval, about wavelengths and signals discipline. During daylight lamps and flags would be as useful as wireless, and much quicker, because ciphering was unnecessary. Flag hoists were still an important part of the navy's communication routines, not least because ships within sight of each other could read them more readily than a Morse message: a black flag for a submarine contact; Q indicating 'all ships to be ready for air attack in seven minutes'; B meaning 'prepare for imminent air attack', and so on. Throughout Pedestal, messages also blinked between ship's bridges, dispatched by yeomen clattering the shutters of Aldis lamps, the words transcribed by their counterparts throughout the fleet.

At 1600 that Sunday evening, 2 August, the two battleships, together with their screening destroyers, sailed from the Home Fleet's base at Scapa Flow, Orkney. The weather, which had been dull and stormy all day, brightened as the great warships gathered way. Churchill had sent a personal farewell message to Syfret: 'All good luck. I am so glad you have this great task to do.' *Nelson*'s commander, George Blundell, censoring seamen's mail, noted that one rating had written in a spirit that would have been mawkish had it not reflected a moving simplicity 'We are bound to do something big, because the spirit of Nelson hovers over this ship.' The cruiser *Phoebe*'s surgeon likewise wrote in his diary: 'There is great speculation about whether the Italian fleet will come out … It promises to be quite historic and quite exciting.' The destroyer *Ledbury*'s pom-pom crew yearned for a chance to fire their guns in earnest: 'Gawd, just to 'ave one of the bastards in my sight,' sighed the gunlayer, an ambition that would be generously fulfilled.

PO Reg Coaker, an ordnance artificer, felt a lingering sense of regret and even guilt that when his wife and mother had come to the railway station to see him off to join the destroyer *Bramham*, he had been too embarrassed to kiss them in front of hundreds of other servicemen: 'You are young, you never think of course that you are going to get killed, but your own mother and your own wife don't share that confidence … They have to do the waiting for a knock on the door, that telegram to say the Admiralty regrets.' He had merely waved the two women a cheerful goodbye: 'I really should have shown more affection.'

At 2000, two hours before dusk, the merchant vessels manoeuvred out of the Clyde estuary in a single column led by *Port Chalmers*, bidden farewell by siren blasts sounding from nearby ships, many of which had heard the 'buzz' about their destination. The crews represented every age group, and included several veterans well into their sixties; the youngest was a ship's boy, aged just fourteen. Three hours later, in full darkness, the immortal cry rang through a score of warship mess decks: 'Hands to stations for leaving harbour! Special sea dutymen close up!' Some great ships, together with many more small ones, sprang into life, then into throbbing motion, sailing in full darkness past Ardrossan and the lovely Isle of Arran. Thereafter Burrough's 10th Cruiser Squadron, with escorting destroyers, set its course for the open sea. Operation Pedestal had begun.

3

Sailing

The battleships and most of the cruisers and destroyers took up stations around the merchantmen north of Ireland on the morning of 3 August – more warships were to join later, some from Freetown. Thompson of *Santa Elisa* said: 'Maybe that escort should have made us feel better, but in a way it made us feel worse. It was so terribly big that the crew realized, for the first time, that we were heading into something deadly.' A Hurricane pilot wrote: 'The Malta run was about the hottest in the book, hotter (in every sense) than the disastrous convoys to Kola Inlet and Murmansk … This was not going to be a mere solemn prank at the expense of the unhappy French' – his squadron had most recently seen action against the forces of Vichy, off Madagascar.

Almost every warship was camouflaged in the Royal Navy's 'Western Approaches' disruptive shaded blue, grey and white diagonals, painstakingly designed to alter its silhouette, except *Kenya*, which was painted in a peculiar 'Mountbatten pink', invented by that officer in his 1941 guise as a Mediterranean destroyer flotilla commander. Every sailor knew, however, that whatever aid such expedients might provide in Arctic or Atlantic murk, under brilliant Mediterranean skies no paintwork would for a moment mask a ship's character from the enemy.

They were ploughing a course that took them three hundred miles out into the Atlantic before turning south, in hopes of hiding their destination from German eyes – postponing the first incoming bombs and torpedoes. Moreover, a week-long passage offered time for exercises, which every course change showed to be essential. They formed four columns, twelve hundred yards apart, each led by a cruiser or destroyer; then began to practise nautical dance steps – zigzagging, emergency turns, closing into two columns. The American *Santa Elisa* experienced special problems, because a clutter of derricks rising high above her deck forward of the bridge impeded the view of the captain and helmsman. Burrough wrote in his subsequent report: 'station-keeping at this stage was naturally very poor, but I was confident that these fine ships could with training be moulded into a well-disciplined team'.

Dudley Mason read aloud to *Ohio*'s crew the Admiralty letter wishing them Godspeed. He then started his own words with a permissible fib: 'You men have been specially chosen for this voyage ... I want no dodgers, no questions asked when an order is given ... I don't expect it's going to be a picnic. But we will have a massive escort. There might be a raid or two, but we're not going to have any trouble getting there.' His remarks prompted a few cheers, but then thoughtful silence. Many men on all the ships said later that, as they watched Scotland disappear behind their wakes, they reflected sombrely that they might never see it again. In a cinema back in Newport, several of *Santa Elisa*'s crew had seen a newsreel of an earlier Malta convoy, 'so we had an idea what we were in for'. A member of *Empire Hope*'s crew said: 'Our first inkling this was no ordinary convoy was a notice outlining the seriousness of the plight of the island and ending with the words "THE CONVOY MUST GET THROUGH".'

Operation *Pedestal*, Cruising Disposition Number 16

Leading destroyers 4,000 yards ahead of leading warships

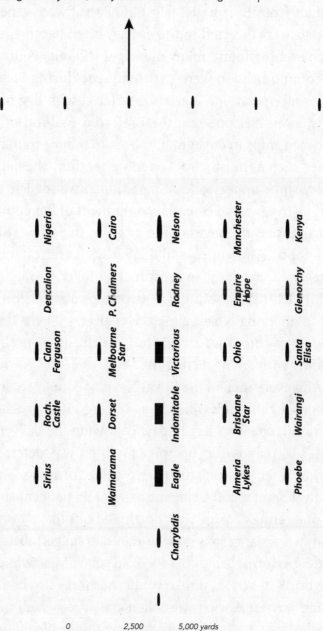

On that first day at sea, it took nine hours to establish wireless links between ships, to get everybody 'on net'. Thereafter, signals from Burrough's flagship *Nigeria* were generally read, but answers directed to the cruiser often became lost. Sunset each evening found many messages still outstanding. Failures of communication were to be endemic during Pedestal, as in all naval operations. Aerials interfered with flag hoists – still now, as in Nelson's era, the preferred method of signalling between ships in daylight, not least to reduce traffic across the ether, such as must attract enemy attention. The flag locker on every ship's upperworks, a partitioned wooden structure five feet by five, was an essential component of her equipment.

In those early stages of the passage, there was still plentiful hot food. And pipes and Player's: cigarettes sold for just sixpence for twenty, so most men smoked prodigiously. And drink: with Plymouth gin at twopence a shot. Phil Rambaut, an engineering officer aboard a cruiser, thought that some of his older comrades drank too much; he himself favoured brandy with crème de menthe.

And every man of the Royal Navy, of course, received a daily ration of rum, unless he chose to forgo it in exchange for an extra threepence a day in pay. Following the daily 1100 hours order to parade for 'up spirits', which was deferred only in action, men queued to receive their formidable tipple, the equivalent of a pub treble measure of 95 per cent spirit, which in one rating's words was 'portioned out like a holy oblation'. Only a few dangerous radicals dared to suggest that men might better perform their duties ice-cold sober. Meanwhile sailors on the brink of action, unlike soldiers ashore, also had entertainment. Jazz and swing were their passions, which meant overwhelmingly American music. Many ships sailed towards battle serenaded by Artie Shaw, Benny Goodman, Glenn Miller.

The weather was changeable, heavy seas sometimes break-ing over the bows of even the bigger ships, which dipped to meet them. The fleet's routines on Tuesday 4 August were simi-lar to those of the previous day: more exercises, destroyers taking oil from cruisers or accompanying tankers. Refuelling was among the least accomplished of the Royal Navy's skills: in the days ahead, senior officers commented unfavourably on several ships' management of the clumsy hoses and couplings, which often parted, especially when a sea was running. Somehow the job was done, however, and sufficient oil passed. That evening *Furious* left the Clyde accompanied by the cruiser *Manchester* and a Polish destroyer, setting a course to intercept the fleet.

The carrier's departure had been delayed by a crisis concern-ing her complement of Spitfires. They were fitted with obso-lete, low-thrust De Havilland propellers: a test take-off by a senior pilot caused his aircraft to plunge within inches of the sea. To make Operation Bellows feasible, smaller hydromatic airscrews, which generated more thrust, were suddenly seen to be essential. Britain was scoured for the necessary equipment, which was brought aboard *Furious* and tested with a successful fly-off. Only then could the old carrier sail to catch up with the fleet, while below decks flight crews laboured around the clock to fit the new props before reaching the Mediterranean.

Geoff Wellum, a twenty-one-year-old veteran of the Battle of Britain, was among the embarked Fighter Command pilots who would fly these fighters. Arriving from makeshift airfields in southern England, they were almost as impressed by the Senior Service's rituals as by its messing arrangements: 'what I didn't cater for is the feeling of pride that one gets when one becomes involved for the first time. Of course I'm not going to tell them that. The RAF is made to feel at home, and tea in the wardroom, all worn leather armchairs and polished mahogany, is a revelation. Loaves of freshly baked bread, bowls of various jams and as far as I can see about a pound of butter for every two people. No rationing, it seems. *Furious* is obviously a happy ship … Our naval hosts display impeccable manners, and I begin to feel better. Whatever is in store for us, this is a damn good team to be with.'

On the evening of Wednesday 5 August, *Nigeria* and *Kenya* parted company, racing ahead to Gibraltar to refuel, and also to exploit the opportunity for Force X's flag officer to hold a final conference with those captains present in its harbour, a historic rendezvous of the British Empire. A Canadian corvette, once challenged by signal lamp as she approached 'Gib', 'WHAT SHIP?', cheekily counter-challenged 'WHAT ROCK?' The destroyer *Lightning* offloaded thirty-nine survivors of a Norwegian merchant vessel, sunk by a U-boat on 25 July, and saved from the sea by the extraordinary good fortune of the fleet's passage within sight of their lifeboats.

Very few people outside an intimate secret circle within the Royal Navy knew much about Admiral Burrough's movements, or indeed about Pedestal. But among those who did was his daughter Pauline, serving in the office of Bletchley Park's naval deputy director, Commander Alan Bradshaw, who was considerate enough to keep her informed about her father's movements. Not that this added much to her happiness: 'I used to worry dreadfully about him – those awful convoys. We were all acutely conscious of the closeness of life and death.' So was her father that August night in the harbour at the gateway to the Mediterranean, which now became a wing of the stage on which he and the fleet were to act out a huge drama. During Burrough's midnight briefing of captains, he was 'his usual charming and confident self', in the words of one of his audience, but Roger Hill thought he also looked 'a little tired and stern'. How could the 10th Cruiser Squadron's admiral fail to show the strain, not forgo some of his accustomed geniality, in the face of the vast responsibility that lay ahead, in which the only certainty was that many men must die before the curtain fell?

Meanwhile at sea, far to the north-west on the morning of 6 August, fog delayed exercises, sometimes reducing visibility to four hundred yards. The fleet made several emergency turns in response to asdic contacts. Wireless-operators – 'sparkies' in Merchant Navy slang – again practised signals procedures. 'Munich' was the radio codeword for an umbrella fighter screen, while 'sober' indicated outgoing aircraft, 'blotto' incomers. There was a constant blinking-light signal traffic between vessels, and the flagship found herself short of Aldis lamps.

Moreover, some telegraphists were frankly incompetent: *Ledbury*'s captain received one signal from *Nigeria* supposedly ordering her to proceed to the Orkneys and Shetlands, an order fortunately ignored. Additional difficulties were caused by the issue of radio-telephoned manoeuvring orders, which were repeated minutes later by encrypted Morse wireless signal, 'thus causing misunderstandings and confusion' in the words of an irritated merchant-vessel captain. From first day to last, the most reliable means of communication between the bridges of Syfret's fleet was for ships to close alongside each other, so that captains could converse through loudhailers.

Almost all the merchantmen had been fitted with electric buzzers to enable Masters to alert gun positions to imminent attacks, but only *Ohio* had telephones. They were warned to expect approaching enemy aircraft to make feints and stage diversions to distract gunners. Naval officers impressed on merchant-ship Oerlikon crews 'the importance of not allowing their vigilance to relax when their side was not engaged, but to expect attacks either synchronised or in succession from all sides', which indeed happened. On 6 August guns were tested, creating an impressively hellish din, though *Rochester Castle* discovered that her Bofors was capable only of single shots rather than automatic fire. Many of the weapons on the merchant vessels were in the hands of seamen accustomed to perform wholly different tasks: *Melbourne Star*'s Oerlikons, for instance, were manned by chief refrigerating officer Charlie Almond and his assistant Bill Dearsley, lamp-trimmer Fred McWilliam and storekeeper Hugh McNeilly.

Even as the merchant-ships reduced speed to twelve knots to conduct their own training, Rear-Admiral Lyster led away his carriers to conduct a three-day programme of flight exercises, which began shambolically. The cruiser *Phoebe*'s captain signalled a friend in a destroyer: 'Keep well clear of me, I am dizzy already chasing this crazy flattop.' It was unprecedented to have so many carriers operating together. There was chaos as planes of four ships' air groups circled, attempting to receive and respond to VHF voice radio direction from their respective parent ships. Both in these exercises and later in battle, the pilots found enemy aircraft hard to spot: the sky is a huge place.

George Blundell, Commander of *Nelson*, who had been born in the reign of King Edward VII, wrote with awe in his diary about all the new-age technology fitted to his ship. His generation sometimes struggled to comprehend radar – RDF, as it was then known. In their ignorance, they were prone to endow its 1942 manifestation with almost mystical powers: 'When an aircraft comes in the director is trained on it, the pip comes along the "Scan" and when the pip is in line with the range mark previously calibrated, the RDF operator presses his trigger. Being down in the office, it is an uncanny feeling watching the scan, hearing the hum of the transmitter valve blowers, seeing a man press a trigger, hearing the six-inch guns go off. Properly calibrated and operated, it ought to be an aircraft every time!'

It was not, of course. Moreover, the technology for vectoring Hurricanes on to enemy aircraft was also highly fallible. *Victorious*'s captain admitted: 'fighter direction is in its infancy'. It had been intended that the carriers' planes would maintain wireless silence in the air, but instead there was a hubbub of chatter, some of it intemperate. Four Fulmars and one Martlet were written off by pilot error. This was a normal rate of loss in carrier operations, inherently dangerous even before the enemy entered the story, but such attrition drained Syfret's fighter strength. Moreover, the low speed and poor rate of climb of *Victorious*'s Fulmars caused Lyster to allocate them the role of low cover, while the Hurricanes and Martlets of *Eagle* and *Indomitable* assumed responsibility for tackling enemy aircraft at high altitude – if they could get there fast enough.

The personalities of one Hurricane unit aboard *Indom*, the Fleet Air Arm's 880 Squadron, must here do duty as typical of the two hundred-odd aircrew of 'the Branch', as the FAA was known, embarked for Pedestal. Most were accomplished pilots, but lacked experience in the skill that mattered most – combat. Hugh Popham had just completed a Cambridge law degree when war came, and he embraced the navy. Although by August 1942 he had accumulated many hours airborne, he had seen no action save against the French. He and most of his fellow pilots detested their huge, red-bearded South African CO, Francis Judd, universally known as 'Butch' – 'the Butcher'. Anger was Judd's default emotion, often lavished on his pilots' shortcomings: 'his temper was volcanic, his ferocity a matter of legend'.

Brian Fiddes, Popham's flight leader, was 'a tall, flexible, rather foppish-looking chap, with strawy hair and protuberant, pale-blue eyes. His voice matched his appearance; it was affected, blasé, the instrument of an amused indifference: "Never a dull, old boy" described everything from a near-miss by a bomb to a raspberry from the CO or a barrier-prang'. Steve Harris was 'a flashy pilot for all his bloodhound-look and almost imperturbable good humour'. Twenty-one-year-old 'Boy' Cunliffe-Owen was 'rich, spoilt, slovenly and usually late, a sheaf of unimportant vices carried with disarming and ruthless charm'. 'Boy' shared Popham's cabin, 'and on the rare occasions that he used a toothbrush or hairbrush, used mine'.

Johnnie Forrest, a celebrated rugby player, possessed 'an indestructible boyishness', forever apparently auditioning to become a school head prefect. Jack Smith was a geologist by inclination, a pilot only by wartime obligation. Dickie Howarth was 'the squadron's licensed clown … he … capered as he pleased'. Just one pilot, 'Bungy' Williams, was a non-commissioned petty officer; thus, shamefully, he alone was denied social access to the wardroom.

Beyond such a miscellany, almost every fighter unit in the world boasted one or two accomplished killers, responsible for most of its collective score of downed enemy aircraft. In 880 Squadron this was twenty-seven-year-old Dickie Cork who, while on loan to the RAF in 1940, had earned a DFC as wingman to the legless fighter leader Douglas Bader. Popham described Cork as 'an immaculate pilot, with the working of the squadron at his fingertips, he moved with the radiance of the head prefect about him, taking the worship of the lesser fry for granted. He treated Butch as if he were a woolly, temperamental housemaster, and no one ever saw him drunk.' Scarcely any of these young aviators would survive the war: the mortality rate among 'Branch' pilots was higher even than that of their shore-based counterparts.

Landing-on was the skill whereby everybody, including off-duty aircrew spectating from the 'goofers' gallery', a cramped weather-deck high on the carriers' islands, judged pilots. Incoming aircraft aspired to succeed each other in touching the flight-deck at ten-second intervals, but a twelve-second pause was thought pretty good. Hugh Popham wrote: 'To achieve it, the aircraft had to back each other up so closely that one was just approaching the round-down as the next taxied over the barrier into the park for'ard. This required careful judgement in the air, quick work by the hookmen in clearing the wire and by the flight-deck engineer in lowering the barrier as soon as the aircraft was safely arrested ... The margin of time between success and failure was about two seconds.'

It was the thrill of danger, much greater than that of a Grand Prix motor race, which drew spectators to carrier flight operations. A key figure was the deck-landing officer, always dubbed 'Bats', whose skill was critical to guiding planes down. A pilot said: 'We were using batsmen waving two paddles, indicating whether one should lower or raise a wing; whether we should lift the whole aircraft higher or lower. He stood at the after end of the flight-deck in one of the catwalks, looking down the wake and looking at our approach, making his judgement as to whether the aircraft was sitting at the correct angle of glide path. They were very gallant ex-pilots who did this job, [for] there were odd occasions when aircraft hit the rundown and removed the batsman in the subsequent crash.'

A plane which failed to catch an arrester-wire with its tail-hook smashed into a collapsible steel-mesh barrier halfway down the flight-deck, sometimes with fiery consequences.

That afternoon's activities concluded with a succession of mock assaults on the convoy, every ship training her unloaded guns in response. At 1715 the air groups demonstrated dive-bombing; at 1730 torpedo attacks; at 1745 low-level strafing; at 1800 medium- and high-level bombing, finally at 1815, a combination of all four. Destroyer captain Peter Gretton recorded that 'one attack by a squadron coming out of the sun caught us completely unprepared, and was a sharp lesson both to radar-operators and to look-outs'. After the flying display, in Hugh Popham's words: 'as dusk drew in, the great concourse of ships turned to the eastward ... The skipper [Tom Troubridge] was in the best of humours, the substantial image of confidence. If he too was visited with qualms like the rest of us, he showed no sign of it; with his tremendous chest and profuse grey hair and invincible bonhomie, he seemed to steady the very ship herself. Whatever demands were made on him, he would meet them with the same vast Olympian calm.'

From 1100 on the 7th, Albacores maintained anti-submarine patrols around the fleet, though there is no evidence that in the Mediterranean they injured, detected or even deterred any Axis submarine. At 1500 that day, the last carrier joined. Aboard her, Spitfire pilot Geoff Wellum felt a surge of emotion: 'Meeting up with the convoy is a moving and unforgettable experience ... history in the making ... the lines of ships all in perfect formation. *Furious* makes her number and we take up position with our faithful watchdog *Manchester* still in position astern. This is truly a mighty force, the greatest concentration of aircraft-carriers the Royal Navy has ever mustered … On either quarter two little destroyers, lithe and brave, press on regardless through the waves, sometimes disappearing from sight in a smother of blown spray, then reappearing on top of a wave crest … There is a certain cheekiness about them and all the time good old *Furious* thunders along steady as a rock.'

On 8 August there were more manoeuvring exercises for the merchant-ships and their close escort, before they were rejoined that evening by the carriers and battleships. Amid noisy demonstrations of the ships' combined anti-aircraft barrages, on flight-decks the leading edges of all the fighters' wings, together with their tails, were painted bright yellow, in hopes of averting friendly-fire tragedies, though events would show that this precaution availed little. The elderly *Argus*, known to the navy as the 'dittybox' because of her inelegant shape and horrible rolling habits in a seaway, parted company from the fleet, having fulfilled her function; she put in to Gibraltar, and played no further part in Pedestal. Five Hurricanes went with her, their pilots having shown themselves too green to be committed to battle.

The cruiser *Phoebe*'s surgeon wrote in his diary: 'this atmosphere of suppressed excitement and speculation is rising … I find it hard to believe there is not another purpose in the operation? The Italian fleet? The wardroom fervently hopes so, but can't believe they will come out and fight.' The general mood was cheerful, upbeat, probably more so among lower ranks than among senior officers, who were most conscious of the blood and tears that lay ahead. Roger Hill wrote: 'The prologue is done, the orchestra has played the introduction; everything possible has been planned and the vast audience are in their seats … The stage lights are on, the curtain rises, and as daylight broke we saw the players, steaming on a calm, deep blue sea, with a clear light blue sky above.' Hugh Popham recorded: 'everyone knew what to expect, and the suspense had a stillness at the heart of it … no one knew what special demands would be made on him in the inevitable battle, what particular contingencies he was going to have to meet. Excitement, fear, suspense were a physical thing, tickling the skin, and danger something to be made light of. The mood on board was cheerful, resolute, taut as a wire.'

That Sunday the 9th, the bigger warships allowed those men off-watch to gather for worship, 'a very simple service', in the words of one of the congregation on *Nigeria*, Commander Anthony Kimmins, 'asking for help and guidance during the next few days. At the conclusion "God Save the King" was sung as I have never heard it sung before.' Forty-one-year-old Kimmins was a privileged spectator. A former career naval officer, he had left the service to become a popular playwright and screenwriter, specializing in light comedy. Recalled to uniform on the outbreak of war, as a fluent broadcaster he was at first permitted, later warmly encouraged, to accompany and report on operations at sea. For Pedestal, he was attached to the staff of Admiral Burrough, becoming one of the most vivid chroniclers of the fleet's ordeal.

George Blundell, on *Nelson*, intervened to prevent the battleship's new chaplain from delivering an eve-of-battle religious homily to the ship's company: 'I can't think of anything more calculated to remind the troops of death.' He considered *Nelson*'s man of God 'wet as a flat fish'. Yet Blundell himself, in the privacy of his own thoughts, and like most of the men embarked upon Pedestal, cherished acute apprehension. That night, he climbed up through the superstructure to stand alone on the otherwise deserted admiral's flag bridge, gazing out into the blackness: 'I felt indeed that some of our party were entering the Narrow Seas on a desperate venture, and prayed to the Ruler of Destiny for his favour.'

On arid, sun-baked Malta that day, RAF fitter Peter Penrose wrote in his diary: 'we are now in a state of emergency again while the convoy is coming in, also working emergency hours as well, 7.30 to 7.30 with no half-days off. Still as long as the convoy gets in, that's all that matters. If this one fails, we shall be in the cart, I'm thinking'.

At midnight on Sunday 9 August, a week after leaving Gourock, Admiral Syfret's ships entered the Straits of Gibraltar, passing between scores of offshore Spanish tunny-boats. Two hours later they met thick fog, which cleared only at 0500 on the 10th. The fleet first experienced a problem that would beset the operation. In poor visibility or darkness, and lacking radar, Merchant Navy captains instinctively sought searoom, a safe distance between ships, in accordance with lifelong experience. Now, however, it was vital they should remain tightly closed up, at the designated intervals of six hundred yards; above all, that they should not reduce speed and straggle. Fifteen knots was fast for merchant vessels, yet painfully slow for a fleet facing constant peril. A naval report later noted: 'the tendency seemed to be for the ships to keep 5 cables' – a thousand yards – 'or more from next ahead, instead of three [cables], and I do not think ships fully appreciate the vital necessity of good station-keeping at all times. The tendency to lag astern was most apparent during dark times.' As Pedestal entered the Straits, *Rochester Castle* found herself two miles to starboard of the convoy, and rejoined only at 0300. By dawn on 10 August, however, and fortunately without mishap, the ships' ranks had reformed. Half the guns' crews were ordered to second-degree readiness, donning helmets, together with white anti-flash hoods and gauntlets.

Syfret had cherished hopes that the poor visibility during the night might shield his fleet's passage from German eyes. He was disappointed, however: Abwehr watchers at Ceuta in Spanish Morocco employed sophisticated, indeed revolutionary, infra-red technology to monitor all British shipping traffic. The Ceuta station was a link in the so-called Bodden Line of German intelligence monitoring posts, built on Spanish territory on both sides of the Straits, which for almost a year had caused frustration and fierce anger to the Royal Navy. Several of the Bodden stations were equipped with bolometers created by Zeiss, and designated *Wärmepeilgerät* 60s, with a range of between six and twelve miles, depending upon weather conditions.

The Bodden Line's posts were linked by radio to the Abwehr station based at the German Embassy in Madrid, which in turn enjoyed direct radio connections to Rome and to Luftwaffe headquarters on Sicily. Back in June, they had monitored and reported the passage of the Harpoon convoy. The Kriegsmarine afterwards emphasized the value of this alert in enabling German forces to wreak havoc.

The Royal Navy persistently urged direct action to put a stop to the flagrant breach of Spanish neutrality which the Bodden Line's existence represented. Churchill concurred, and in January 1942 Special Operations Executive supplied two anti-Franco Spaniards with explosives which they used for Operation Falaise – demolition of the Abwehr's Tangier monitoring station. This did the navy little good, however, since other posts remained operational. Both the Foreign Office and SIS – SOE's bitter rival intelligence organization – opposed further sabotage or commando operations on Spanish soil, on the grounds that these would escalate tensions with Madrid. They preferred diplomacy, backed by the implicit threat of a

British blockade. Fear of being deprived of fuel and food imports, on which Spain's existence depended, was indeed the principal reason why its dictator Gen. Francisco Franco had prudently declined to become the open ally of Hitler.

On 27 May 1942, British ambassador Sir Samuel Hoare had an ill-tempered personal confrontation with Franco at which he demanded the dismantling of the Bodden Line. The Caudillo responded blandly that German civilian technicians were merely assisting the construction of his own coastal defences, that the Bodden Line comprised Spanish and not Nazi installations. Under further British pressure, his ministers agreed that the German personnel should withdraw, but in the event Admiral Canaris' intelligence officers continued to man the observation posts, and the bolometers, on both sides of the Straits. Only in 1944, when the outcome of the war was no longer in doubt, did the Spanish belatedly withdraw some of the facilities they had for so long accorded to the Third Reich.

On the night of 9/10 August 1942, the Ceuta station detected and reported the passage of Pedestal, though it failed to provide accurate details of the fleet's composition. Axis commanders believed that USS *Wasp* was at sea with Syfret, and also much overestimated the size of his cruiser force. On the 10th more warships joined or rejoined the fleet through the first miles of its Mediterranean passage. Most had been refuelling at Gibraltar – the carriers *Indomitable* and *Eagle*, the cruisers *Cairo*, *Charybdis* and *Manchester* and a bevy of destroyers.

Syfret's force now assumed its ordained cruising formation. The destroyers formed an outer screen, sailing ahead and on the flanks. The average age of the crews of these escorts was twenty-four, which helped to explain the informality that prevailed aboard most of them, by comparison with the big ships. *Ledbury*'s captain Roger Hill joined the navy in 1927 as a fifteen-year-old cadet, and overcame a reputation for indiscipline in harbour, if not at sea, to assume command of the little Hunt-class destroyer eighteen months before Pedestal sailed.

He was a notably eccentric, colourful, impulsive, brave, irrepressible figure – Eddie Baines of *Bramham* described him, not unaffectionately, as a 'ragamuffin' – who liked to do things in his own way. Hill was fond of tinkling a piano presented to his ship by the famous comedian Leslie Henson, and was accompanied at sea by a tiny grey and white terrier bitch. Hill wrote: 'sailors are lonely and sentimental creatures, and lavish affection on any kind of pet, so that she was thoroughly spoiled'. He dubbed his three twin four-inch gun turrets Pip, Squeak and Wilfred, after cartoon characters of the day. Yet the ship's doctor said long afterwards: '[Hill] wasn't very popular with the ship's company ... [They] called him "Phyllis" and it wasn't a term of endearment. He used to get tantrums, swearing at the ship's company and the gunnery officer. That doesn't go down very well.'

Escorting another convoy less than a week before joining Pedestal, Hill joined several members of his crew who hurled themselves into the sea in an attempt to rescue the crew of an RAF Sunderland flying-boat, which crashed nearby, after being hit by friendly fire. Accustomed as were the ship's company to the eccentricities of their captain, most recognized that his personal rescue bid was a foolhardy act by the man responsible for the direction of their ship, together with the lives of all who sailed in her. The war had already cost the Hill family Roger's twin brother Christopher, killed ten months earlier when a U-boat torpedoed the battleship *Barham*. His former first lieutenant had recently been relieved of duty after being diagnosed with mental health problems, and Hill had little confidence in his replacement. He mustered *Ledbury*'s crew before they joined the fleet and told them: 'You know what happened on PQ17; as long as there is a merchant ship afloat, we shall stay with it.'

The main body of the fleet was arrayed in five columns, led in the centre by Admiral Syfret on the flag bridge of *Nelson*, followed by *Rodney*, the four carriers, with *Charybdis* bringing up the rear. No writer has improved upon Winston Churchill's description of battleships at sea as 'gigantic castles of steel', their bows dipping as they advanced in stately procession, 'like giants bowed in anxious thought'. Everything about these behemoths was on a heroic scale – the nine vast sixteen-inch guns mounted on each ship forward, the 120 tons of fresh water consumed each day by her crew of 1,360 men, the twenty-five tons of fuel devoured each hour by her engines. Where for centuries the Royal Navy had fought its wars behind what it called 'wooden walls', now men watched, ate, slept, prepared to fight in echoing, unyielding, yet astonishingly vulnerable masses of painted steel, softened only by a few wooden after-thoughts of furniture, canvas and planking, together with the flesh of men. Before the war ended, *Rodney* had steamed 160,000 nautical miles in the service of the King, a remarkable record. However, the strategic utility of most of this great odyssey remains an open question.

To port and starboard of the battleships steamed two columns of merchantmen, each led and tailed by a cruiser. Three more destroyers kept station behind the fleet, watchful for submarines that sought to dog its course, and the ocean tug *Jaunty*, designated as a rescue craft, though it was already apparent that she was too slow to keep up. Two oilers followed behind, receiving destroyers two at a time, passing to each perhaps a hundred tons of fuel in a couple of hours.

A *Daily Express* war correspondent, Norman Smart, gazed in wonder from the cruiser *Cairo* at the assembly of ships ten miles wide, stretching 'almost to the blue bowl of the horizon'. He mused lyrically, yet by no means fancifully: 'When this war is a misty memory in the minds of old men, they will still talk of the convoy for Malta which entered the Mediterranean in August 1942.' Syfret dispatched a Churchillian signal to all ships of his fleet: 'The garrison and people of Malta who have been defending their Island so gallantly against incessant attacks by the German and Italian air forces are in urgent need of replenishments of food and military stores. These we are taking them, and I know every officer and man in convoy and escort will do his utmost to ensure they reach Malta safely. We may be sure that the enemy will do all in his power to prevent the convoy getting through and it will require every exertion on our part to see that he fails in his attempts ... Malta looks to us for help. We shall not fail them.'

The rest of that day, 10 August, passed amid continuous tension, with every British and American ship at the highest pitch of expectation. The Axis gained their first pinpoint on the composition and course of Syfret's fleet not from its own aircraft, spies or U-boats, but instead from a Frenchman. It is sometimes forgotten how deep was the hatred many of his nation, in those years, nursed towards the British, intensified by the Royal Navy's bombardment of the French fleet at Mers El Kebir in July 1940, to prevent it from falling into the hands of Hitler. Churchill observed irritably during the 1941 campaign to occupy Syria that he wished the French Army had fought as vigorously against the Germans the previous summer as it did against Wavell's advancing forces.

Only a few months before Pedestal, Admiral Syfret had himself commanded naval operations to pre-empt the Japanese by seizing French Madagascar, where again Vichy's forces fiercely resisted the British on land, at sea and in the air. Now, around 1700 on the afternoon of 10 August, the pilot on a routine Air France shuttle flight from Marseilles to Algiers, Commandant Marceau Méresse, radioed from his flying-boat above the Mediterranean that he could identify below at least thirty-two ships, including two battleships, two carriers and two cruisers, of which he obligingly provided the position and course.

This intelligence, which added some precise detail to the earlier Abwehr sighting report, electrified the German and Italian high commands when it was forwarded to them by the Vichy Armistice Commission. At first, both Axis headquarters were so impressed by the size of Syfret's fleet that they believed it could not have entered the Mediterranean merely to supply Malta. Was this to be an allied invasion of French North Africa? Or perhaps another attempt to move reinforcements to Egypt by the direct route, rather by the tortuous transfer of men and tanks around the African continent?

Gen. Ugo Cavallero, C-in-C of Italy's Comando Supremo, feared an allied amphibious landing, and cancelled a scheduled trip to Libya. Both the Germans and Italians expected the carrier aircraft to launch pre-emptive strikes against Axis airfields on Sardinia, though in truth such an initiative far outreached Admiral Lyster's remit or capability. Admiral Luigi Sansonetti argued that, with so many carriers at sea, the British might simply be intending a mass fly-off of fighters for Malta.

What was certain was that something big was coming. To meet any of the above contingencies, every squadron and warship in the theatre was ordered to readiness. Following the late July intelligence alert about a British naval thrust, seven German and Italian U-boats had been ordered to form a patrol line in the western Mediterranean, while a further twelve Italian boats established a second line further east, around the Skerki Banks. South of Cap Bon, more than a score of German and Italian fast torpedo-craft were to prepare a night ambush. On 10 August the German 3rd E-Boat Flotilla's commander, Kapitänleutnant Kemnade, flew to Rome to plan the forthcoming battle – he had consulted with Rommel, in North Africa, a few days earlier. Captain Loycke of the German naval staff directed five boats to Empedocle, a Sicilian port. Two boats at Mersa Matruh, Libya, were ordered to prepare to move east to attack a second British convoy, thought to be sailing from Alexandria.

Access to most of the Mediterranean shores gave Kesselring enviable flexibility to move his strike aircraft at short notice between bases. As soon as the British were confirmed to be at sea, a Stuka dive-bomber wing was transferred from Libya to Trapani, Sicily. Twenty-eight of the Luftwaffe's formidable Junkers Ju88s, some of them torpedo-bombers, were redeployed from Crete and the Aegean to Sicily. More Ju88 bombers and torpedo-bombers were moved from southern France. An Italian pilot who watched the arrival of his German counterparts wrote sympathetically: 'No time-outs were on offer for them. Many had been flying since the outbreak of war, some had experienced Spain, the Battle of Britain, North Africa and Russia, their only breaks short leaves or hospital time. They were assuredly very experienced, but also very tired.'

Meanwhile Mussolini's Regia Aeronautica redeployed from the Italian mainland scores of bombers and escorts to airfields on Sicily, Sardinia and Pantelleria. Fighter pilot Pier Paolo Paravicini landed at Gela with his squadron, earmarked for operations against the British fleet. He was nonplussed when a mechanic immediately asked if he had brought spare spark plugs for his engine – they wore out very fast. Of course Paravicini had not – pilots assumed that vital equipment of this kind would be provided by the logistics staff: 'This makes me tremble for the future of the war effort of my country on which all my hopes are pinned.' In total, more than six hundred Axis reconnaissance aircraft, fighters, bombers, dive-bombers and torpedo-bombers were being arrayed beside a score of runways across the region, armed, bombed-up and fuelled, committed to confront Pedestal.

Only the role of the Italian surface fleet remained uncertain. Throughout the ensuing three days, much of Mussolini's impressive navy stood at readiness. Three heavy and three light cruisers, with an escort of twelve destroyers, were ordered to readiness to attack Syfret's ships south of Pantelleria: only the Italian battleships, lacking fuel, were to stay at home, though the British were denied this knowledge, which would have much relieved their apprehensions. Italy's naval chiefs hereafter anguished, as they had anguished since the war began, about whether to regard the confrontation that now beckoned as offering them a historic opportunity or a mortal threat.

4

First Blood

1 HUNTERS

Paul Theroux has written: 'The very word Mediterranean signifies sunny skies and balmy weather, and for thousands of years these shores have been a kind of Eden, fruitful with grapes and olives and lemons.' Lawrence Durrell provided a similar reflection: 'The Mediterranean is an absurdly small sea. The length and greatness of its history makes us dream it larger than it is.' This narrow expanse of water between two continents has provided one of mankind's foremost settings for trials by battle – Salamis, Actium, Lepanto, the Nile and innumerable lesser encounters. Its bed is strewn with more wrecks of war, accumulated over millennia, than any comparable space on the planet.

Tuesday 11 August dawned as blue and brilliant as only the Inland Sea at high summer can be, with funnel gases wafting a slight haze across the horizon. After a week of forging through Atlantic breakers under troubled skies, the glassy calm of the Mediterranean seemed to Syfret's crews unreal, and certainly unhelpful to their prospects. For miles over the water on this almost windless morning, men could hear fighter engines warming aboard the carriers. Thereafter, on every flight-deck four planes stood at continuous readiness, pilots strapped in. Below, in the hangars, more aircraft were crowded wing to wing. A pilot wrote of these nether regions: 'The air has a dead, flat taste, and stinks of oil and petrol; and this smell, a sickly cloying smell, seems to condense onto the metal surfaces of the deck and the aircraft and the tools in a tacky, black film.'

On the merchant vessels, Bofors and Oerlikon guns' crews fell in before first light, and would not be stood down until darkness fell once more. Lifeboats were swung out over the sea on their davits, ready to be lowered within seconds if a ship was hit. Above warship superstructures, radar scanners pulsed, invisibly alive. 'Submerged beneath the surface inaction,' wrote an officer, 'men pored over their sets, listened intently to the crackle of their headphones, peered through their binoculars with unblinking, rapt vigilance.' Every man knew that 'sooner or later the peace would be shattered; and nerves, jumping at every pipe, at every change in course or revs, screamed out for it to happen and be done with'. During the first two days in the Mediterranean, much energy and air effort were wasted chasing shadowers which failed to identify themselves by transmitting radar signatures, and turned out to be British – Hudsons or Sunderlands flying from Gibraltar. Syfret wrote crossly later: 'It cannot be emphasized too strongly that the RAF were a definite hindrance to the operation.'

There was no doubt that the enemy would come, uncertainty only about exactly when, where and in what guise – by air or submarine or surface warship. On the previous day photographic reconnaissance photos of Taranto, landed and processed at Malta, showed a tanker alongside an Italian battleship, and later the huge vessel was spotted being towed towards the harbour mouth. There seemed a real possibility that she was being prepared to come out and fight.

Bletchley Park provided the navy with plentiful general information about the Luftwaffe's deployments, but could offer little that assisted fighter-direction officers on the carriers. The official historians of wartime intelligence wrote: 'There was a great volume of Enigma evidence about the high priority the German Air Force gave to stopping this convoy, but very little of it was of immediate operational value.' There is a twenty-first-century delusion that Bletchley's codebreakers gave the allied high command an ever-open window upon the motions of the enemy. This is untrue. Ultra indeed provided critical information, but its output was patchy and often behind the curve of events.

Syfret's fleet was informed, for instance, that while Italian submarines were stationed in the western Mediterranean, the only German U-boats in the region were in transit to and from the eastern sector. In reality, three German as well as five Italian submarines had been deployed in anticipation of its coming. One of the former, *U-331*, was damaged by an RAF Hudson on 8 August, and returned to La Spezia to effect repairs and take off two wounded men. Though the boat sailed again four days later, it failed to engage the British ships. The remaining seven submarines, however, were ready and waiting. Indeed, an Italian boat had already made the first attempted attack, of which they were oblivious: at 0340 on the 11th, Lt. Gaetano Targia's hydrophone-operators on *Uarsciek* reported heavy propeller noise, which grew steadily closer and louder. Just before dawn, at 0442 sixty miles south of Ibiza, Targia launched three torpedoes, intended for *Furious*.

The submarine captain reported the carrier sunk, and this dramatic news was broadcast on Italian radio that same evening. In truth, *Uarsciek*'s torpedoes missed, but false claims would become a commonplace of the coming battle, from both sides. Targia claimed afterwards to have been heavily depth-charged before he and his boat escaped from the area, but no British ship logged such activity, though a corvette reported hearing explosions. The probability is that Targia made some sort of long-range attack, which in the half-light went unnoticed by lookouts or listening watches aboard Pedestal's screen warships.

Peter Gretton of *Wolverine* wrote: 'it was hard to keep the asdic operators alert during long uneventful passages. The noise of the "ping" ... is a natural soporific and my sympathies were with the men who had to keep awake. I found that only by continual "briefing" by the Officer on the Bridge was it possible to keep the operators interested and up to the mark.' 'Briefing', that morning, meant impressing upon duty ratings the certainty that U-boats were lurking, groping for gaps in the defensive screen. Captains urged on their men that the difference between success and failure in anticipating an attacker was a matter of life and many deaths.

Roger Hill said: 'The ping from the bridge loudspeaker was the background to one's thoughts and senses. I could hear [it] in my cabin, and if there was an echo on the end of the ping, I was on the bridge before the officer of the watch had called me.' Yet the western Mediterranean was a bad place for asdic. The intermingling of its warm, relatively shallow waters with those of the chill Atlantic produced hydrophonic anomalies that created false echoes. In the hours and days ahead there would be repeated alarms of submarines ... and also failures to detect predators that delivered death.

While sea power played a critical role in the Second World War, gunfire clashes between rival fleets, such as characterized the naval conflicts of history, did not. In European and Pacific battles alike, most outcomes were decided by the success or failure of air and submarine attacks upon warships and merchant vessels, which would dominate the Mediterranean clash of August 1942. Undersea warriors of all nationalities lived and died in a squalor unique among sailors. Yet most accustomed themselves to the privations, even took a perverse pride in them. 'Serving in submarines is like being a member of a Service within a Service. Its members are bound together with a camaraderie unequalled elsewhere,' wrote a skipper.

Moreover, the thirty-odd German, Italian and British boats deployed across the Mediterranean in August 1942 were to play roles in the saga of Pedestal out of all proportion to their size. Lt. Alastair Mars, captain of *Unbroken*, then at sea off Sicily, gloried in the might of his damp, stinking, six-hundred-ton steel tube, its maximum underwater speed just eight knots. He wrote of his first docking at Gibraltar: 'We passed the towering sides of *Renown*, *Rodney* and *Eagle*, bos'ns' pipes shrilling in salute, smug in the knowledge that despite their comparative enormity, we could destroy any one of them with a minimum of effort. They, on the other hand, might never find, far less sink, us. Without their escorting destroyers they were powerless against a submarine – even against the under-sized *Unbroken*. For a young lieutenant, it was a satisfying thought!'

Italian submariner Giuseppe Roselli Lorenzini likewise sought to imbue with romance the experience of war under-water for the benefit of prospective recruits to his service: 'The struggle possesses beauty: the submarine unleashed at full speed on the billow of the immense ocean, in search of an enemy striving to escape her; a handful of men closeted in those steel hulls thousands of miles from the distant Homeland. Contrast this means of making war, against fellow-warriors, with that employed by others who seek to attack [from the air] their distant homes. The submariner's experience has a manliness that should rouse the enthusiasm of every right-thinking youth.'

If this seems Latin bombast, many young men of all nations indeed responded to the challenges of war with the same enthusiasm as Roselli Lorenzini – until the hours when terror, pain or mere relentless discomfort displaced it.

In 1942 submariners shared with bomber crews the knowledge that they were more likely to die than to see old age. Of forty-two thousand men who served in the wartime German U-boat arm, twenty-eight thousand perished and another five thousand became prisoners. Mediterranean service was especially hazardous, because in those clear, shallow, relatively confined waters even submerged boats were readily spotted from the air. Dönitz's submarines suffered severely in the theatre during that late summer of 1942: on 4 August *U-372* was pinpointed by an RAF Wellington which dropped flares; then in daylight it was forced to the surface by two destroyers' depth-charges and scuttled. North of Alexandria on the same day, *U-97* was badly damaged. Later in the month, two other German submarines were crippled.

U-boat crews received almost double the pay of Germany's other sailors, together with special leave facilities. They were provided with the finest food available, though given the chronic wetness in which they existed, within days of starting a patrol fresh vegetables began to moulder and rot. Italian menus sounded more exotic than they tasted: *antipasto* of meats from the Po valley, prepared with tinned ham; *pasta asciutta* with seafood sauce, created from anchovies; *médaillons à la russe* – tinned meat mushed into meatballs, accompanied by tinned sauerkraut. Mussolini's submarines put to sea provided with a generous supply of Tuscan or Piedmontese wine, while German crews were issued with *Kujambel*, a fruit drink made from lemons, sugar and water that was popular as well as healthy.

Sergio Parodi of *Delfino* said: 'On patrol the most important order shouted by the first lieutenant was "*Fondo alla pasta!*"' – 'throw the pasta!', a command to start the supper, familiar in every Italian family kitchen. This call, delivered as soon as a submarine surfaced, provided a marker buoy in the latter part of each twenty-four hours: the first period, throughout daylight, was often silent and passive, while the submarine lingered submerged, propelled by its battery-powered electric motors, with much of the crew resting.

During the second phase, the submarine patrolled on the surface in darkness, its diesel engines roaring, bridge lookouts striving to pierce the night. 'Of course the *pasta* order was given only if no enemy was suspected near the submarine … In a high and confused sea, hot food inevitably gave way to the famous *caponata*. Into the big pot went biscuits steeped in water – quite possibly sea water – to which was added tuna in oil, onions (if there were any), anchovies, olive oil, capers and a drop of vinegar.'

Yet Italian submariner Livio Villa wrote ruefully in his diary: 'It is scarcely a nice war that we fight, unable to move about, prisoners of the boat, deprived of freedom to act or fight alone, to run away or even to hear [above the din of machinery]. We must wait upon prey that needs to approach close to us, and we are ourselves vulnerable to ambush and unimaginable perils. Every moment carries the risk of proving our last, each manoeuvre of the boat must be executed with meticulous care.'

The background orchestration of submarine life was common to all navies, as described by a skipper: 'the dominant clatter of the engines, an occasional voice, the scrape of a boot against an iron ladder, the apologetic shuffle of slippers through the wardroom. Percolating through them all, the angry moan of the sea as it swishes past our sides.' He noted also 'the strange smell peculiar to submarines: the curious combination of oil and damp and gas that dries the mouth and lines the throat'.

Admiral Karl Dönitz's favoured Type VIIB U-boats, such as were now deployed in the Mediterranean, had an external length of 211 feet, but an internal one of just 142. Half the fifty-man crew lived in the forward torpedo space, double-bunking and hammocking amid the spare torpedoes. Every two days it was necessary to clear the tubes, check propulsion systems, then reload. Men yearned for action, to liberate a few precious square feet of extra living space, by expending a salvo of torpedoes. They ate sitting on the bunks, used fresh water only for drinking and tooth-cleaning. Foul-weather clothing – *Ubootpäckchen* – was issued to the watch officer and three lookouts before they climbed the conning-tower when the boat was surfaced, likewise the superb Zeiss 7 × 50 binoculars which were the most prized booty of allied sailors who boarded a foundering submarine. Wet clothes were dried beside the diesel engines, but even there condensation dripped from bulkheads. The stench noted above was known to Dönitz's men as *Ubootmief*. Most noticed it only when they went ashore, and masked it with drenchings of 4711 or Kolibri colognes, especially when they hoped to meet girls.

Batteries, to power the electric motors for submerged movement, ran under the deck plating for most of its internal length: a U-boat carried 124 large cells. A submarine obliged to remain for a protracted period underwater might move as slowly as three knots, to extend precious battery life. In the twenty-foot-

long control-room were crowded the wheels, levers and tillers
to adjust rudder and diving planes, the engine-order telegraph
and gyro-compass, blow and vent valves for the ballast and
other tanks, the navigational plotting table and the periscope
viewfinder. Dials and gauges filled every inch of wall space –
bulkheads in naval terms. In the centre of the control-room
stood the ladder up which men ascended through the conning-
tower to the bridge when a boat was surfaced. To have a chance
of escape in an emergency, it was necessary to flood the entire
space while wearing personal breathing apparatus, then open
the hatches. Once the hull of a submerged boat was fractured,
few men profited from such an opportunity. Most often, the
detonation of a mine or depth-charge precipitated a swift death
for all aboard, as within seconds the sea burst through every
compartment.

Aft of the control-room were the two big 1,160-horsepower
diesels, requiring an air feed through an induction pipe from
the conning-tower, and thus usable only on the surface. There
was also a distillation plant to make fresh water, mainly to
supply the battery cells. The sternmost compartment of all
housed the electrical room, with its two 375-horsepower
motors, together with clutches that determined which propulsion system was linked to the propeller shafts according to the
surfaced or submerged condition of the boat. German submarines were also fitted with a stern torpedo tube. British, Italian
and American boats differed from German ones in detail, not
in general layout.

In an underwater world bereft of privacy, men spent much
time awaiting access to the toilet – the 'heads' in British
parlance. They accepted the necessity to defecate in front of
their mates. Italians, notably sensitive about anal hygiene, used
a bucket that did duty for a bidet while squatting naked in the

boat's central passageway. The almost comical apology for an officers' wardroom – recreation and messing space – measured around nine feet by seven on most boats, of which two feet was in the central passageway, through which men passed continually. At night in the warm Mediterranean, when captains believed themselves safe from aircraft, usually at night, some trimmed their boats with the hull awash, then allowed the crew to clamber in shifts on to the casing, to splash luxuriously and clean themselves with salt-water soap.

For men off-watch in British boats, there were games of solo, cribbage and uckers. They ate, slept, read, occasionally quarrelled, and wrote interminable letters home, more than a few of which were never posted. They speculated and reminisced, 'and all the time the air becomes heavier and more sour, tempers grow short and nerves are frayed, and you curse the day you ever volunteered to serve in these caricatures of sardine tins'. After a day submerged in the Mediterranean, this submariner wrote, 'the boat was an oven'. When men were not suffering diarrhoea, they were almost assuredly constipated, given the absence of scope for exercise.

Discipline was unfussy yet taut, because the lives of every crew depended upon rigorous observance of procedures. In the U-boat service, a lookout's failure to spot a threat – enemy aircraft or warship – prompted a formal investigation, should the defaulter survive it. Flushing the toilet was a complex technical procedure: at least one German boat was lost because of a flush accident that allowed seawater to flood its batteries. Use required permission from the watch officer, because waste ejection dispatched bubbles to the surface, which might be spotted by an alert ship's lookout. It was necessary to operate correctly a five-position lever, and to calculate how much air pressure to build up, to overcome external sea pressure. If the lever was set to 'Blow' with insufficient strength, the user suffered the indignity of 'getting your own back'.

Major Tony Frank was an army officer who spent many days aboard a British submarine in the Mediterranean before being landed to launch a commando operation. He was impressed by the experience. The crew, he wrote later, 'look very scruffy and dirty, but all know their jobs well. Discipline is rather different to ours [in the army], but effective.' As the days wore on, he noted 'the strained, eager eyes, shaggy beards, filthy clothes, voices calling – odd cries – but somehow it all works out'. He appreciated the excellence of the food, such as no soldier saw on a battlefield.

After long hours submerged, however, Frank noted that most men suffered headaches and nausea. When at last the boat surfaced and the hatch was opened, a rating held the captain's feet on the conning-tower ladder, amid the rush of foul air released, 'rather like being in the neck of a champagne bottle when the cork is drawn'. On reaching the bridge, both the captain and first lieutenant were promptly sick, and the crew took some time to recover from confinement. Exhaustion or outright mental collapse were significant drains on the manpower of every nation's submarine service, as some sailors found themselves unable to endure further captivity and boredom interspersed with spasms of terror.

Hitler's submarine fleet was more successful than that of Mussolini. This is partly explained by the technical superiority of the former's boats – they were faster, stronger and had much better fire-control systems. Although U-boat crews included few Nazi fanatics, one of Dönitz's captains who criticized the conduct of the war was reported for sedition by his first lieutenant when their boat returned from patrol. Oskar Kusch was promptly tried and shot.

While the Kriegsmarine's underwater branch recruited some of the Reich's most technically proficient young men, more than a few of their Italian counterparts were illiterate. Luciano Barca, a young midshipman, described how Gaetano, a former fisherman in his crew, begged him to write on his behalf to the parish priest back at home in Maratea, to ask for the year's lunar cycle. This touchingly simple youth explained that the moon influenced when and where he and his companions at sea should cast their nets. Barca declined: 'Frankly, I couldn't face writing that letter which harked back to the magic numbers of Etruscan priests or wizards.'

Barca's batman had served a twelve-year sentence for theft, armed robbery and rape, though he proved an excellent servant. The man eventually deserted, but left as a parting gift for his officer a fine edition of Dante's *Divine Comedy*. Given the limitations of some of the human material in the Italian underwater service, it is remarkable how much its boats achieved. Mussolini's submariners proved themselves among the most skilled and courageous members of his armed forces, as some would demonstrate in the course of Pedestal.

It is sometimes forgotten that in 1942 submarines were less than half a century old as effective weapons of war: they had little more tradition than heavier-than-air flight. The tools used by both sides in the contest between undersea hunters and their intended victims were still relatively primitive. Asdic detectors, connected to probes fitted beneath escorts' hulls, often lost contact with their target several hundred yards before reaching the point at which depth-charges were to be launched. In the 'dead time' between sonic blindness and firing, U-boats could carry out violent evasive manoeuvres.

Destroyer captain Peter Gretton wrote: 'the depth-charge attack, though stimulating to carry out, impressive to watch and terrifying to receive, was not nearly as lethal as had been hoped and expected [before the war] ... If they exploded more than a few feet from a superbly engineered U-boat hull, they might spring leaks and traumatize the crew, without sinking the boat.' Hydrostatic pistol settings on the cylindrical explosive charges were decided by hasty guesstimates on a destroyer's bridge or in her adjacent asdic cabin, followed by adjustments to the charges, made with keys carried by the ratings manning the ship's stern launchers.

Torpedoes – 'eels', as German crews knew them – were beyond human intervention once they left their tubes. Crude computers, 'fruit machines' in British slang, mounted in submarine control-rooms helped captains to judge deflection when taking aim at a moving ship. But much depended on the skill and luck of the officer peering through his periscope sight, and above all on his courage in defying the enemy escorts to creep to within point-blank range – a few hundred yards – before giving the order to fire.

An overwhelming proportion of the torpedoes loosed by aircraft, surface ships and submarines during the Second World War missed their targets, some because of technical malfunction, but most because they were fired at too great a range. Caution improved the firer's prospects of survival, but diminished his chances of achieving a hit. When twelve Royal Navy Albacore aircraft attacked the battleship *Tirpitz* on 9 March 1942, all missed. The after-action report concluded that pilots had delivered their torpedoes well beyond the thousand yards laid down in fighting instructions: 'the attack was badly carried out', despite little defensive fire from the battleship.

Many times in the coming August days in the Mediterranean, submarine torpedoes would miss for the same reason. U-boat officer Wolfgang Lüth said: 'as a submarine commander you were ultimately on your own. Peering through the periscope you made the decisions all on your own. During an attack you held the sole responsibility for the entire crew that was stuck with you inside that iron coffin. And so every hit you scored was a kind of vindication of yourself.' He added of commanders, 'if a man is a success, his men will follow him even if he is a fool'. A British officer wrote: 'Commanding a submarine was one long series of frustrations and anti-climaxes, punctuated only occasionally by the thrill of positive achievement.' This last was now to become the reward of one of Dönitz's captains in the western Mediterranean.

2 THE NEMESIS OF *EAGLE*

That morning of 11 August, as on the days preceding and following, Albacore biplanes from Lyster's carriers maintained anti-submarine patrols, but none located an Axis boat. The principal responsibility of the three great vessels, carrying between them seventy-two fighters, was to sustain a defensive umbrella against bomber attack. To fly off their own aircraft, they were obliged to diverge from the course of the fleet, to increase wind speed over their flight-decks, while the cruisers *Sirius*, *Phoebe* and *Charybdis* provided anti-aircraft protection respectively for *Victorious*, *Indomitable* and *Eagle*. None of the fighters could comfortably patrol below five thousand feet, because their engines overheated in the summer sun.

At 0815 the first German 'snooper' aircraft were detected by radar, and thereafter a procession of intruders approached the fleet, which Hurricanes were scrambled to intercept. They failed: though they claimed one shot down, the damaged Ju88 in reality made it home. None of the naval fighters could climb fast enough to catch reconnaissance aircraft at twenty and even thirty thousand feet. Successful interceptions might be made only if fighters took off ahead of radar warnings of 'bandits', thus enabling them to reach high altitude before the intruders arrived.

Moreover, the mere routine of carrier operations imposed steady attrition upon Syfret's strength. That morning of the 11th, one of *Victorious*'s two-seater Fulmars ditched in the sea when a wing caught fire, and a Hurricane from *Indomitable* did likewise after being damaged by a Luftwaffe twin-engined bomber. Later in the day, a Martlet ploughed into the Mediterranean when ammunition exploded in its wing. Yet another ailing aircraft narrowly escaped the same fate: after landing, it was found that its engine had been wrongly assembled. In the course of the morning, there were more submarine alarms, probably false, because no Axis captain reported making approaches at those times. Although *Uarsciek* had failed to sink a British ship, the Italian boat continued to trail the fleet. At 0839, it surfaced to radio a sighting report which reached Rome at 0930, prompting an alert to all Italian air and naval units, of which a British decrypt reached Syfret at 1055.

The sun blazed down, causing many men on ships' upper-works, dressed in tropical whites and shorts, to become painfully burned, including Peter Gretton, who paid the price for a sensitive skin and constant duty on the open bridge of *Wolverine*. Before the passage was completed, there would be a striking contrast between the bronzed or reddened deck officers, gunners and lookouts and the pallor of engine-room, communications and technical staff, cooks and suchlike, sweating below without benefit of air conditioning or even effective ventilation. Some very young men who did not think much, like seventeen-year-old midshipman Mike MccGwire on *Rodney*, had begun to treat the voyage as something of a lark. MccGwire, a few months after passing out of Dartmouth top of his class, had yet to see action, was still to see men die: 'everyone had been acting as if they were just going on leave, as discipline was relaxed, there was not much work, and there was the general carefree air of a holiday crowd about the ship's company'.

The first significant event of the day was the start of Operation Bellows, the subsidiary phase of Pedestal, the take-off from *Furious* of thirty-seven Spitfires for Malta. The carrier and her escorts moved out to port of the fleet. Hundreds of off-watch sailors clambered up from below, to spectate from neighbouring ships: Spitfire take-offs, without benefit of cata-pults, were a novelty for the Royal Navy. The RAF pilots, with no previous experience of operating from a carrier, were obliged to become airborne within two hundred yards of releasing their brakes. Once full power was applied, there was no scope for a change of heart: death was the nigh-certain penalty for faltering. They were now sixty miles north of Vichy-controlled Algiers, 584 miles from Malta, rather more by the dog-leg southern route appointed for the Spitfires, inland across Vichy-owned Tunisia, keeping as far as possible from Axis airfields.

Once airborne, pilots were ordered to maintain an altitude of at least twenty thousand feet, preserving radio silence and flying in four loose formations, each of nine or ten aircraft. The fighters had been painted in desert camouflage, with big Vokes tropical air filters fitted under their noses, and ninety-gallon fuel overload tanks slung beneath their bellies. Wooden wedges were inserted in the wings to hold the flaps as high as possible during take-off – these should fall away once in the air. Most of the pilots' logbooks already recorded hundreds of flying hours, but their six-hundred-and-fifty-mile dash with no margin of error, over terrain with scant scope for crash landing, was no joke even for the most practised.

They were dismayed by the sight of armourers removing the linked ammunition belts from their guns' magazines to save weight. Cannon shells were replaced in the wings by cigarettes, in short supply on Malta. The reasoning was simple: even if the Spitfires encountered enemy aircraft, they lacked fuel to fight. They were told that their personal kit would follow them to the island, by submarine or transport aircraft, and curiously enough it did. During the hours of preparation for the fly-off, *Furious*'s Royal Marine band played on one of the ship's lowered flight-deck lifts – popular numbers that included 'Happy, Tell Yourself You're Happy'. The carrier worked up to full speed, generating a thirty-five-knot wind on the flight-deck, to offer the aircraft every possible ounce of lift for take-off. A flier muttered: 'the old girl is really gathering up her skirts'.

Each pilot in turn primed his engine with six strokes on the Ki-Gass pump before start-up. Then at 1229 the flight-deck officer signalled the first Spitfire forward, raising both arms before showing a green flag. A thumbs-up from the cockpit, then brakes off, full power, three thousand revs and a pilot's muttered imprecation 'come on, little Spitty, build up speed'. Flaps down, braking effect, flaps up, wedges fell away and the fighter surged into the sky. Glancing below and behind, *Furious* suddenly looked tiny: 'Fancy having to land back on that thing for a living!' thought an RAF pilot, thankful to have escaped sea service in the Branch. Haze shrouded the mountains beyond the Algerian coast ahead.

Back on the carrier Geoff Wellum, scheduled to fly with the second formation of Spitfires, had just started his engine when he saw men on the flight-deck staring and pointing towards some spectacle across the sea. He was momentarily distracted, then a radioed word of command caused him to focus on his task: 'discipline and purpose' took over. Wellum gunned the Spitfire, rose steeply away from the ship, throttled back to 1,950 revs for the three-hour flight, then cursed before the burning glare as he realized that he had left his sunglasses below decks.

Far below the gathering formations of fighters, every man on the fleet's decks was stunned by the spectacle that distracted the flight-deck crew on *Furious*. Even as the Spitfires lifted away, Hurricane pilot Douglas Parker was making a downwind approach to land back on *Victorious*. To his horror, he glimpsed torpedo tracks below in the water, streaking towards *Eagle*. He had no time to radio a warning; instead he watched, impotent, as four German 'eels' struck successive giant hammer blows against the port side of the carrier. It was 1315, and the first disaster of Pedestal, a triumph for Hitler's Kriegsmarine, had begun to unfold.

On 4 August, when Kapitänleutnant Helmut Rosenbaum sailed from La Spezia in *U-73*, one of Dönitz's 750-ton Type VIIBs, he had not hitherto enjoyed an especially successful war. During a year in command, he roamed widely across the Atlantic and sank five vessels, but none of significance. On a recent milk run to Tobruk, carrying urgent tank spares for Rommel, he was depth-charged by a British aircraft which damaged his boat so badly that it was unable to dive. He was obliged to make the thousand-mile passage back to Italy on the surface. Having survived this trauma, for three months repairs consigned *U-73* to a dockyard. When the twenty-eight-year-old Rosenbaum received orders to embark on a new patrol, he remained troubled by persistent technical problems aboard, including a faulty direction-finding aerial, leaking bilge pump, leaking periscope and exhaust cut-out. Nor was he altogether happy about his crew. A curiosity: his senior watch officer, twenty-three-year-old Horst Deckert, was the son of German-American parents who lived in Chicago. The rest of the men, most very young, were drawn from all over Germany.

Following so many setbacks and frustrations, Rosenbaum wrote, 'I am longing to go out!' He started emergency diving exercises as soon as they were at sea, but was dismayed to find that by midnight four of his crew were experiencing high fevers, while others complained of stomach cramps and diar-rhoea – it was hard to sustain men's health under extreme heat. Next day, in the Gulf of Genoa, he recorded difficulty in receiv-ing wireless signals from his shore headquarters, but pressed on towards the designated patrol area, eighty miles north of Algiers.

He was west of Sardinia when, on 10 August, he was alerted that the Abwehr's watchers had identified *Eagle* and the cruiser *Cairo*, sailing from Gibraltar at 0300. Rosenbaum wrote: 'I will take absolutely any chance I get at them.' In darkness early next day, the 11th, south-west of the Balearics, he was told that a large British fleet was heading towards him. He spotted two unidentified twin-engined aircraft, probably German, and recorded that one-third of his crew was stricken with the boat's stomach bug: the accustomed squalor of existence aboard every submarine became significantly worse in *U-73* that morning. Though castor-oil treatment had some success as a panacea, few of Rosenbaum's men were feeling their best as the boat's climactic hours unfolded

The first indication of the approach of the Pedestal fleet came when the submarine's hydrophone-operators detected propeller noises to the west. Thereafter several ships passed above, before the boat slipped unnoticed between two screen destroyers, little more than eight hundred yards apart. There followed a nerve-jangling 150-minute approach as the submerged *U-73* crossed one column of ships between *Ohio* and *Santa Elisa*. It was an essential reality for all attacking submarines that they must position themselves in the path of warships steaming at speed – they could never hope to catch up with them from astern. Converging upon a prospective target demanded the nicest adjustments of course and depth to achieve a firing position while remaining undetected.

When next Rosenbaum raised his periscope, it was to behold a thing of beauty for a hostile submarine captain: *Eagle*, fast approaching his position. Even as he peered through the lens, the carrier changed course to zig away, but he understood that well before she reached him she would revert, zagging back towards him, into the path of her doom. Dropping the periscope, he took his boat deeper for a couple of minutes, then rose once more. When he scanned the sea at 1302, he saw first a destroyer three miles away, then *Eagle* approaching almost bow-on, last ship in the starboard column of the fleet. Swinging the periscope in a complete circle, he also noted the cruiser *Charybdis* and eight merchant vessels. He could have attempted a shot at any of them, but his orders demanded priority for the richest prize – the carrier. 'If only I had one more *röhre*' – torpedo tube – he lamented, he could have fired at the cruiser as well.

He slowed from eight knots to three. Perhaps surprisingly, he allowed himself a shot of brandy before starting the attack. Then he returned to periscope depth, saw the carrier again 'looking like a giant matchbox on a pond', registered even the fighters on her flight-deck. He ordered the torpedoes to be set to run at a depth of twenty feet, which should strike their huge target in her engine-rooms, the very guts of *Eagle*. By a curious chance, he himself knew the ship quite well: as a naval cadet in 1933, he had been among a party that went aboard the carrier, at anchor off Qingdao. There was no sentiment now, however. At 1305 Rosenbaum ordered his torpedoman, a nineteen-year-old who was among the youngest members of the crew, to fire four torpedoes in a narrow pattern, at intervals of 2.5 seconds. The boat shuddered as the salvo left the tubes, then Rosenbaum ordered as steep a descent as he could contrive, hastened by making every available crewman scramble forward, to compensate for the lost weight of the expended torpedoes. He recorded in *U-73*'s log 'Crash-dive, v. loud sinking and breaking-up noises'. He knew that he had only a minute or two, at most, to put depth and distance between his boat and avenging British escorts.

The pre-war *Eagle* was a notably happy old ship. By August 1942, with a crew overwhelmingly composed of 'Hostilities Only' ratings, the atmosphere had changed, and many men were learning their jobs while they performed them. Veteran petty officer Roy Northover observed that at least half the ship's company did not even know where their abandon-ship stations were. When the torpedoes struck, by good fortune an unusually large number of men were topside, 'in the nets' beside the flight-deck, because *Furious* had never before flown off Spitfires, and a crowd gathered to spectate as the first aircraft left the nearby carrier.

From the moment of the four thunderous detonations against her hull, there was no doubt of *Eagle*'s fate. The carrier took an immediate list to port, which steepened swiftly as thousands of tons of water poured into her machinery spaces.

Every stoker in B boiler-room perished, along with all the greasers in the port engine-room, and a hapless defaulter confined in the ship's punishment cells, together with the carrier's senior engineering officer, forty-one-year-old Irishman Geoff Mandeville. A pilot in the very act of taking off sought to claw his Hurricane into the air, but instead found the aircraft sliding unstoppably towards the sea, a spectacle never forgotten by those who witnessed it. A more fortunate flier was in the midst of lunch. 'My soup plate flew up and hit me in the face,' wrote Mike Crosley, 'and all the lights went out. A second later the reason burst upon my senses in three or four shattering explosions. Every one of us in the wardroom was lifted off our seats and thrown onto the floor.

'There was a small, grey shaft of light coming from one of the scuttles. By its light I could see shadows leaping over the tables and making for the door ... The slope of the deck was increasing by the second, so I joined the surge aft. Smoke was coming up from the open hatchways. The smoke smelt strongly of cordite and it was black as hell. The old girl was obviously not going to last long. As we shuffled along in silence we could hear dreadful sounds coming from below, crashes and shouts.'

On reaching the upper deck, Crosley saw the forward ladders jammed solid with men. A fellow airman warned him to stay off the flight-deck, because he himself had narrowly escaped being crushed beneath a toppling Hurricane. Beside the door of the aircrew ready-room, the pilot glimpsed his cherished portable gramophone, which had entertained them through hours of standby boredom, lying desolate beside a heap of smashed records.

Against the cant of the ship, attempts to lower boats failed. Roy Northover saw ghastly sights as some men leapt from the

high side towards the sea, and instead smashed themselves on
the exposed keel. Arthur Thorpe, a thirty-eight-year-old war
correspondent, scrambled up a ladder to the upper deck as the
ship listed over almost to her port rails. The sea, normally ten
feet below, was surging ominously a bare two feet below them.
Thorpe found himself beside the ship's first lieutenant, who
was inflating a lifejacket, as he did also. Hundred-pound
six-inch shells cascaded out of the racks on the tilted starboard
side above them, rolling and bumping past, threatening life
and limb. Journalists have a reputation for asking foolish questions, which Thorpe sustained by demanding: 'is she going?'
The officer nodded. Some ratings knotted a heavy line to the
rail, then slithered down it into the thick oil already coating
the sea. Thorpe followed suit; then, 'with all the poor swimmer's dread of deep water, I splashed and kicked clear of the
ship'.

For the first few moments after the impacts on the hull, air
mechanic George Amyes speculated absurdly that *Eagle* had
struck a school of whales. 'Then the ship started to list, I saw a
pair of seaboots hurtle through the air, followed by other
debris. I thought "aye, aye, we're in trouble here".' As he slithered down the side of the hull, he heard two men shouting
frantically that they could not swim, which prompted an
officer to seize one in each arm, saying 'now's your chance to
learn!' Amyes was dragged twenty feet underwater, before
thrusting himself upwards towards the sunlight, to find 'there
was bodies all the way round, rubbish and oil'.

Mike Crosley was tormented by memories of the sounds he
had heard on the stricken ship only a few moments earlier, 'the
screams and pitiful shouts of men's voices echoing up the
engine-room ventilators as they lay trapped below in darkness'.
Now, amid hundreds of heads bobbing around him in the

water, he was astonished to find some sailors singing. A passing senior officer, still wearing his brassbound cap, urged the choristers to silence, since 'they would need their breath later'. Roy Northover hauled himself up the steeply tilted flight-deck by clutching an arrester-wire, and was cutting free a Carley raft lashed to the carrier's superstructure – her island – when he found himself four feet above the sea. Abandoning the raft, he had swum a few yards when a voice cried 'she's going!', and 'lo and behold she went right over, so that her bottom came up'. As the great ship turned turtle, Mike Crosley was appalled to see one of her four bronze screws impale a ship's boat, 'tipping those inside her into the boiling white foam as if they had been rag dolls'.

Arthur Thorpe was fortunate enough to gain a hold on the side of a raft to which several men were already clinging, gazing for a few moments in disbelief upon the red expanse of the carrier's bottom. 'Then came a mighty rumbling as the sea poured into *Eagle*, forcing out the air. Water thrashed over her in a fury of white foam and then subsided.' Some men walked upright round the bilge keel as she turned over, but few of these survived its submersion. Crosley nonetheless thought that suction killed fewer than might have been expected. He said sentimentally: 'She went carefully. She never meant to hurt a soul.'

Thorpe suddenly felt a shock at the base of his spine, prompted by detonation of the first of a succession of depth-charges from warships hunting the U-boat responsible for their fate. These impacted on men in the water 'like a sledge-hammer hitting you in the back', and killed several. Crosley wrote: 'every time an underwater explosion concussed our vitals I could hear the most dreadful oaths coming from the bobbing heads, new words which I had never heard before'.

The destroyers' actions seemed to the struggling men to be thoughtlessly callous. Yet it was vital to deter the U-boat from making a further attack, even if they could not sink her.

Captain Lachlan Mackintosh had commanded *Eagle* for just two months. A colourful Scot and 29th chief of his clan, he often donned a kilt and played the bagpipes as his ship entered harbour. Now, in the sea, still affecting a uniform cap, even oil-streaked features did not conceal his identity. Some of his crew pulled him aboard a raft, crying 'Up the *Eagles!*' Roy Northover and others pushed every injured man they could find up on to another raft, while they themselves clung to its attached floater nets.

Mike Crosley spent two hours in the water, increasingly desperate as successive destroyers steamed by unheeding: 'I was by now camouflaged with oil and beginning to look like any other piece of flotsam, and not like a human being at all.' At last, however, the fleet's 700-ton tug rescued him. 'There was a feeling of both relief and sadness aboard *Jaunty*. Those of us who were able tried to do something useful to help the injured and encourage those who were pale and motionless. Others were scanning the oily water and pointing out floating objects, asking the tug's lookouts, who had binoculars, to look to see if they were human.' Rating Henry Rathbone attempted to clean and dry himself on a big blue flag he extracted from a signal locker.

The destroyer *Lookout*, also searching for survivors, threw down heaving lines which men struggled to grasp in oil-clogged hands. Roy Northover said: 'There was me and 250 men being pulled in, and I was almost dead then ... Anyway they eventually put a line round me and pulled me up. On deck the ship's doctor, tending the injured, demanded brusquely: "help me throw this body over the side". As he did so, the *Eagle* man was shocked to recognize an artificer whom he knew well, married with two children, who had not a mark on him, 'and must have been caught with the depth-charges'. Among the throng of survivors crowding the deck, a friend recognized Northover himself only by his broad Dorset tones – his features were entirely concealed by oil.

George Amyes paddled for an hour on a wooden spar, until he saw men on a nearby raft pointing eagerly to a ship approaching. Then he recognized *Jaunty*, which had been built in the dockyard at Hull, whence he himself came: 'I thought well, if this isn't like coming home I don't know what is.' Ropes and scrambling nets were trailing the ship's side, but the exhausted man missed them as the tug pulsed slowly past: 'I thought "this is it. I'm going to be left behind after all".' He grabbed the last rope he could see, which tore skin and flesh from his hands as he clung to the hemp. Then somebody slung him a looped line, by which 'I eventually got pulled up just like a ruddy hooked fish.' Once on the deck, 'I was tossed among a heap of writhing and retching humanity … Quite soon I wished I was back in the water, where it was nice and comparatively peaceful.' Men on other ships were astonished by the spectacle of the tug, her upperworks and even rigging rendered invisible by five hundred blackened and half-naked survivors.

No man with that fleet ever afterwards forgot where he was when he witnessed or learned below of the abrupt extinction of *Eagle*. A spectator on *Nigeria*, war correspondent Anthony Kimmins, wrote: 'it was so sudden that it almost took one's breath away. A great brown column of water rose to three times the height of her topmast … For a long time everyone was very quiet and subdued.' Hugh Popham was at readiness in the cockpit of his Hurricane on *Indomitable* when the flight-deck engineer waggled the fighter's ailerons to attract his attention. He was stunned to see *Eagle*, half a mile to port on a parallel course, with steam and smoke pouring out of her.

He heard the series of muffled explosions, as the doomed ship swung in a circle before turning over: 'For a few seconds her bottom remained visible; then with a vast gust of steam and bubbles she vanished. All that remained was the troubled water, a spreading stain of oil, and the clustered black dots of her ship's company.' Even when Popham was flagged to take off, a sense of shock clung to him in the cockpit as climbed into the sky. He saw the pattern of ships below alter course as they zigzagged, the sea heaving as escorts bombarded the German submarine's presumed position.

At 1330 Helmut Rosenbaum recorded almost contemptuously in *U-73*'s log the sounds of thirteen depth-charges exploding in distant water: 'hopeless allied submarine tracking'. Another nine charges fell during the next hour, without troubling the Germans. A technical error by an engineer, which must have prompted a stream of obscenities from the man's mates, caused a haemorrhage of oil which the U-boat skipper was fearful that warships would spot, but they did not. At 1815, with the British fleet already fifty miles to the east, at last he thought it safe to surface and report his success in sinking *Eagle*. This prompted a flood of congratulatory signals including 'ROSENBAUM BRAVO!' The captain scribbled euphorically 'everything seems to be going my way today'. It continued to do so: soon afterwards, he received a signal reporting that he was to receive the *Ritterkreuz* – Knight's Cross – customarily awarded for a major U-boat success – prompting 'a yell of delight from the crew'. Germany's new hero also had the good fortune to have aboard a supernumerary from a

Propaganda Company – a war correspondent. Thus, there was a bard to sing *U-73*'s song to its nation's press.

There were also war correspondents on the other side, to describe *Eagle*'s tragedy. Among these, Arthur Thorpe was doubly fortunate, because he had already experienced one giant warship sinking, that of the carrier *Ark Royal*. Now, he wrote fancifully that his fellow survivors on *Jaunty*, thrilled to be alive, 'reminded me of a train in a London station crowded with happy schoolchildren bound for a day out in the country'. In truth, however, most men were sobered, indeed awed, by the destruction of *Eagle*, which made mock of the fleet's anti-submarine screen. *Nelson*'s commander – executive officer – wrote bleakly: 'What a tragic failure this convoy has been! The first anybody knew about it was to see *Eagle* listing over … No noise, nothing was heard. I've never before seen such a thing. It makes one tremble. I saw Skinner [a naval construction expert on board the battleship] looking like a man who'd seen a horrible nightmare, he was sweating and white and I heard him say "they couldn't have had anything closed".'

Deep below decks in the transmitting station of *Furious*, manned in action by members of the ship's Royal Marine band, news of *Eagle*'s fate was telephoned from the Fighting Top to the musicians. Every man serving in the bowels of every ship was conscious of his special vulnerability: a bass drummer named Arkinstall attempted black humour, saying in a heavy West Country accent: 'Now then lads, don't forget it's alphabetical order out of here.' Seaman Chris Gould wrote from *Lightning* to his fiancée Vera: 'I'm not very happy darling as things have started a lot too soon this time. We have been at action stations since one this afternoon when we had a concentrated attack by submarines. It was B[loody] awful; a carrier was hit and sunk in less than 8 minutes. I saw it go. She just rolled on her side and disappeared. What an awful feeling it gives you in the pit of your stomach … We stand a much better chance in a destroyer, but that doesn't stop you being frightened.'

When a sailor loses his ship, he forfeits his home and possessions. After *Eagle*'s commander had come aboard *Nelson*, his counterpart on the battleship wrote: 'Poor old St. Croix was very shattered, having lost all his belongings. I lent him my best uniform and my best pipe, but he seems to resent that my ship has not been sunk too!' On the destroyer *Bramham*, petty officer Reg Coaker said: 'It was quite clear we were going to be in for a fair old dusting-off.'

The fly-off of Spitfires from *Furious* had been interrupted by the need to make emergency turns, to put distance between the ship and Rosenbaum's assumed position. Geoff Wellum's formation was still circling above the fleet when *Furious*'s operations room broke R/T silence to warn: 'Keep clear, keep clear of the carrier. Air attack.' A formation of Ju88s had been detected approaching, at extreme high altitude, 32,000 feet. The enemy aircraft ignored the Spitfires, and surprised the British by making no attempt to bomb: they were merely conducting reconnaissance. Fighters scrambled from the carriers failed to reach altitude before the enemy flew away, causing Syfret to observe wearily in his report, 'it will be a happy day when the fleet is equipped with modern fighter aircraft'.

By 1430 every Spitfire, save one which technical failure obliged to put down on *Victorious*, had set course for the hot, sweaty flight to Malta. The island at last loomed ahead looking, in Wellum's words, 'like a brown leaf floating on water'. Once on the ground, the moment their engines stopped, armourers scrambled on to the fighters' wings to replace cigarettes with cannon shells. The lean, earnest Air Vice-Marshal Park met Wellum, whom he had encountered in 1940 during the Battle of Britain, and warned him to say nothing publicly about the loss of *Eagle*. 'I don't want word to get around just yet to the Maltese if it can be avoided. Their morale has taken enough knocks for the moment … They'll hear soon enough.' As indeed they did. Thirty-six fighters from *Furious* landed safely, in itself a considerable gift of fortune and a feat of airmanship.

A letter later written in England to the father of Mike Crosley, a surviving pilot from *Eagle*, by the young man's grandmother, merits quotation because it reflects the moving naivety of elderly civilians amid a world in flames: 'I was so very pleased to hear about Michael; as his ship has been sunk, he will I expect have a good rest and it will be fine if he goes to America; the time he has gone through must be a very great strain on him, but he is young and keen; he always has been a dear, good boy; I do hope he will be spared to be a great help and blessing to others.' There would be no American dockyard holiday for Crosley, such as crews of damaged British warships sometimes enjoyed; instead there would be many, many more fighter sorties with the Royal Navy before his war, and everybody else's, was done with.

5

'Stand By to Ram'

The men of Syfret's fleet were severely shaken by witnessing the fate of *Eagle*, and it is fair to assume that the same was true of the admiral. Every commander must set a brave face upon losses, which are inseparable from the experience of battle. But even naval veterans find it shocking to see a huge vessel, created by the expenditure of boundless resources and human ingenuity, abruptly vanish beneath the sea as if it had been a plaything drowned in the bath. It was a bitter business to lose a capital ship before the fleet had even entered the most perilous waters. Moreover, the August day was far from ended. That afternoon, the immense formation of ships executed almost continuous emergency turns to evade further alleged submarine threats. So violent was one such manoeuvre that *Nelson*'s anti-mine paravanes were carried away.

The admiral received a signal, based on decrypts of Axis wireless traffic, warning him to expect an air attack at dusk. He ordered the destroyer screen to prepare to meet this, with some escorts taking station six thousand yards out as pickets to provide early warning, while the Hunts closed in astern of the big ships. At 2000, pilots aboard the remaining carriers were hanging around flight-decks, watching the red and orange hues of sunset herald the darkness that would relieve them from imminent action. Then the tannoy crackled, 'Scramble the Hurricanes! Scramble the Hurricanes!'

On *Indomitable* fitters, in temporary occupation of the four aircraft at readiness, pressed starter buttons, before pilots leapt over the wings to take their places. The carrier turned into wind, and within minutes the fighters were climbing at full boost, heading northwards to search the sky at twenty thousand feet, seeing nothing, descending to twelve thousand, then returning to twenty, where suddenly they glimpsed below a formation of twenty-seven Ju88s. Each fighter in succession peeled off, dived to intercept. Hugh Popham and his wingman gave an enemy aircraft a burst each from extreme range, then pursued the bomber away from the convoy until they felt obliged to turn back, after failing to catch up. Once more, they cursed the relative sluggishness of their own aircraft.

This first air skirmish, which began at 2045, was a German improvisation: Italian commanders, having rushed aircraft to Sardinia, opted to hold off using them until the following day, when they would launch a coordinated operation. The Luftwaffe bombers used tactics that would become familiar – synchronized assaults at varied heights, from differing attitudes, to split the tremendous British anti-aircraft fire of Oerlikons, Bofors, pom-poms, 4.5-inch, 5.25-inch and 6-inch guns, and even the battleships' sixteen-inch monsters. Hurricane pilot Douglas Parker saw *Rodney* and *Nelson* fire their thunderous main armament, eighteen guns each hurling across the sky more than a ton of explosive and shrapnel, set to detonate in the path of the attackers. Every aircraft within hundred of yards was thrust upwards by the violently displaced air: 'when these large shells burst close, the sea boiled up in the face of the incoming torpedo aircraft. An awe-inspiring sight,

and quite often [the enemy] broke off their attacks immediately after receiving the heavy blast.'

On one wing of the flag bridge of *Nelson*, Syfret's Royal
Marine orderly manned a Bren light machine-gun which did
no discernible damage to the enemy, but gave much satisfaction
to its firer, and even to the admiral himself, clad in impeccable
whites with a pipe clamped in his mouth, who directed its aim.
The Ju88 twin-engined bombers made shallow attack runs
from the port side of the convoy, while three Heinkel 111
torpedo-bombers approached at extreme low level. The
warships' barrage seemed to daunt the latter, which released
their weapons at long range, with little conviction, and
achieved no hits.

Most of the Ju88s showed no greater enthusiasm. Their
bombs fell harmlessly into the sea, though two near-missed
Victorious, which took violent evasive action. Another stick of
bombs fell close to one of the tankers astern of the convoy, and
yet another German aircraft attacked the tug *Jaunty*, returning
from transferring her cargo of survivors to Gibraltar-bound
destroyers. The little craft proved too slow to catch up the fleet,
and was ordered to quit the operation and retire to the Rock.
Her loss proved significant, leaving the fleet without a rescue
and towing vessel.

'The noise was deafening with all our armament in action,'
wrote Ron Stockwell, an ammunition number on *Nigeria*.
'Busy at my job I sneaked a look to port and wished I hadn't
– the torpedo-bombers were coming in low and straight, with
streams of tracer going out to meet them.' Gerhart Suppiger,
on *Santa Elisa*, thought the enemy planes resembled dragon-
flies. Warships of every navy responded to a torpedo attack by
swinging towards the tracks, to present the narrowest possible
profile. To counter this tactic, Axis aircraft were briefed to

approach from different angles simultaneously, as they did through the days that followed.

Many ships harboured pets aboard, which reacted variously, and pathetically, to the sound and fury. *Nigeria*'s ship's cat emerged from below on to the gun deck carrying three kittens, which she began to feed. By contrast Minnie, the destroyer *Laforey*'s pet monkey, was discovered, in the words of an officer, 'cowering in a dark corner, her teeth chattering and on the verge of hysteria. A tot of rum administered to steady her nerves led to severe alcohol addiction, eventual DTs and a drunken death.'

Anthony Kimmins wrote: 'The din was terrific, but through it all you could hear the wail of sirens for an emergency alteration of course ... and the answering deep-throated hoots of the merchantmen ... Then suddenly a cheer from a gun's crew, and away on the port bow a Ju88 spun vertically downwards with both wings on fire, looking like a giant Catherine-wheel.' The only damage suffered by the fleet was that the Walrus amphibian in the hangar of the cruiser *Manchester* was riddled with bomb splinters. A naval officer aboard a merchantman, *Rochester Castle*, wrote: 'the attacks were not pressed home, the enemy showing marked inclination to avoid the barrage'. One of her sister ships reported the wireless antennae shot away by friendly fire. As the aircraft engines receded and darkness became complete, some men thought: 'if this is the best they can do, maybe it won't be too bad'.

Relatively small warships moving at thirty knots, and even freighters making fourteen knots, offered poor targets for high-altitude bombers – dive-bombers had by far the best chances of success, as the US Navy found in the Pacific. On *Ledbury*'s bridge, Roger Hill eavesdropped on the relay of Hurricane radio chatter: 'We were listening to our pilots, who were short of fuel, but stayed up … The darkening sky was lit in all directions by shell explosions and tracer; I saw a great sheet of flame as a fighter crashed as it landed on a carrier. Then it was dark, the firing had ceased and we listened to the last fighters, with petrol gauges showing nought, landing on.'

In the dying light four Hurricane pilots approaching *Indomitable*, their own carrier, were at first fascinated and even charmed by the spectacle of the ships 'enclosed in a sparkling net of tracer and bursting shells, a mesh of fire. Every gun in fleet and convoy was firing, and the darkling air was laced with threads and beads of flame.' The fliers searched the vicinity for the German aircraft which they assumed were the guns' targets; spotted nothing, for the enemy was gone. It was dark enough now for the airmen to see the blue haze of each other's exhausts. They circled above, waiting for the shooting to stop; after an hour in the air, they were watching their fuel gauges with increasing unease. Yet as they became obliged to descend, they found themselves pursued by fire from wildly overexcited ships' gunners.

They radioed for instructions, and heard controller Stewart Morris's calm voice say, 'Stand by, Yellow Flight. Will pancake you as soon as possible.' Brian Fiddes, their leader, called somewhat irritably 'if you'd stop shooting at us, it would be a help', and received no response. Almost every ship heard a pilot shouting over the radio: 'for Christ's sake stop that pom-pom. He's knocking my fucking tail to bits.' Bursts of fire persisted, however, even when the Hurricanes switched on their navigation lights. Recklessness of this kind by ships' gunners was a feature of the air–sea war from first to last – even a year later, a frightful toll of allied aircraft was exacted by 'friendly fire' during the invasion of Sicily. Before entering the battle zone Ramsey Brown, Master of *Deucalion*, had expressly told his ship's gunners to fire on any plane which approached within fifteen hundred yards, without awaiting an order.

Tom Troubridge, *Indom*'s Falstaffian captain, took the huge risk of turning on the ship's deck lights as he turned into the wind, then steamed a straight course at twenty-six knots, to help his fighters to get down. Yet the shooting persisted. There was a small tragedy on Troubridge's own carrier: the much loved deck-landing officer, Geoff Pares, was mortally wounded when a shell from a nearby 4.5-inch gun exploded prematurely above his head. Pares' illuminated signalling bats were also shattered, so that as the Hurricanes came in, his deputy was obliged to improvise, guiding them down with a torch in each hand and a third held between his teeth. 'Butch' Judd, the choleric South African CO of 880 Squadron, was fired on by one of *Indom*'s pom-poms even as he landed. Leaping out of his cockpit he ran across the deck to the lieutenant in charge of the gun and seized him by the throat. 'You bloody useless bastard!' he roared, shaking him like a dog. 'You brainless oaf! Don't you know a Hurricane when you see one!'

Hugh Popham found himself last man aloft, darting repeatedly towards the carrier's flight-deck, only to meet renewed shell bursts or red tracer, which obliged him to climb away. Then his radio went dead, and he saw that he had only ten gallons of fuel, less than fifteen minutes' flying. Anticipating a ditching, he jettisoned the cockpit hood and unclipped his parachute harness. Down to five gallons, at 2130 in desperation he approached a ship that was unidentifiable in the darkness, then sheered away on seeing a funnel looming ahead of his nose. Next astern, he thought that he glimpsed a flight-deck, lowered his undercarriage and suddenly glimpsed a pair of lighted bats, guiding his approach. He had no hope of making a second attempt, though the ship beneath him suddenly swung to starboard. Despairing, he crammed down his nose, cut the throttle and tried to match the flight-deck's turn.

The wheels banged down on the invisible plating so brutally that the plane's undercarriage collapsed. It skidded on its belly towards the island of *Victorious* – which the ship proved to be – precipitating a scream of metal upon metal and a cascade of sparks. As the Hurricane abruptly halted, Popham glimpsed a tongue of flame beneath his cockpit and hurled himself out. Seconds later the plane erupted into an inferno visible to half the fleet. Fire crews doused the blaze before Jumbo, the flight-deck crane, heaved the remains over the side. A voice close beside Popham demanded: 'did anyone see the pilot? Well, either he must have made a ruddy quick getaway, or the kite must have landed on by itself.' The man with no right to be alive said mildly across the darkness: 'It's all right. I was the pilot.' *Victorious*'s captain described it as a miracle that all the fighters made landings, 'only achieved at the expense of several crashed aircraft and the exposure of many lights and flares. They had done no good, as the enemy held their attack till it was too dark for our pilots to see them clearly.'

Admiral Lumley Lyster reported later: 'It is evident that the dusk attack[s] must be met with either anti-aircraft guns or fighters. Both cannot operate together … That the gunfire is a deterrent to the attacking torpedo aircraft is evident, and perhaps it may deter some of the more timid bombers and dive-bombers. The fact remains, however, that the dividend paid for a tremendous expenditure of ammunition is remarkably low. I have the strong impression that not only is it inaccurate, but also undisciplined.' Admiral Cunningham had written to the Admiralty months earlier that 'no AA fire will deal with the simultaneous attacks of 10–20 aircraft'.

Furious, having completed the Spitfire fly-off for which she came, had started back towards Gibraltar at full speed, escorted by five destroyers, at 1830. The sea was so placid that, before they set forth, some ships closed alongside each other, to transfer *Eagle* survivors to those returning westwards. That night – a sad night, given lingering memories of the carrier – *Wolverine* was on the starboard beam of *Furious*, passing a few apparently uneventful hours en route to Gib, when the telephone rang in Peter Gretton's cabin: 'Radar contact bearing green two-0,' said the officer of the watch. Gretton sprang from his bunk and up the ladder to the bridge. The contact could only be a fishing boat, such as the watch officer on their outbound passage had mistaken for a submarine and come close to ramming … or a U-boat. The destroyer leapt to full speed ahead, altered course to close.

At a range of a thousand yards the watch officer spotted a shape in the darkness, then a glistening white bow wave. At three hundred yards Gretton ordered illumination of the ship's bridge searchlight. Yes, there it was, the low, sinister, thrilling form of a submarine's upper hull. Gretton ordered a signal to be made to his flotilla leader, then broadcast to his crew 'stand by to ram'. For whatever reason, he made no attempt to use his Hedgehog anti-submarine projector, which threw a pattern of charges forward. A signaller reminded the captain to ring the ship's alarm gongs for 'Crash Stations', then fifty yards from the enemy the destroyer switched on her big fighting-light.

It was 0100 on 12 August. Gretton said 'a five-hundred tonner, I think', then at twenty-six knots the destroyer smashed into the aft end of the conning-tower of the Italian submarine *Dagabur*. The British officer wrote exultantly: 'long hours of practice had reaped their reward and our dreams had come true'. They ignored the screams of drowning men from the submarine, a callousness probably influenced by the recent memory of *Eagle*'s fate at the hands of another such enemy. Gretton said briskly later: 'we thought they were Germans'.

On *Furious*, the starboard lookout sang out: 'Light Flashing Green 30. *Wolverine*, sir. "Have-rammed-and-sunk-enemy-submarine"'. Yet success – the boat foundered with all hands – came at a price. The destroyer's first lieutenant reached the bridge from his bunk to blurt tactlessly to his triumphant captain: 'Oh Lord, what have we hit now?' The chief stoker had been hurled across the wardroom so brutally by the shock of collision that he broke several ribs, though this did not prevent him from donning diving gear to explore the flooded bow and direct temporary repairs. A pipe had been fractured, causing a rush of men to the upper deck desperate to escape from the engine-room, which was filling with scalding steam. All power failed for a time, the asdic receiver was wrecked and the ship's bow was twisted into a tangle of steel that required months of dockyard work to repair.

After an hour, however, the chief engineer reported the ship capable of making steam for one engine, which enabled her to reach Gibraltar at eight knots. Escorts not infrequently rammed submarines when opportunity allowed, a tactic that highlighted their captains' lack of confidence in depth-charges, and for that matter in guns. Admiral Syfret afterwards expressed frustration that precious destroyers should be crippled to sink submarines when the fleet's foremost priority was to keep his screen intact – to hold the enemy off the big ships, rather than to 'kill' U-boats. *Furious* and her escort reached the safety of Gibraltar a few hours later, without further incident.

*　*　*

Thus ended the first day, opening act of the supreme melodrama of Pedestal. From the viewpoint of commanders, the loss of *Eagle* was a bodyblow, eliminating one-seventh of the Royal Navy's entire fleet-carrier strength, and one-fifth of Syfret's fighter support. Sixteen Hurricanes went to the bottom with the ship, though four more that had been airborne at the time of the sinking landed on other carriers. Among the twenty thousand men sailing eastwards, the blow to morale, or perhaps more appropriately to confidence, was severe. Helmut Rosenbaum's success emphasized the fallibility of the destroyer screen, though it was immensely powerful by the standards of the day. The U-boat's achievement was attributed by escort commanders to the freak asdic conditions which prevailed in the western Mediterranean. Yet there would be plenty more such carnage inflicted by submarines as they sailed closer to Malta, when the same explanation could not apply.

The truth was simply that in 1942 the Royal Navy's technology for locating and weaponry for destroying enemy submarines were relatively primitive. Moreover, even when a U-boat was attacked before it struck, it was deemed impossible to detach two or three escorts for the hours of successive depth-charge runs necessary to offer a fair chance of sinking it. The objective of the destroyers must be merely to hold the ring, to frustrate attacks. The priority was to maintain the integrity of the chain of warships around the capital ships and cargo carriers, to defend them against the air attacks which, in the days ahead, were bound to come in far greater strength than on 11 August.

Berlin Radio not unreasonably trumpeted the success of Helmut Rosenbaum. Its English-language bulletin concluded: 'This carrier was part of an exceptionally heavily escorted convoy which is heading eastward towards Malta. Operations against this convoy will continue.' A German press report described *Eagle* as 'one of the few aircraft-carriers which England still has'. It mistakenly asserted that the ship's aircraft were designed to reinforce the defences of Malta, 'which will now wait in vain … Kapitänleutnant Rosenbaum and his brave crew have thrown a big new spanner into the wheels of the English.' Back in Germany, the U-boat skipper's proud father told reporters that his son 'always wanted to join the Navy – from the age of ten he kept an album with photos of ships he cut out of newspapers'. It was a measure of Italian scepticism about all things German that Count Ciano, Mussolini's foreign minister and son-in-law, expressed doubts in his diary about the truth of Berlin's claims to have sunk the *Eagle*.

Even as Syfret's ships steamed on eastwards other British sailors, airmen and even a few soldiers, hundreds of miles beyond their field of vision, sought to assist their progress by impeding that of the enemy. A diversionary convoy sailed from Port Said at dusk on the 10th, composed of three merchantmen escorted by three cruisers and a dozen escorts, which were joined at sea next day by a single merchantman and two cruisers from Haifa, the whole force commanded by Rear-Admiral Sir Philip Vian. At 1700 a U-boat radioed a sighting report of this group, 155 miles west of Haifa. At first, Axis staffs took a serious view, judging that 'the operations in progress aim at something more than merely supplying Malta'. Their own shipping movements in the Aegean were temporarily suspended, and Italian cruisers put to sea.

But the diversion failed to persuade the Germans to leave their most experienced anti-shipping squadrons in Crete rather than shift them west to Sicily. On the night of the 11th, as planned by Admiral Harwood in Haifa, Vian's ships

dispersed and returned to their respective bases. This retirement caused momentary alarm in Rome next day, when reconnaissance aircraft were unable to locate them. However, it
quickly became apparent to Kesselring, and to Mussolini's
Supermarina, that Pedestal, and Pedestal alone, was the big
event. The diversionary force was too weak, its motions too
feeble, to achieve the intended British purpose.

The RAF also did its best, albeit a poor best, to maul the
enemy's air forces on Sardinia. At dusk on the 11th, even as
Ju88s bombed the fleet, nine twin-engined Beaufighters from
Malta strafed with cannon the Axis airfields at Elmas and
Decimomannu. On the outbound flight, they spotted below
three Italian submarines – *Axum*, *Avorio*, *Otaria* – leaving
Cagliari to join the assault on Pedestal, and duly reported these
sailings to Valletta. Three more boats – *Cobalto*, *Dandolo* and
Emo – also left port that night. The Beaufighters' raking of
Elmas achieved little beyond destruction of one Savoia torpe-
do-aircraft, but they did better at Decimomannu, wrecking
four Savoias and a Cant, damaging eleven more aircraft and
exploding two torpedoes awaiting loading. The Italian history
notes: 'the attack was especially effective because it caught
several units in the midst of refuelling aircraft'. Two soldiers
were killed, four more wounded.

On the fliers' passage home, they wirelessed the electrifying
news that they had spotted two light cruisers and accompanying destroyers of Admiral da Zara's 7th Cruiser Squadron leaving port. This confirmed an earlier Ultra intelligence report
that the eight-inch cruiser *Trieste*, already on passage from a
Tyrrhenian port, would soon enjoy the company of the six-inch
cruiser *Muzio Attendolo*, which had sailed from Naples at
0930, and the eight-inch cruisers *Bolzano* and *Gorizia*, which
had left Messina that morning on a circuitous course around

the north shore of Sicily. The Beaufighters orbited the warships until shortage of fuel obliged them to turn for home. They were replaced by a Wellington from Malta, the crew of which had been hastily summoned by a message flashed on to the screen of the cinema where they were relaxing: 'Members of Special Duty Flight, report to Luqa'. When the airmen duly rose to comply, in a moving gesture of tribute the Maltese civilians in the audience clapped them.

At 2355, the crew of 'O-Orange' reported radar sightings of four cruisers and eight destroyers moving east. At 0130 the Wellington dropped flares to illuminate the enemy warships, provoking a heavy but wild barrage of flak. While the flares still hung in the sky, the bomber crew released eight 250-pound bombs without effect on the squadron, and indeed perhaps without expecting any, then turned for home. One of its crew shrugged: 'As far as we were concerned, this was the end. We [had] done what we were sent to do.' Yet the little sortie prompted a more important sequence of events than the lone bomber crew recognized: the British had their first sightings of major Italian warships at sea. These could pose a deadly threat to Burrough's cruisers and destroyers, once the British capital ships turned away west, as the planners intended, that very evening of the 12th. Conversely the Italians, neurotically apprehensive about the preservation in being of their fleet, knew that the British knew that their cruisers were on passage.

During the night of the 11th/12th also, two RAF B-24 Liberators of Middle East Command attempted to bomb Decimomannu, though their loads fell harmlessly on empty fields. Two other Liberators returned home without attacking, having failed to find the target on this moonless night. A lone Wellington meanwhile bombed Catania. Air Vice-Marshal

Park expressed bitter disappointment that he had not been allocated more heavy bombers for attacks on enemy airfields. If the RAF had been able to mount intensified sorties against the planes crowded on to every strip on Sardinia and Sicily, he wrote afterwards, 'the story of the convoy might have had a happier ending'.

There was one more British operation, almost pathetically feeble. Around midnight a six-man party of commandos from Lord Jellicoe's Special Boat Section was put ashore near Catania by the submarine *Una*. They had sailed from Valletta two days earlier on a mission to sabotage Axis aircraft on the ground. Their first act on landing was to blow up a power pylon linking Catania and Syracuse, which awakened every Axis soldier in the region. A British NCO was captured by two members of the tax and customs police, and early next morning the rest of the disheartened group were discovered by local fishermen: the soldiers surrendered before they even approached their airfield target. If these operations were unimpressive in scale, they represented the best that British forces could contrive with the meagre resources at their disposal.

Britain's prime minister received the day's news from the Mediterranean while attending a grisly Moscow banquet with Stalin, who taunted him brutally about the failure of PQ17: 'Your navy runs away!' The message from Pound to 'Tulip' – the incongruous code designation for Britain's leader while in Russia – read: '"PEDESTAL" according to plan, except that EAGLE was sunk by U-boat south of Balearic Islands. To-morrow Wednesday is crucial day. Thirty-six SPITFIRES landed Malta from FURIOUS. Nothing else of naval interest except one Italian U-boat sunk.' Churchill gave orders to his staff that future messages from the Admiralty about the

Mediterranean operation should be upgraded in priority to 'most immediate' and delivered to him personally, no matter what was the condition of his conferences with Stalin.

Syfret and Burrough could take heart, first by reflecting that despite the rapidity of *Eagle*'s sinking the overwhelming majority of her ship's company, 930 men, had been rescued; only 136 perished, most in the engine-rooms. For those obliged to entrust themselves to its waters, the warm Mediterranean was merciful. Meanwhile Operation Bellows, the fly-off to Malta, had been a complete success. The first Axis bomber attacks had been delivered without conviction: if this was the best the enemy could do with his immense air strength, the fleet could meet him with fair assurance. The merchant-ships of the convoy were conforming to orders, and to formations, with better discipline than naval planners had expected.

Yet that night they were still more than four hundred miles from Malta. It would be hard to overstate the strain upon Syfret, Burrough and their captains: they bore responsibility for a fleet which was hideously vulnerable, upon a sea exposed to the enemy, who could operate from naval and air bases close at hand on every side. After witnessing the fate of *Eagle*, many men thereafter took such pathetic precautions as they could, in anticipation of being obliged to abandon their own ships. Chris Gould of *Lightning* told his fiancée: 'I've got my wallet in my pocket with £4.15 and a door key in. Also I carry this pen around with me; of course I always wear my watch.' On all the ships of both the Royal Navy and Merchant Navy, many men below decks now wore lifejackets in their hammocks as well as in action. The merchant vessels were sailing with lifeboats slung out on their davits, poised for lowering, a prudent rather than jittery precaution.

Nelson's George Blundell was nonetheless unjustly pessimistic when he wrote in his diary that Pedestal was already 'a failure'. Britain's directors of war, foremost among them Churchill himself, had understood from the outset that the dash for Malta must be a bloody business, in which ships would be lost – perhaps many ships – and men must die. The loss of *Eagle* was a mere payment on account of the grimly anticipated 'butcher's bill'. As Pound told 'Tulip' in his signal to Moscow, the next day, Wednesday 12 August, when Syfret's fleet would approach within almost point-blank range of enemy shores, enemy airfields, enemy warships of all kinds, would be crucial in determining the fate of Pedestal.

6

The Twelfth

1 'SHALL I BE KILLED TODAY?'

During the night of 11/12 August, there was a lull. While the ships of Pedestal maintained a rigorous black-out, below decks lights blazed amid a relentless clamour of activity: men clattering up and down steel ladders, through mess and machinery spaces, opening and clanging closed the great steel hatches that locked compartments, a process which caused passage through a warship at action stations to take three or four times as long as the same movement in harbour. The previous day's events made it unnecessary to remind any man of the importance of preserving the integrity of watertight doors. It was for this reason that bigger ships distributed dumps of food, chocolate, cigarettes: during prolonged periods in action, it was almost impossible for men to move between compartments.

Guns might be silent in the darkness, but nothing else was. Screws turned; machinery roared, vibrated, hummed by night as by day; asdic, bridge and radar watches were meticulously sustained. Most men on most ships received a hot meal, snatched some sleep amid the arduous business of reammunitioning – shifting shells from magazines to the guns' lockers and ready-racks, to replace expenditure during the dusk attack and persecution of the unfortunate Hurricanes. Spare hands on both warships and merchant vessels were employed refilling cannon-shell clips and drums. Endless maintenance work had to be performed: on *Ledbury*, for instance, an ordnance artificer worked all night to repair a broken piston on one of the destroyer's forward guns; some Bofors and Oerlikons were malfunctioning in consequence of exuberant gunners' continuous firing, in place of the three- or four-second bursts they were trained to use.

The ships of Syfret's 6th Destroyer Flotilla basked throughout the war in the generosity of a rich American widow named Mrs Hanrahan, whose husband had served with the US Navy's 6th Battle Squadron at Scapa Flow in 1918. 'Aunty May', as she was known to the grateful crews, dispatched consignments of socks, vests, pullovers and balaclavas, both before and after her own country had joined the struggle. While the warm clothing was little use on Pedestal, her gifts of records certainly were: ships broadcast Aunty May's jazz contributions to the mess decks during lulls in the action.

Captains thought themselves fortunate if they closed their eyes for two hours. 'It seemed as if the great ships were stealing secretly through the sea,' mused Roger Hill, contemplating the dim shapes of cruisers and carriers while sipping cocoa on *Ledbury*'s bridge at 0430, amid the tangle of bulbous nineteenth-century voice-pipes that mingled with twentieth-century electronics around his high chair: 'What do you think on a morning like this? Do you think, "A lot of people are going to get killed today," or even "Shall I be killed today?" No, not a bit, not once the game is on … With all the things to attend to, all your young men around you and dependent on you, and all the other ships in sight, you never think anything will happen to you.'

More than a few crews, early that morning, reflected that they would enjoy the reassuring power and presence of Syfret's capital ships for only another fifteen hours before they turned back for Gibraltar at the entrance to the Sicilian Narrows.

The breeze was a light easterly, aligned with their intended direction, which meant that the carriers were spared from making a drastic change of course before flying off fighters. An officer described the scene aboard *Victorious*: 'The standby squadron was all set on deck, with the aircraft armed, fuelled and waiting and the pilots in their cockpits gazing upwards and perhaps munching a biscuit. Men stood by the lanyards which were secured to the wingtips of the aircraft – others lay by their chocks and yet more men sat astride their starter motors. The flight-deck officers fiddled with their flags and the Commander Flying nursed his microphone. There was a tenseness in the air, expectancy and waiting, all waiting, for those vital seventeen seconds which would follow the Boatswain's Mate's call "Fighters stand to". The mad scramble to get the aircraft off and then the eighteenth second should see the ship returning to her station, with her fighters airborne over the sea.'

At 0620, a German plane radioed a position report on the British fleet, and soon afterwards Ju88s and Cants began to circle. Two Fulmars and two Hurricanes took off from the carriers at 0630, and the standing patrol was reinforced to number twelve fighters. Such a force was thereafter maintained aloft, and occasionally exceeded, for most of the day – but the rotating flights of defenders were nonetheless always outnumbered by the attackers. On the 12th, the pilots' experience was frustrating from the outset because, as on the previous day, they found themselves lacking speed to drive off any enemy 'snooper' for long.

Yet on returning from a patrol Hugh Popham never lost his delight in the first glimpse of his beloved home carrier, 'the white water curving away from her hull, her attendant destroyers to port and starboard. For all her lopsidedness, her proportions are so rakish, the strength of her lines so exact, each fresh sight of her from the air has the same compelling excitement. The squadron breaks up, the [landing arrester] hooks go down. For a moment or two, as one pushes back the hood and lets in a blast of cold air that sends the dust in the bottom of the cockpit swirling, the pleasure of flying, the strangeness and the wonder, obliterate anxiety. This is something to have done, to have flown so easily over a great ship on a glittering sea, to have this liberty. This is our everyday, our humdrum marvel.'

Here were sensations such as inspired a host of young men of all nations to brave extreme hazard.

Those hours brought the British a small success, out of sight of the fleet: they exploited the enemy's preoccupation with Syfret's ships to slip out of Malta two empty merchantmen, survivors from the ill-fated Harpoon convoy. These were barely seaworthy after suffering mine damage and were protected by two equally leaky escorts. Their decks were painted in Italian colours to deceive questing eyes. Several enemy aircraft were indeed dispatched on search sorties, after Axis eavesdroppers intercepted their wireless transmissions, but none found them through the day of the 11th. That night *Matchless*, one of the escorts, exchanged ineffectual fire with an Italian destroyer inside Vichy French territorial waters in the Sicilian Narrows. The *Lanzerotto Malocello* and an accompanying MAS fast torpedo-boat were engaged in minelaying, and declined to be distracted to engage Force Y, as the little convoy was designated. This was fortunate for the British, who were in no state to fight. Next morning, several Axis planes overflew the west-

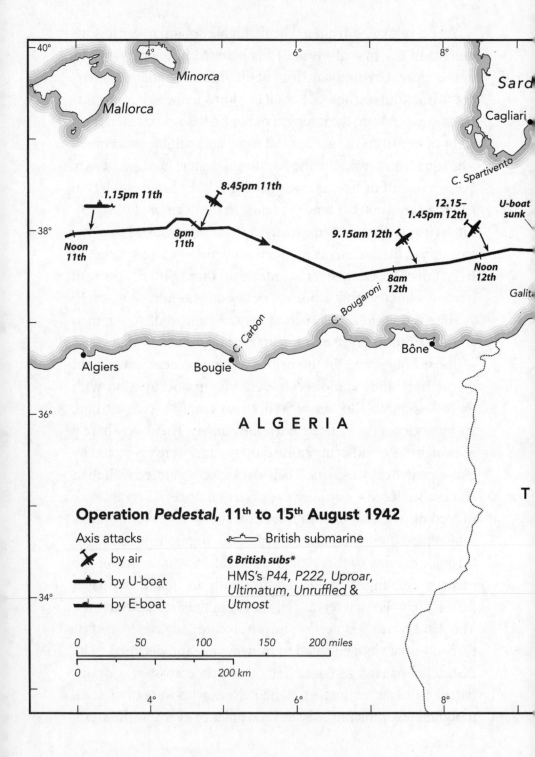

40°

4°   6°   8°

Minorca

Sard

Mallorca

Cagliari

C. Spartivento

1.15pm 11th   8.45pm 11th   12.15–
1.45pm 12th   U-boat
sunk

9.15am 12th

38°   8pm
11th

Noon
11th   8am
12th   Noon
12th

Galit

C. Carbon   C. Bougaroni

Bône

Algiers   Bougie

36°   A L G E R I A

Operation *Pedestal*, 11th to 15th August 1942

Axis attacks   🚢 British submarine

✈ by air   ***6 British subs****

⚓ by U-boat   HMS's *P44, P222, Uproar,*
Ultimatum, Unruffled &

34°   ⚓ by E-boat   *Utmost*

0   50   100   150   200 miles

0   200 km

4°   6°   8°

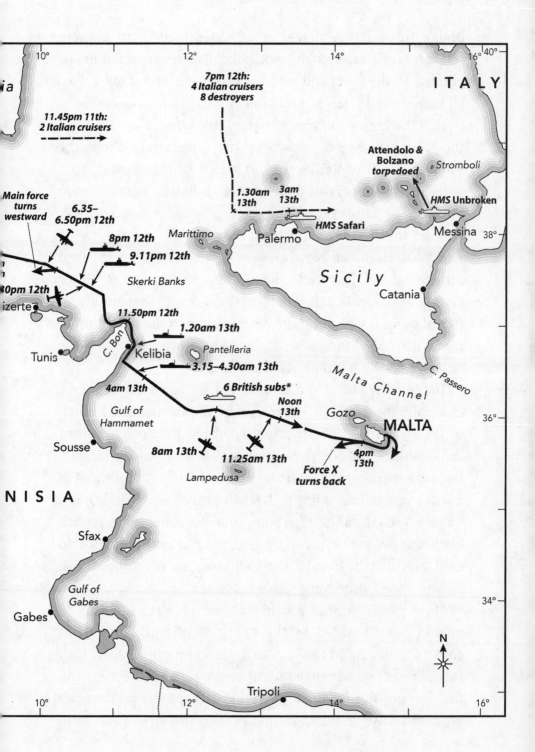

ITALY

**7pm 12th:
4 Italian cruisers
8 destroyers**

**11.45pm 11th:
2 Italian cruisers**

**Attendolo &
Bolzano
torpedoed**

Stromboli

**1.30am
13th** **3am
13th**

HMS **Unbroken**

HMS **Safari**

Messina

**Main force
turns
westward**

**6.35–
6.50pm 12th**

8pm 12th

9.11pm 12th

Marittimo

Palermo

S i c i l y

Skerki Banks

Catania

30pm 12th

izerte

11.50pm 12th

1.20am 13th

Pantelleria

Tunis

C. Bon

Kelibia

3.15–4.30am 13th

Malta Channel

C. Passero

4am 13th

*Gulf of
Hammamet*

6 British subs*

**Noon
13th**

Gozo

MALTA

Sousse

8am 13th

11.25am 13th

Lampedusa

**Force X
turns back**

**4pm
13th**

NISIA

Sfax

*Gulf of
Gabes*

Gabes

Tripoli

N

bound ships. Rome now correctly identified them as mere fragments of Harpoon, not worth the diversion of attention or bombs. To the relief and perhaps surprise of the ships' crews, all four arrived safely in Gibraltar two days later.

That Wednesday morning of the 12th, Syfret, Burrough and the Admiralty were gnawed by the uncertainty about the intentions of the Italian fleet. It was a British nationalistic nonsense of that era to pretend that all Italian warriors were bunglers or fainthearts. The truth was more complicated. Though the shortcomings of most – not all – of the Italian Army had been laid bare in 1940–1, Mussolini's navy ranked fifth by size in the world, being bigger than that of either Germany or Russia. It had entered the war with six battleships, two of them faster and more heavily gunned than any ship of the British Mediterranean Fleet; twenty cruisers; sixty-one destroyers; over a hundred submarines.

The Italians revealed a notable gift for special operations, especially those employing frogmen. The British made much of their own success in crippling three Italian battleships through the Royal Navy's November 1940 air attack on Taranto, but six superbly brave Italian swimmers achieved almost as much a year later by mining the battleships *Valiant* and *Queen Elizabeth* in Alexandria harbour, so that neither played much further useful part in the war. Italian submarines and mosquito craft, of which the Royal Navy would see much in August 1942, also achieved some remarkable successes.

Many of Mussolini's warships, however, were beautiful to behold but ill-suited to the cruelties of war. Their light construction caused them to struggle in rough weather; they lacked radar, good rangefinders, anti-aircraft protection and flashless powder for night operations. There was no naval air arm, and bitter mutual antipathy with the air force. Ciano wrote acidly in his diary on 13 July 1942: 'The real war at sea is not between us and the British, but instead between our own navy and air force.' Japan's naval attaché in Rome was of the same opinion. He expressed respect for Italy's armed forces, but added: 'there is no collaboration between the navy and the air force. Everyone is fighting on their own. This seems absurd to me.'

Italy's warships were short of officers, and a bitter class divide – much deeper than in the Royal Navy – separated ratings from the men who commanded them. Kesselring attributed many of the Italian armed forces' troubles to a lack of human sympathy, exemplified by starvation rations issued to Mussolini's troops in the field contrasted with the relatively sybaritic messing arrangements of their leaders. While it would be foolish to idealize the relationship between officers and men on British and US warships, in the Italian Navy the divide was absolute. A stoker said: 'The officers were a separate caste who would never speak to a humble sailor except to give him an order.' Once when this man ventured on to the upper deck to smoke, he sought to strike up a conversation with an officer standing nearby, 'but he acted as though he had not heard what I said'. When the sailor's own light cruiser was sinking, he felt nothing but darkly amused contempt for the captain's cries of 'Long live the Duce! Long live the King! Long live Italy!'

In an early wartime action, the heavy cruiser *Fiume* ignited smoke floats on her fo'c'sle to lay a protective screen, but these set fire to the wooden decking, causing both sides to assume that the ship had been hit. A young midshipman on a gun position below the bridge was bewildered to observe that while this emergency was being addressed, above his head the admiral, his chief of staff and his second-in-command conducted

an angry squabble about which should take precedence entering the cramped command position.

A senior naval officer, Giuseppe Pighini, said that Italian
warship design emphasized speed and looks at the expense of
armoured protection and indeed hull strength. Fifty-three-
year-old Admiral Alberto da Zara of the 7th Cruiser Squadron
was by far the most respected Italian commander, and enjoyed
an additional reputation as a wit and playboy who loved horses
and women. His comrades cherished a belief, possibly well-
founded, that he had seduced the Duke of Windsor's wife
when she was still Mrs Wallis Spencer, living in China in
1924–5. Da Zara now laboured under the handicap of a poor
relationship both with the Italian Navy's pre-war chief of staff,
Admiral Domenico Cavagnari, and with his successor Arturo
Riccardi, who granted him little latitude for pitting his cruisers
and talents against the Royal Navy.

Naval headquarters in Rome – the Supermarina – was
notorious for the sluggishness of its thinking and decision-
making. Again and again, opportunities came and went before
the admirals and their staffs had decided how to address
them. Before the war Admiral Sir Dudley Pound, then based
on Malta as C-in-C Mediterranean, enjoyed frightening
visiting Italian naval chiefs by taking them for breakneck
drives around the island: 'just as well to get the measure of
one's possible enemies', he said gleefully. Now that war was
upon them, these same Italian commanders were terrified of
having ships sunk, because they could not be replaced. They
were repeatedly worsted in actions with British fleets even
when themselves superior in numbers. Captains were never
permitted to exercise initiative; instead they were always
obliged to ask themselves 'what will the admiral say?', even
'what will the Duce say?' Admiral Angelo Iachino remained

fleet commander despite his repeated refusals to grapple with the British.

Mussolini promised that he would commit his ships at 'a decisive moment'. Hitler's commanders were sure that this would never come. Because Italian warships never exercised under realistic conditions, Kesselring dubbed Mussolini's fleet 'a fair-weather navy', with a superabundance of admirals and shortage of almost everything else. A pre-war Italian study of the maintenance of sea communications between the home country and its Libyan colony in the event of hostilities concluded that the problem was insoluble. A paper from the Rome Supermarina stated that 'if we persisted in the attempt [to sustain a shipping supply line between Italy and North Africa], we should end by losing our entire Navy and merchant fleet'. If the final outcome was not quite that bad, Italian pessimism proved well founded.

When 'ABC', Admiral Sir Andrew Cunningham, relinquished his post as commander-in-chief of Britain's Mediterranean Fleet in April 1942, he wrote in a farewell message to his command: 'the enemy knows we are his master on the sea, and we must strain every nerve to keep our standard of fighting so high that this lesson never fails to be borne in upon him'. This was an overstatement: the Royal Navy was paying a terrible price in lost ships to keep the seas in what Mussolini yearned to make 'an Italian lake'. But Cunningham's words were justified in that many Italian admirals feared to go head-to-head with the British.

And as influential as all the above in determining the Italian Navy's conduct in the Mediterranean was lack of means to move its ships. The Axis powers experienced immense problems in shipping oil from Ploeşti in Romania, through the Black Sea and the Aegean and across the Adriatic. This was German oil, because Hitler now owned Romania. He promised generous supplies to Mussolini, but his pledges went unfulfilled: herein lay an important element in the Duce's grievance about German perfidy.

In 1942, whenever Berlin suggested that the Italian fleet might put to sea to engage the British, Rome shrugged that its capital ships, especially, lacked fuel to do so. In June, the Italian Navy received only twelve thousand tons, one-fifth of the volume consumed by convoys to North Africa; the balance had to be drawn down from reserves. Battleships' bunkers had to be almost drained to provide oil for convoy escorts. This situation still obtained when Pedestal was launched. Yet at this date – and a vivid indication of the still-uncertain outcome of the war – Cavallero and his colleagues cherished hopes of receiving succour within months. The Wehrmacht seemed poised to capture the oilfields of the Caucasus, and Hitler promised Mussolini his share of this priceless booty.

On the morning of 12 August, the British were doing everything possible to deter an Italian fleet sortie: eight of the Royal Navy's submarines in the central Mediterranean, most patrolling south of Pantelleria, were ordered to take the risk of allowing themselves to be glimpsed on the surface by the enemy in hopes of frightening Supermarina. If the Italians still appeared likely to launch an assault on Pedestal, however, a difficult decision would be required. Burrough's squadron, if left alone with the merchant vessels, would be heavily outgunned by the enemy's battleships and cruisers. *Nelson* and *Rodney* had been sent to sea to enable Syfret to meet a sortie by capital ships on equal terms, or to see off Italian heavy cruisers. If, however, the British battleships remained overnight with the convoy in the Sicilian Narrows, instead of turning back as planned, they would be exposed to mortal risk from submarines, mines and air attack.

At that moment, it was indeed the Italian Navy's intention to fight. Da Zara, who had sailed the previous night from Cagliari with two cruisers and had been reported by the RAF's Wellington reconnaissance aircraft, held their course towards a rendezvous planned for 1900 on the 12th just north of the eastern tip of Sicily with the heavy cruisers *Gorizia* and *Bolzano*. Another cruiser, the *Muzio Attendolo*, was proceeding south from Naples, expecting to be joined from the north by *Trieste*. This combined force of three heavy and three light cruisers, escorted by eleven destroyers, was to steer an interception course through darkness to fall upon whatever remnants might be left of Pedestal early on the morning of the 13th off Pantelleria, sixty miles south-west of Sicily.

The previous day, both Luftwaffe and Regia Aeronautica planes had been landing on this little island, barely half the size of Malta, in readiness to join the attack. Italian torpedo-bombers were obliged to make nervous descents in a crosswind, weapons already slung beneath aircraft bellies. A squadron of German single-engined fighters, making their first landings there, were oblivious of the hazards of its rough strip. Several Messerschmitt Bf109s approached too fast and were damaged or wrecked. One skidded into the side of a hill beside the runway, hurling its pilot from his open cockpit on to the rocks, a dead man.

Yet, while Kesselring prepared to commit prodigious air power to attacks on Syfret's ships, the Italian naval high command was preoccupied with securing cover for its cruisers. The admirals knew that on Malta a force of RAF Beaufort torpedo-bombers was standing by to meet an Italian sortie. Thus they demanded German fighters to sustain an umbrella over Da Zara's squadron. In truth, it is unlikely that the RAF could have done much harm to the Italians, even had every plane available been committed, including Lyster's Albacores. The pilots who would have had to execute the attack regarded it as a suicide mission: one of them, Wing-Commander Patrick Gibbs, wrote later: 'the role of torpedo squadrons was never enviable; their crews were often kept in suspense by the hourly possibility of a call which seemed more likely to result in death than in glory. [The aircraft] were immobilized for weeks on end by some target that would not come out and fight, yet the attack for which they waited might annihilate the squadron.'

Even as Da Zara and his cruisers headed towards their rendezvous, argument raged at three successive meetings of the Italian chiefs of staff about whether they should be permitted to fight. All that Syfret and Burrough knew on the morning of the 12th, however, as they approached the waters most dangerous to the fleet, was that the coming days might bring any combination of surface action, submarine attack and massed air assault. Dawn placed Pedestal within a few minutes' flying time of Axis airfields on Sardinia, and fifty miles north of Bône, in neutral Vichy French Tunisia. Ahead lay a new patrol line of enemy submarines, awaiting their coming. In addition U-205, U-73 and the Italian Uarsciek were still trailing the fleet, delivering regular course and position reports. In the first two hours of the day from 0550, British ships logged twelve reports of alleged submarines or periscopes seen and of asdic contacts, or at least a semblance of them.

Every ship's crew grew accustomed to hearing the thuds of exploding depth-charges, to seeing great spouts of white water and spray that rose sixty feet in the sky astern of an attacking escort. Such activity continued all day, even during air battles. At 0741, the cruiser Kenya took hasty evasive action after sighting three torpedo-tracks closing on her port side. During the hours that followed, destroyers carried out a series of attacks on a contact later identified as the Italian submarine *Brin*, which escaped unscathed: flotilla leaders recalled the escorts, rather than allow them to stray from the screen in pursuit.

RAF aircraft from both Malta and Gibraltar strove to assist the anti-submarine battle. At 0934 that morning a long-range Sunderland flying-boat from the Rock spotted the Italian *Giada* seventy miles north-west of Algiers and delivered a depth-charge attack. The damaged submarine crash-dived and escaped destruction, but began to leak so badly that it was obliged to surface. Four hours later, another Sunderland caught the boat in plain sight and inflicted further damage with a pattern of four depth-charges. The slow, lumbering aircraft then closed for a second run, to be met by a hail of Breda machine-gun fire from the Italians, which caused it to crash into the sea and explode. *Giada*'s crew, astounded at their own achievement, gave a riotous cheer. The boat proved able to limp into neutral Valencia that night carrying one man dead and eight wounded, but without further incident.

Even as that battle was being fought – a small epic in its own right for both airmen and submariners – at 1135 Franz-Georg Reschke's *U-205* closed on Pedestal from the port side. An alert asdic-operator on *Pathfinder* immediately detected the threat. Within minutes the destroyer was dropping depth-charges as two deafening blasts on *Nigeria*'s siren, audible for miles across the sea, prompted the fleet to make an emergency turn, such as would be often repeated during the hours that followed. An engine-room artificer aboard a convoy escort said of the experience of serving below in such circumstances: 'you cannot see the submarine, you only have a mental picture of her. My reaction was always the same. When the attack started I would become aware of my heartbeat … it was heavy

but normal ... When we increased speed, the prelude to dropping depth-charges, my heart would increase speed until it
raced with mad excitement.' On that morning of 12 August,
following detection Reschke – at thirty-two one of the older
U-boat skippers and among the less successful, with a final
record of one sinking in twelve patrols – broke off his attack
and set about escape, pursued by more depth-charges from
Zetland. Reschke and his crew survived, but the escorts
achieved their purpose, for the fleet left the submarine behind.

2 DOGFIGHTING

The day's first assaults by a visible enemy came from the sky.
Ultra intelligence had informed Syfret on 9 August that no
Luftwaffe units were based on Sardinia, while the Italians had
130 planes there. Thereafter, however, the Germans shifted
scores of strike aircraft and fighters to the island. By that morning of the 12th, 189 bombers and fighters and sixty-seven
torpedo-carriers were available on Sardinia, while a further
thirty-six torpedo-carriers and 132 other aircraft were standing by on Sicily. Throughout the battle, Kesselring's squadrons
repeatedly switched bases between Sardinia, Sicily, North
Africa and Pantelleria, to conform to the easterly progress of
the British ships. From airstrips across the theatre, the Germans
now deployed 456 aircraft – 328 dive-bombers, 32 high-level
bombers and 96 fighters. A further twenty Ju88s, anti-shipping
specialists, flew into Sicily from Crete on 11 August, and eight
more the following day, having completed an escort mission
for an eastern Mediterranean convoy, supplying Rommel's
forces in North Africa.

The Italians had an additional paper strength of 328 aircraft:
ninety torpedo-bombers, sixty-two bombers, twenty-five

dive-bombers and 151 fighters. Immediate availability throughout the Pedestal battles was around half the above numbers. The Regia Aeronautica, like Mussolini's navy, owned many handsome machines which were hampered by functional shortcomings. The tri-motor Savoia-Marchetti Sparviero or Sparrowhawk, many of which the British would encounter in the August battles, was nicknamed by its crews *il gobbo maledetto* – 'the damned hunchback' – because of its curious profile. Sparrowhawks were outclassed by other nations' bombers, and lacked good bombsights, navigational aids, armour-piercing bombs. The Italians' dive-bombers were so hopeless that they were obliged to beg Stukas from the Germans. Only Italy's torpedo-bombers were superior to those of the Luftwaffe.

The weakness of the entire air fleet was that its designers favoured manoeuvrability over speed and firepower. Fighter pilots loved stunting the open-cockpit biplane Fiat CR42, but scores died in it. Mussolini's best fighter was the Macchi Thunderbolt, fitted with a German Daimler-Benz engine, which the Royal Navy often met over the Mediterranean. In the course of the entire war, however, Italy produced a mere 13,253 aircraft against Germany's 72,030, Britain's 92,034, America's 163,049. British factories produced as many planes in a month as did Italians in a year.

Nonetheless, if few of the Regia Aeronautica aircraft now deployed against Pedestal were in the same class as those of the Luftwaffe, there were so many of them that they represented a formidable threat. For at least the next twenty-six hours, Syfret's fleet would be sailing within 150 miles of enemy airfields on Sardinia, Sicily and Pantelleria. In the Sicilian Narrows, between the island and Cap Bon, after the capital ships turned back Burrough's convoy must pass through a stretch of sea only a hundred miles wide, heavily mined by the enemy.

The fleet's radar picked up most incoming enemy aircraft at ranges of between eighteen and twenty-five miles, six or seven minutes' flying time, and corresponding warning. In the short but game-changing history of radar, British warships had become so addicted to its powers that the regular interruptions of scanning, to change alternators in the Type 281 gun-ranging sets, caused 'ten minutes' acute anxiety', in the words of *Charybdis*'s captain. Admiral Lyster later expressed dismay that the instinct of both pilots and fighter-direction officers was to concentrate upon giving close protection to the fleet. But 'the only successful interceptions are those well out, giving the fighters a chance to make a proper attack before the enemy breaks up into attacking groups'. Lyster may well have been wrong about this: fighters could perform invaluable service by distracting enemy aircraft as they attacked: causing them to miss their aim was more useful and more important than shooting them down.

At 0907 radar heralded a formation of enemy aircraft, which proved to be composed of nineteen Ju88 bombers, approaching the ships head-on. Their pilots craned in their cockpits for a first sight of British fighters, especially coming out of the sun: 'our eyes burned and streamed from the exertion, our necks grew sore with continual twisting in every direction, the harness cut into our skin as we turned in our seats'. They were met by sixteen Hurricanes, Fulmars and Martlets, which engaged them in fierce dogfights, while every gun in the fleet opened up.

A few of the German aircraft were sufficiently discomfited to jettison their bombs, but most pressed on. It became a new frustration for the carrier pilots that even when they hit enemy aircraft their .303 machine-guns lacked killing power: again and again, the big, robust Ju88s suffered damage yet flew on. Despite these disappointments, however, this first air battle of the day was an unequivocal victory for the British. Of nineteen enemy attackers, at least two failed to return – both the fleet's gunners and the pilots claimed far more, but that was to be expected. Much more important, not a single bomb hit a ship. The raid was over in six minutes.

As the fighters landed back on the carriers, wildly excited pilots jumped down from cockpits, exulting in their successes. Moreover, soon afterwards two Fulmars executed a carefully planned ambush of an Italian Savoia-Marchetti 'snooper', pouncing on it from out of the sun, so that the plane fell in flames close to the North African coast. The Fulmars did less well against a similar aircraft shortly afterwards: Italian gunners inflicted so much damage during a long chase that the fighters were obliged to break off and hasten back to their carrier to seek repairs. Aboard the destroyer *Penn*, petty officer Les Rees 'like a fool' expended an entire roll of precious personal camera film, taking pictures of attackers with his little box Brownie. Only much later did he realize that, however impressive was the air battle in the eyes of beholders, the planes – never closer than five hundred yards – showed up as mere specks on Rees' 'holiday snaps'.

For more than two hours thereafter, there was no serious air attack. Outer screen destroyers occasionally loosed off at a 'snooper' which came too close but, until after midday, an uneasy tranquillity prevailed. The fleet drove steadily onward, captains conscious that while they were now eighty miles closer to Malta than they had been at dawn, they were also within seven hours of Syfret's departure and loss of his carriers and priceless fighters.

At noon, the Italians made their big air effort of the day, a carefully coordinated operation from bases on Sardinia, scarcely a hundred miles to the north. Seventy strike aircraft, heavily escorted by Thunderbolt fighters 'which should suffice to clean the sky', in the cocky words of their commanding general, closed in waves upon the fleet, as in response the carriers launched every plane they could put into the air. First lieutenants on the warships took up action stations aft, distanced from their captains so that they might survive to assume command if, as sometimes happened, the bridge officers were wiped out by a bomb or shell.

The enemy's first wave was composed of ten tri-motor Savoia bombers, equipped with parachute-retarded mines which they dropped in the path of the convoy from a height of ten thousand feet: so-called *motobomba FFs*, innovative self-propelled, radio-controlled 900-pound torpedoes, with a 240-pound warhead and a mercury automatic device that activated an electric motor on hitting the water, starting the weapon on a nine-mile circular course. Chiefly to distract the defences, old Fiat Falcon biplanes simultaneously attacked the British escorts with small bombs at low level.

The three Hurricanes first engaging the attackers, which were escorted by fourteen fighters, shot down two and damaged several of the remainder. The Italians released their *motobombas* two miles ahead of the fleet before turning for home. Syfret gave the signal for his ships to make an emergency turn to evade the weapons – which the British characterized as mines – all of which exploded harmlessly a few minutes later. Meanwhile Martlets engaged the low-level Italian biplanes, only three of which pressed their attacks, one of them near-missing a destroyer. Although this attack failed, it deserves notice that both the Germans and post-war Americans were much impressed by the *motobomba* technology, which the Luftwaffe later developed into a highly effective weapon.

The afternoon attacks had scarcely begun, however. After a brief lull, at 1242 two new formations of Italian torpedo-bomb-ers droned towards the fleet – thirty-three Sparrowhawks and ten Savoia SM84s, escorted by Reggiane fighters. The torpe-do-carriers had taken off in low spirits. Martino Aichner kept unhappily repeating to himself a statistic avowed by his squad-ron, that a pilot might hope to survive no more than three attacks on enemy shipping. For Aichner and several of his comrades, this was the third. He was also dismayed that the attack would be led by a novice, Captain Manfredi. 'We release all together from too great a height to have a chance of a hit before the ship takes evasive action. We see it make a high-speed turn to port, churning up a great cascade of water.'

Admiral Syfret's action diary, kept by his secretary, who stood behind him on the flag bridge of *Nelson*, conveys a vivid impression of the intensity of the action:

> 1222 – Aircraft [falls] over *Indomitable*'s side, destroyer [ordered] to pick up pilot bearing 210.
>
> 1225 – [Signal] from *Nigeria*. About 10 mines [dropped ahead]
>
> 1226 – Turn together to 076
>
> 1229 – Bomb dropped bearing 110 between 2 destroyers on screen
>
> 1234 – Explosion bearing 120 … Mines going off [exploding]
>
> 1236 – *Manchester* opens fire
>
> 1238 – 6 parachute mines dropped ahead of *Fury*
>
> 1240 – Torpedo-bombers coming in on port bow. *Rodney* firing 16″ barrage
>
> 1241 – *Derwent* reported circling torpedoes

1242 – [Torpedo] dropped abreast *Rodney*. Another off *Rodney*

1244 – Aircraft splitting. *Nigeria* [signals] torpedo ahead

1245 – Executive signal [to fleet turn] 45 degrees to port. Starboard side – torpedo-bombers on beam

1247 – 4 more torpedo-bombers starboard quarter. 2 flying towards Green 110. 3 torpedo-bombers port quarter

1252 – Still one or two torpedo-bombers near carriers on starboard quarter. Turn together to 075 executed.

1253 – Parachute coming down bearing 050

1254 – 2 mines bearing 320

1256 – ? Aircraft shot down

1257 – Tall black column of smoke starboard quarter

1258 – Torpedo-bombers on port quarter

1259 – 4 torpedo-bombers on port quarter are opening out.

1301 – *Wishart* dropping depth charges in position C

1302 – [Signal] from *Kenya*: torpedo-bomber in sight bearing Green 45. Explosion ahead.

1304 – 1 coming in from port quarter

1306 – Emergency turn to starboard

1307 – *Wishart* drops depth charges bearing 030

1308 – Enemy fighter ahead

1309 – *Vansittart* investigating [asdic] contact port beam. Emergency turn to 075

1315 – *Nigeria* opened fire starboard side. Bombs dropped.

This snapshot, from an almost random page of the admiral's action diary, covers less than an hour of 12 August: similar activity and flag signal exchanges continued almost uninterrupted through that longest day, amid a cacophony of gunfire, aircraft noise, bomb and depth-charge explosions. Roger Hill,

The architects of Operation Pedestal:
First Sea Lord Sir Dudley Pound and the
prime minister, photographed together at sea.

The devastation of Malta, including the opera house that was once its pride.

The governor Lord Gort on his daily round (above), RAF chief Keith Park (left) and the man charged with Malta's deliverance, Neville Syfret (below, centre).

SAILING: following Harold Burrough's briefing of the merchant ship captains, aboard *Nigeria* he offers a parting handshake to Dudley Mason of *Ohio* (above) and discusses the plan with Drew of *Manchester* (below, left). The mustachioed little figure (below, right) is Wren, of *Rochester Castle*. Note that the Merchant Navy as well as Royal Navy officers are clad in their Sunday best.

(Above) An Italian aerial reconnaissance shot showing a fraction of the Pedestal fleet at sea. (Below) *Indomitable* with (right) her captain, the doughty Tom Troubridge, wearing his invariable tiny binoculars.

THE FIRST DISASTER: a supremely dramatic image of *Eagle*, listing and about to turn turtle, snapped from one of the escorts. The cargo liner in the foreground is *Glenorchy*. Notice that her lifeboats are all slung out, ready for a hasty evacuation, as they were on every merchant vessel throughout Pedestal.

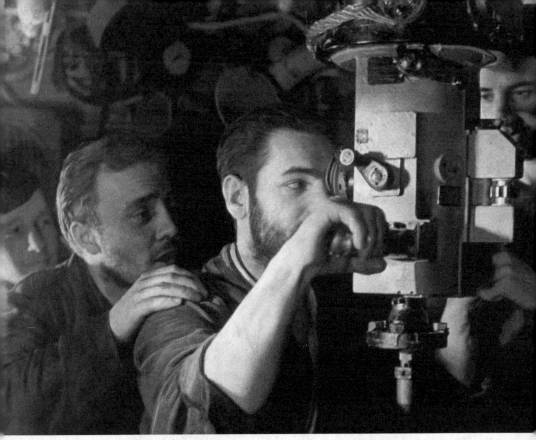

(Below, right) Helmut Rosenbaum, *U-73*'s skipper, awarded the Knight's Cross for sinking *Eagle*. The other images show different boats, but emphasise the enforced intimacy in which crews lived and fought – though the men below look much cleaner than most submarine crews on a long patrol.

(Above) *Eagle* survivors being transferred to a destroyer – a scene often repeated during Pedestal. Notice the scrambling net up which so many men clambered from boats.

(Below) *Eagles* at Gibraltar being filmed for a propaganda newsreel in fresh whites and cheery smiles. In the oil-soaked Mediterranean a day earlier, these same men would have been unrecognizable. (Right) The carrier's captain, clan chief Lachlan Mackintosh.

THE ENEMY: (above) Kesselring with Rommel; (below) Giacomo Metellini, an Italian fighter pilot with film star glamour and (right) Admiral Alberto da Zara, best of the Italian admirals, seen during an Asian interlude in the 1920s when he allegedly conducted an affair with Mrs Wallis Spencer, later Duchess of Windsor (shown here).

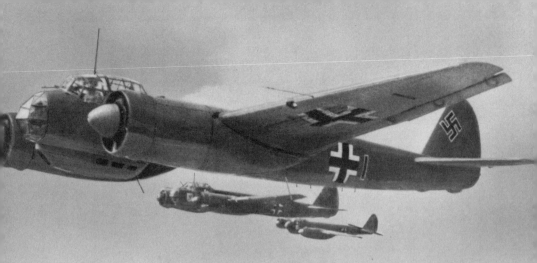

(Above) German Ju88s, superb multi-purpose bombers that inflicted heavy losses on Pedestal. (Right) An Italian Sparrowhawk torpedo-bomber makes an attack. (Below) The Ju87 Stuka dive-bomber which both the Germans and Italians used for many attacks on Syfret's ships.

A depth-charge explodes behind the stern of a destroyer.

A rating makes a signal. Throughout the battle, lamps clattered almost continuously on ships' bridges – the fastest and most secure means of communication.

Destroyers rammed two Italian
submarines during the battle.
(Above) The crippled *Cobalto*,
moments before she sank.
(Right) Some of her disconsolate
survivors, huddled aboard *Ithuriel*.
(Below) Peter Gretton stands at the
centre of his triumphant crew of
Wolverine, which sank *Dagabur* with
all hands.

THE AIR BATTLE: (above) rearming a Hurricane and (below) the carriers in line astern, an Albacore visible in foreground, wing folded. (Right) Burrough on his bridge, though during the battle he wore whites.

Some Pedestal fighter pilots, including Popham (top left), Carver (with bandaged arm, bottom right), Cork (in his cockpit, bottom left) and Crosley (in naval cap, top right).

The flight-deck crew race to take charge of a Martlet landing on HMS *Indomitable*.

Anti-aircraft armament: (left) a 20mm Oerlikon cannon, of which every ship carried several. (Below) A 2-pounder multiple pom-pom, of which every warship carried at least one, and the bigger ships several batteries. The anti-flash capes worn by the crews protected against burns, at the cost of being infernally hot to wear in action. Many men preferred to accept the risk of fighting half-naked.

on *Ledbury*, reflected that, though he had often suffered air attack, this was the first battle in which he had actually glimpsed bombs descending through the air towards his own and other ships. Senior officers knew that the U-boat threat was ever present, even as aircraft attacked and engaged in dogfights overhead.

David Scott, a wide-eyed eighteen-year-old midshipman at the anti-aircraft plotting table of *Rodney*, wrote in his diary: 'Lt. Campbell in the director-tower told me he had actually seen a periscope at six hundred yards, then the tracks of bubbles coming towards us.' The enemy need only get lucky once, amid so many torpedoes launched and bombs dropped, and he could achieve another devastating success, such as the previous day's sinking of *Eagle*.

British planes broke up some of the midday raids before they reached the screen, and gunfire from the destroyers increased the enemy's disarray. Several attackers released their torpedoes impossibly far out, at eight thousand yards. Those that came closer did no harm. One gallant Savoia pilot approached *Nelson* with both wing engines on fire, vainly releasing his torpedoes just before crashing into the sea. Italian losses were relatively light – only four aircraft failed to return to base, though several others were damaged – but the defending fighters had fulfilled their foremost duty sufficiently to distract attackers, to spare the fleet from harm, though the early-afternoon attacks caught most on decks, refuelling. The returning bombers' crews informed Rome that they had sunk two merchantmen and a destroyer, and damaged a battleship, three cruisers and two further merchantmen. In truth, they had merely killed or stunned an impressive number of fish. The Italian torpedo-bombers landed at Elmas on Sardinia, which had been attacked in the early hours of that morning by

Beaufighters. A pilot noted that 'some burnt-out shells of aircraft are still smoking on the strip'.

When the early-afternoon wave of attackers turned for home, the fleet had been in action for eighty minutes. Already gun positions were heaped with empty shell and cannon cases. In the steel boxes of every cruiser's six-inch gun turrets, the crews of twenty-seven men apiece laboured in almost intolerable heat, while below decks a further twenty-two men suffered worse, serving in each turret's shell-room and magazine. Pom-pom loaders' hands were bruised from banging in fresh clips, each weighing twenty pounds. A hazard throughout the operation was that blast from ships' main armament disrupted reception on their radar equipment.

Sunburn was extracting its toll from hundreds of the men fighting for their lives under the blazing orange orb of a Mediterranean August. Some captains, especially on the big warships, rigorously imposed in action the discipline of wearing helmets, coveralls, white asbestos anti-flash hoods and long gauntlets – George Blundell on *Nelson* reflected wryly that his chief coxswain 'looks just like an old crusader' in this rig. The logic was plain: bereft of skin protection, men could suffer terrible burns amid a sudden explosion or outbreak of fire. On smaller ships, however, many captains allowed their men to accept this risk and fight half-naked, wearing whatever they chose. *Ledbury*'s pom-pom crew affected shorts and football shirts, to which one man added a Sherlock-Holmesian deerstalker.

Scheduled meals were abandoned: chests of biscuits and corned beef were instead laid open for crews to snack when they could be spared. On most warships, each man in rotation was permitted to absent himself from his action station for thirty minutes, except during an attack, to perform essential bodily functions. Buckets were meanwhile set around the decks, for those caught short. Yet mortal peril affects different men in different ways. Jack Harvey, a naval signaller on *Rochester Castle*, became chronically constipated, unable to achieve a bowel movement for three days, until the battle ended.

Merchant seaman Patrick Fyrth wrote: 'the difference between the cinema and real life is that [movie] directors always show naval and air battles as an ever-increasing cacophony of noise with tracer and other missiles coming and going from every direction at once. In real life the noise doesn't increase at a regular rate. It rises and falls, sometimes being deafening and then falling almost to silence. Once or twice the noise ceases altogether, and one is conscious of the water lapping the ship's side, and the fact that the sun is bright and hot.'

A light-cruiser officer observed that, for the men serving below decks in engine- and boiler-rooms, conditions during an air attack 'resembled the inside of a giant's kettle against which a sledgehammer is being beaten with uncertain aim. Sometimes there was an almighty clang; sometimes the giant in his frustration seemed to pick up the kettle and shake or even kick it.' The gunnery officer on *Eskimo*, John Manners, said: 'Most of the time, one did not know what the hell was going on.'

Dickie Cork, the twenty-five-year-old Fleet Air Arm veteran who had been credited with destroying five enemy aircraft during the Battle of Britain, was that day flying from *Indomitable* the only Hurricane armed with 20mm cannon. He had his first success of Pedestal at 1230 when he shot down a Savoia-Marchetti bomber. Then, flying further south off the coast of Tunisia, he accounted for one Junkers Ju88 and shared

in the destruction of another. Later he shot down a
Messerschmitt Bf110 and another Savoia-Marchetti, thus
claiming five enemy aircraft in that single day, the 12th. As
always, a handful of exceptionally skilful pilots accounted for
a wildly disproportionate number of successes. A critical
element in Cork's deadliness was an experience of air fighting,
which most of his comrades lacked.

'Danger is relative', wrote a destroyer captain, 'and listening
to the pilots reporting their sightings and actions with enemy
aircraft – sometimes we could hear the guns firing – made the
odd bombs falling around us seem quite safe compared to
their repeated engagements.' Thus far, though explosive near-
misses had prompted moments of terror for a few men on a
few ships, most had found the day more thrilling than daunt-
ing. Hereafter, however, it became progressively grimmer. A
cruiser's surgeon wrote ruefully: 'I would like to lash to the
deck of an a/c carrier, going to Malta, those politicians and
brasshats who say that the dive-bomber is a useless machine.'
Yet hate was absent from most men's feelings, even amid terrifying experiences. Engineer Tom Brunskill, a veteran of almost
twenty years at sea who was now aboard *Glenorchy*, reflected
later: 'I did not then, nor at any time subsequently, feel any
personal animosity towards the chaps that were trying to sink
us and I cannot remember anyone expressing any. We were all
doing our job ... it seemed so dispassionate, and it was that
which seemed to me so frightening.'

The Italians, who had begun the day's attacks, were now
replaced by Sicilian-based German Ju88 bombers, some of
which dropped parachute-retarded mines, while others
attacked merchant-ships, for the first time in earnest, at very
low level. At 1320 three bombs near-missed *Deucalion*'s starboard side, two more exploded off her port side and yet another

pierced a derrick and drove straight down through the deck and her No. 5 hold before emerging below the waterline and belatedly exploding. Alfred Holt's Blue Funnel liner, named for the Greek hero, son of Prometheus, who took his family aboard an ark during a great flood, was carrying eight thousand tons of stores and munitions. Bearing cargo from New Zealand she had docked at Glasgow only on 3 July, after completing a seven-month global circumnavigation.

Her Master, Ramsey Brown, was reckoned among the most forceful personalities in Pedestal's Merchant Navy contingent. The column of water thrown up by one of the bombs drenched gunners on the fo'c'sle and bridge, 'all of whom remained at their posts and behaved in a most exemplary manner', in the words of the captain. Able seaman Bill Lawley, manning the ship's helm, was a veteran of sixty-two: ashore or in the Royal Navy, he would have been deemed far too old for active service. Here and now, however, he kept his post. The Master could see cargo falling out into the sea from a huge hole in the starboard side while a torrent of water coursed into the hull. He ordered the engines stopped so that the damage could be examined. One of the ship's holds soon flooded, and all power was temporarily cut off. His crew gave heartfelt thanks that their 1,500 tons of aviation spirit and TNT had failed to explode. The destroyer *Bramham* stood by and lowered scrambling nets, while Brown ordered the lifeboats lowered to deck level, in preparation for a possible Abandon Ship.

Damage control was a vital art, in which every navy's proficiency increased with experience. Thus by 1945 both the British and Americans were saving grievously wounded ships such as perished in 1942. The Royal Navy developed special skill in controlling blazes: at action stations, every warship deployed strong Fire & Repair parties, commanded by responsible officers, ready to leap into action the moment a vessel was hit. Meanwhile men of the Merchant Navy sometimes performed remarkable feats to save their ships: many such emergency responses would be required in the course of the Pedestal saga. But merchant vessels had much smaller crews than warships, less rigorous discipline and training in the skills of battle. Strength of will could play a critical part in saving bombed or torpedoed ships: some crews had it; others did not.

A British Medical Research Council study found that the most dangerous time for survivors of a bomb or torpedo hit was between its impact and sailors reaching a boat or raft. Seventy per cent of merchant-ships that were destined to sink went down within fifteen minutes of being attacked; barely one in ten stricken merchant vessels remained afloat for more than an hour. Crews were thoroughly aware of these realities, of the need to beat the clock. Here was evidence to justify a headlong dash for escape, such as often took place after a hit. Even as *Deucalion*'s officers were assessing the damage, without orders or permission terrified greasers and assistant stewards lowered two boats, pulled away from the ship and sought refuge along-side *Bramham*.

Ugly scenes followed, as the fugitives attempted to ascend the destroyer's nets to reach her deck. In the words of her captain, Eddie Baines, 'we jumped on the men's knuckles … invited them to choose: either they could get back into their boats and stay there, or go back to the *Deucalion*'. The twenty-nine-year-old Baines was no respecter of persons. An ebullient character who had commanded his ship for only three months, he was described by one of his officers as 'in some ways as crazy as Don Quixote, and the windmills he tilted against were usually those of Authority … Pomposity and arrogance were targets he could not resist.' Now there were shouted exchanges of obscenities between his bridge and the boats before the reluctant merchant seamen pulled back to their own vessel. This episode, and the altercations, caused *Deucalion* to remain hove to for an hour while the rest of the fleet steamed on eastwards. Ramsey Brown belatedly got his ship under way again, having been ordered by Syfret at 1349: 'if you can steam, steer inshore and follow coastal route' rather than trail the convoy, already some twenty miles ahead. Hereafter, *Deucalion* was on her own.

3 MINES, *BOMBAS*, TORPEDOES AND A CANARY

The leading vessels of the fleet were now 355 miles from Malta. Admiral Syfret's log recorded a twenty-minute lull, between 1357 and 1417, before he received renewed warnings of both approaching aircraft and lurking submarines, swiftly followed by the sounding of warships' alarm gongs and a new onslaught. First came an incongruous little aerial comedy of which the British, far below on their ships, were at the time oblivious. The Italians unveiled a radio-controlled aircraft, designed to crash on to a warship, laden with two tons of explosive. So that its accompanying guidance plane could more easily track its progress to the target, the converted SM79 *bomba* was painted yellow, and thus nicknamed *il canarino* – the canary. It took off at 1300, in the hands of courageous pilot Mario Badji, who baled out at six thousand feet before his aircraft approached the British fleet.

The device's inventor, Gen. Ferdinando Raffaelli, was personally manning the radio guidance transmitter aboard an accompanying Cant escorted by two fighters. The *canarino*, however, stubbornly refused to respond to his signals. The Cant pursued its pilotless companion southwards, while

Raffaelli struggled in vain to divert its course towards the British fleet. As the yellow bird crossed the Algerian coast, the general abandoned the mission. When the *bomba* ran out of fuel, it ploughed into the side of a mountain in the Atlas, 150 miles inland, precipitating a spectacular explosion. The Vichy authorities, baffled by the absence of human remains in the wreckage, declared the crew to be presumed 'completely carbonized'.

The next attack on the fleet was made by two Reggiane fighters, which streaked low over *Victorious* as she was landing on Hurricanes. The Italian planes, mistaken for British fighters, each released a hundred-pound fragmentation bomb before escaping amid a storm of flak. One bomb broke up on impact, showering the flight-deck with fragments that killed two men and wounded two more. The other exploded in the sea off the carrier's bow.

Then came another wave of Heinkel torpedo-bombers, escorted by Messerschmitt Bf110 twin-engined fighters. Allan Shaw, aboard *Ohio*, said: 'The sky seemed to be full of planes. How the hell they ever got through the spider's web of tracer being fired at them, I'll never know.' Yet even the most impressive barrage, which dotted the airspace above the fleet with smoke puffs of many colours, left a remarkable amount of empty sky through which attackers flew on undamaged. Les Rees, on the destroyer *Penn*, watched with numb fascination as a formation of torpedo-bombers passed in front of his own vessel, and a few hundred yards astern of the destroyer ahead, with every gun on both ships firing at them: 'The gap was no more than five hundred yards wide and one would think they would never get through because a good shot with a catapult should have been able to hit the nearest one, but not one of those aircraft fell into the sea.'

The defending Hurricanes of 880 Squadron suffered severely. Two enemy fighters shot down one pilot as he was attacking a Heinkel. Two planes forced-landed on *Indomitable*, badly damaged. The squadron CO, 'Butch' Judd, was hit by a Heinkel's rear-gunner, who blew a wing off his Hurricane, causing the rest of the plane to cartwheel into the sea from five hundred feet.

It was an oddity that Judd was detested by most of the squadron, and especially by Hugh Popham. Meanwhile the men of the fleet, who never met him, had formed a warm admiration for the huge South African, formed through eaves-dropping on the VHF relay of the air battles, in which he was credited with shooting down three enemy aircraft. Roger Hill's first lieutenant wrote: 'gradually we were getting to know some of [the pilots] by voice. One in particular we came to know and admire, though to us he was only Black Leader' – this was Judd, and the sailors grieved when he ceased to respond to Control. Finally another pilot reported him shot down while making an attack on a German bomber from directly astern, such as he repeatedly warned his pilots not to attempt, because the enemy's gunners were too good.

Ledbury dashed to pick up an airman parachuting into the sea, in hopes this might be Judd. There was dismay when the man instead proved to be German, prompting callous cries of 'Chuck him back!', influenced by what the fleet was enduring at Luftwaffe hands. It is often suggested that there was little hate between British and German fighting men, yet the ratings of another destroyer, many of them from Chatham which had been devastated in the Luftwaffe's 1940–1 blitz, 'almost rioted' in the words of an officer rather than rescue a ditched German pilot. *Ledbury*'s captive was thereafter employed carrying sandwiches and lime juice to the guns' crews. Dickie Cork succeeded Judd in command of their Hurricane squadron.

That afternoon seemed painfully long and slow as the fleet crept by the island of Galita, off the Tunisian coast: 'we began to wonder if we should ever pass it'. Hugh Popham, having crash-landed on *Victorious*, was now a mere spectator of the air battles: 'I was depressed and disappointed and – some cowardly little voice insisted – rather relieved.' He dutifully badgered the ship's Commander Flying for an aircraft to fight, yet was grateful that none was found for him. All that afternoon, the killing continued. Three Fulmars from *Victorious* shot down an Italian aircraft which landed in the sea, where its crew were fortunate enough to be retrieved by a fishing boat. But two British aircraft hit by the bomber's gunners also came down, and their pilots went to the bottom with the fighters.

At 1515 Syfret warned Burrough that the merchant vessels were 'observed to be straggling', a loss of discipline probably influenced by witnessing the mauling of *Deucalion*. Through the afternoon, the carriers' Hurricanes and Martlets engaged again and again, as successive waves of Ju88s attacked the fleet. Hugh Popham, aboard *Victorious*, wrote: 'the pilots snatched meals when they could and rushed back on deck to take their places on the revolving wheel of readiness. In the hangar the maintenance crews worked like men possessed to make the aircraft serviceable as they were struck down. They were coming on now with battle damage to be repaired as well as the normal troubles of oil-leaks, coolant-leaks, sprained oleo-legs and whatnot. The hangar itself was a shambles as aircraft were ranged, struck down, stowed, refuelled and re-armed at top speed; and the hangar-deck became more and more slippery with oil.'

One episode especially shocked and grieved the flight-deck crew of *Indomitable*: Robert Johnson, CO of 806 Squadron, sought to land a damaged fighter without benefit of flaps and thus hopelessly fast, while himself badly wounded by German guns. He caught an arrester-wire which tore off his hook; then slewed off the deck and on to the catwalk. There followed a tragedy which lasted only seconds, yet to witnesses seemed to unfold in slow motion: Johnson tried in vain to pull himself out of the cockpit, then gave a brave, despairing wave as his Martlet toppled over the ship's side, fell on its back and disappeared. A fellow airman wrote after that episode: 'it became more and more evident that we were not on a picnic; we had not got to the easiest bit for the enemy yet, either'.

Ensign Gerhart Suppiger, commanding the US naval armed guard aboard *Santa Elisa*, became concerned that the black-and-yellow chequerboard recognition panel painted atop an RAF rescue launch being borne to Malta on the ship's deck was giving attackers an aiming point. The chief mate hastily covered the crash boat with a tarpaulin. Many men on both warships and merchant vessels, committed to roles away from the guns, envied the close-range-weapons crews, especially Oerlikon gunners: 'they could do something – they could almost see the whites of [the enemy's] eyes'.

Seaman Charles Hutchinson wrote: 'I found that the worst thing of all is the waiting for what you think will be a certain hit, and watching everything around you go up, and wondering who will be next ... but when actually in action firing, you forget it all and just crack away. At times I felt rotten, but it was soon over when there was anything to do, in fact when we were in local firing and I was laying on one of the torpedo planes and firing the gun myself, I didn't worry two hoots what else was coming at us, I was enjoying myself.'

Midshipman Mike MccGwire wrote: 'The noise was positively exhilarating. There was something about the wild rhythm of the pom-poms that got in one's blood.' On *Glenorchy*, engi-

neer Tom Brunskill repaired a broken weapon, then begged its
gunner to be allowed a turn firing at enemy planes. Brunskill
was rewarded with a refusal laced with expletives. At the opposite end of the spectrum, two Swedish survivors aboard the
cruiser *Phoebe* were invited to make themselves useful by
refilling Oerlikon drums: they refused, solemnly invoking
their rights as neutrals.

And then it was once more the turn of the Italian Navy's
submarines, which unleashed a wolfpack ambush, detected by
both asdic contacts and surface sightings from 1515 onwards.
Through the hours that followed, destroyers reported repeated
skirmishes and depth-chargings. Commanders were conscious
that every escort which left the formation to attack a real or
imagined submarine exposed an opening for another enemy.
At 1559 Admiral Syfret sent an irritable signal to *Intrepid*: '3
destroyers are <u>not</u> required to hunt a contact.' At Admiral
Burrough's suggestion, escorts began dropping a depth-charge
every ten minutes until nightfall, as a deterrent. At 1616
Pathfinder threw a pattern of five around a contact, later identified as *Granito*, which she drove off. At the same moment, on
the fleet's starboard side, torpedo tracks were spotted from the
Italian *Emo*. The attack failed; no British ship was hit. *Emo*
escaped the escorts' attentions that followed, not least because
they were distracted by spotting the periscope of *Avorio*.
Energetic depth-charging by the screen kept both this boat
and yet another, *Dandolo*, beyond effective torpedo range until
the fleet had passed – the last-named was sufficiently damaged
to return to base.

Cobalto was less fortunate. Its captain, Raffaello Amicarelli,
afterwards gave his own account of the afternoon's events. He
first sighted the fleet at 1345, he said, when he estimated his
own position to be thirty miles north-east of Galita. He saw

many aircraft overhead, and a succession of destroyers, and descended to 160 feet to put himself out of sight until the screen had passed and the British heavy ships came within range. He told the crew they would shortly be attacking and ordered the torpedoes to be set at readiness. He returned to periscope depth several times, heard depth-charges, but so distant that he assumed that he himself was undetected. He was wrong. He returned cautiously to periscope depth, while moving slowly northwards to reach an attack position. Then, at 1629, he glimpsed the most terrifying sight a submarine commander could encounter: the bow wave of a destroyer, *Pathfinder*, three hundred yards distant and heading straight for him.

Amicarelli dragged down the periscope and ordered a deep dive, but they had descended only fifty feet when the first depth-charge exploded above: 'The conning-tower shudders as though it must split open, and the entire vessel is forced downwards.' A ghastly ordeal followed. Many men doubted that their boat would ever rise again. *Cobalto* checked at four hundred feet, and the helmsmen wrestled to regain control. The rudders appeared jammed, the boat unnaturally heavy, stern down. Most of the lights were gone, so that only dim emergency illumination remained effective. They struggled in vain to rebalance the tanks. For the crew, 'the pressure within the hull was almost unendurable': blood oozed from men's ears. Amicarelli claimed that his crew remained calm throughout this ordeal, but if this was true it can only have been a paralysis of terror. At last, despairing, he ordered the boat's vents blown. To their vast relief *Cobalto* began to rise, first very slowly, then with uncontrollable speed, while listing heavily to port.

On the flag bridge of *Nelson*, Syfret's secretary logged the events of those minutes in the admiral's action diary, which vividly evokes the chaos and some of the successive dramas of the afternoon:

1622 – PATHFINDER investigating starboard side
ITHURIEL bearing Green 40, investigating contact on port quarter

1626 – PATHFINDER still investigating

1629 – PATHFINDER attacking

1634 – TARTAR [signals]: Torpedo in sight starboard side
LOOKOUT [signals]: Submarine in sight. Fired depth charges. 3 patterns. Confirmed contact

1637 – Emergency air attack signal
LOOKOUT dropping more depth charges starboard side

1638 – Force F Turn together to 106 [flag, VHF and wireless signal to the fleet]

1639 – TARTAR [signals] T[orpedo] starboard

1641 – [Burrough signals his squadron] Turn together to 061
Hunt in progress (TARTAR and LOOKOUT)

1651 – Firing on port quarter – cruisers

1652 – Depth-charge dropped near INDOMITABLE

1654 – Disabled submarine on surface 160 degrees.
Destroyer firing at it. ? destroyer ramming

1656 – Submarine brought to surface – fired on by ?
ITHURIEL on port quarter

1702 – 2 destroyers attacking [signal to ITHURIEL] Well done – rejoin as soon as you can.

This last incident unfolded as *Cobalto* surged back towards the surface. Above the submarine, an able seaman manning the rangefinder on the destroyer *Ithuriel*, named for one of Milton's angels, suddenly cried out: 'periscope green seventy!', and sure enough the bridge crew glimpsed the feather of its wake midway between themselves and *Indomitable*. The ship's captain, lean, elegant thirty-one-year-old David Maitland-Makgill-Crichton, a flamboyant lifelong bachelor known throughout the service as 'Champagne Charlie', barked a string of orders: 'Hard astarboard, twenty-four knots; make the alarm signal; depth-charges ready; press the U-boat alarm.' *Ithuriel*'s wireless-operator messaged hasty warnings to neighbouring ships. The periscope briefly vanished, but the ship's asdic was gaining echoes from below. A seaman pulled the lever to release a pattern of depth-charges which exploded astern seconds later. Every man on the destroyer's upperworks craned in vain for a glimpse of wreckage.

Ithuriel was turning in readiness to make a second attack when suddenly there was a roar from the deck: 'There she is!' As *Cobalto* wallowed on the surface, its hull awash, the destroyer's guns opened fire, almost immediately tearing a gash in the side of the conning-tower. For the submarine's crew, now bent only upon survival, there was a new crisis: they found the hatches jammed. A midshipman, Stelio Romito, wrestled desperately with the upper access until at last it yielded, revealing a thin shaft of merciful sunlight amid a clatter of British machine-gun fire striking the steel. In the submarine's control-room the captain followed a gunner in a dash up the ladder, where they were confronted by *Ithuriel*'s bow towering above them, yards away. Amicarelli yelled down to his crew to abandon the boat.

On *Ithuriel*, Crichton called: 'Full ahead both engines. Prepare to ram!' The 1,900-ton warship raced towards the submarine as bedraggled figures appeared from below and scurried on to her casing. Crichton belatedly realized that he might wreck his own ship unnecessarily: the submarine was already doomed. He ordered his engines full astern: 'You don't need to use a hammer on a boiled egg'. The collision, at 1702, was nonetheless shattering – 'a delightful scrunch', in Crichton's facetious words – abaft the conning-tower on the starboard side, hurling several Italians into the sea. For a few seconds the two ships remained locked together, and four of *Ithuriel*'s sailors jumped down on to the conning-tower. There was no hope of salvage, however – the engine-room was already waist-deep in water.

A minute or two later, the Italian boat heeled over and sank. Despite the initial congratulatory signal to *Ithuriel*, on learning more of what had taken place Crichton's senior officers were dismayed that he had severely damaged the destroyer, necessitating her retreat towards the dockyard. An exasperated Syfret described the ramming as 'completely unnecessary' when *Cobalto* was already obviously doomed. The admiral was even less pleased when Crichton spent the ensuing hour lowering boats and picking up *Cobalto*'s survivors, in all three officers and thirty-eight crew, at a time when an 'imminent air raid' warning was already in force. Amicarelli gave an effusive, perhaps extravagant account of his treatment once aboard the British ship: 'I am received with military correctness. I am kept at a distance from my crew and put in an officer's cabin where I receive the visit of the second officer who delivers a note from the captain. It is a very chivalrous note with expressions of congratulations for the attack attempt and condolences on the loss of my ship and of my men. With his crew, I must think of myself as a guest, not a prisoner.'

Syfret's crews had now been at action stations for over twelve hours and were growing weary. They had experienced a surfeit of thrills and terrors. Thus far, as evening approached, the British could feel that they had had the best of the day. Two submarines had been sunk, the attacks of a pack of others beaten off. Scores of enemy aircraft had done their worst, and succeeded only in damaging a single freighter. George Blundell of *Nelson* wrote: 'bombs fell all over the place ... The barrage put up by the fleet was aesthetically one of the weirdest and most wonderful and beautiful I have ever seen. People who had seen it all had a look on their faces as if they'd seen a vision – the sort of look a man would have on his face after he'd looked on the Almighty. It was the purple sea and black sky, the red in the West, and the pearls and rubies of the tracer necklaces, putting up this miracle display ... Our AA fire has much improved since last year.' *Victorious*'s captain took a harsher view: 'the number of aircraft shot down by what must have been a vast expenditure of ammunition was disappointing, though a great deterrent to the faint-hearted among the enemy'.

He was unjust: the barrage had an undoubted effect on the accuracy of the attackers' bombing. Moreover the carriers' pilots, most of them inexperienced in air fighting and flying outclassed aircraft, had held at bay far superior enemy numbers, inflicting more losses than they suffered. Of the 117 Italian and fifty-eight German aircraft that had attacked thus far, at least eight bombers, a reconnaissance aircraft, a torpedo-carrier and a fighter had been shot down by fighters or guns for no significant result – only a glancing hit on *Victorious* and damage to *Deucalion*. For those who spoke in the language of the football or cricket field, as even admirals of that generation liked to do, they had successfully defended their goal, their wicket, all through that fierce and dangerous day. How unjust it was, then, that even as the sun dipped towards the horizon, entirely new enemy teams arrived on the pitch. These inflicted punishment which brought Pedestal within a hair's breadth of failure.

7

Cruel Sea

1 'I BELIEVE THEY'VE BUGGERED US!'

The battered *Ithuriel*, limping alone behind the fleet after her encounter with *Cobalto* and rescue of the submarine's survivors, was attacked by five enemy aircraft at 1749. The destroyer's bent bow reduced her speed to twenty knots, but she successfully evaded their bombs. At 1813, Admiral Syfret sent out a general signal informing all ships that the fleet would turn west for Gibraltar an hour thereafter; this, although forty minutes previously warning had been received that yet another large Axis aircraft formation was heading its way.

Victorious had been suffering difficulties with her aircraft lifts which slowed flight-deck operations. By a supreme effort, however, the carriers now put into the air twenty-two fighters to meet over a hundred Stuka dive-bombers, Ju88s and Savoia 79s, both Italian and German, escorted by Bf109 and 110 fighters. Lt. Rodney Carver, commanding 885 Squadron, took off with one arm in plaster. Michael Hankey, a survivor from *Eagle*, had already flown so many sorties from *Victorious* that he was visibly exhausted and was advised to stand down. He insisted on taking off, however, and shot down an enemy aircraft before himself being bounced by a twin-engined

Messerschmitt that he failed to see behind his tail. Weariness
led inexorably to carelessness, fatal to a pilot in action. Hankey
and his aircraft vanished without trace beneath the surface of
the Mediterranean.

Giacomo Metellini and two other pilots of his Reggiane
fighter group took off from Chinisia in Sicily, and met the
bombers they were tasked to escort over Castelvetrano. Many
of the fighters failed to make the rendezvous, however, and
Metellini described the bomber leader's decision to press on
regardless as showing 'great courage and some recklessness ...
What we had expected happened: a swarm of Hurricanes and
other British types from the carriers descended on us the
moment we approached the convoy ... Not all the bombers
returned to base.' Metellini wrote of 'absolute chaos' in the sky
that evening, with a dozen different varieties of British,
German and Italian aircraft milling over the convoy, 'constant
explosions from the defensive fire of the warships'.

He claimed that he and his two wingmen 'hit many enemy
aircraft', but admitted that they landed 'without discovering
their fate. I remember seeing a [Savoia] flying low over the sea
on a mad dash to reach a position where it could launch its
torpedo without being hit by AA fire, then a Hurricane getting
on its tail. I dived towards the enemy, opened fire but almost
immediately my guns jammed ... I was hit by a long burst, felt
an impact on the port side of the canopy, saw a hole where a
bullet had passed through, missing me by a few centimetres,
only because my head was bent forward, trying to clear my
guns.'

Then Metellini found his engine missing beats: the temperature gauge warned that it was overheating, probably with
coolant pipes severed. He hastily weighed the options of baling
out, perhaps to fall into British hands, or instead pressing on.

He decided to make for Sardinia, climbing as hard as he could, to win altitude before the engine expired. He kept transmitting his position by radio, but received no response. 'After twenty minutes that seemed an eternity, I found myself above the base at Alghero when the engine finally seized.' He glided down to a safe landing, to discover that one of his two wingmen had been shot down, while the other made it home to Sicily. His group claimed three Hurricanes shot down.

Meanwhile sixteen torpedo-bombers, led by the famous Italian flying star Captain Carlo Emanuele Buscaglia, began their approach at 1836, even as sirens demanded an emergency turn by the entire fleet, which failed to save the little destroyer/minesweeper *Foresight* from being hit by one of Buscaglia's Savoias, so that she was left almost immobilized. Eight Stukas peeled off at ten thousand feet to make their screaming descent to bomb, sometimes attacking as low as a thousand feet. Ed Randall, American chief engineer of *Santa Elisa*, said: 'circling maybe five thousand feet above us, the dive-bombers look casual, almost contemptuous. Then, two or three of them peel off and dive, usually towards the carriers and other warships. They figure they can wait around like vultures and pick off the cargo ships at leisure. As they roar down, you can make out the swastikas … At first they look like as they're floating, distant, impersonal, harmless. Then they start zooming right at you, and you realize they're [dropping] 500-pounders.'

Bombs exploding close to merchant vessels often disabled their generators and thus the electrically powered gyro-compasses. In the engine-rooms, observed Randall, 'every near-miss sends tools crashing to the floor as the ship lurches. You keep your ears cocked for the telephone order from the bridge to abandon ship, because you don't want to get caught like a rat in a sewer.' Lt. Tony Hollings of *Ledbury* wrote that, while on deck in action, men were so absorbed by their own duties that they were sometimes oblivious of the deafening clamour: 'down below it sounded like a thousand hammers all hitting the ship at once' – water displaced by blasts struck hulls with savage force.

A bridge lookout aboard the 1,920-ton destroyer flotilla leader *Laforey*, on the port bow of the convoy, shouted: 'Enemy aircraft bearing green 10!' – five Ju88 bombers. The gunnery officer wrote later: 'no echo showed on our primitive radar to reveal their range so everything depended on Able Seaman Daft manning the director-rangefinder. "Range, Daft," I shouted.' Captain Reginald Hutton rang the bridge telephone to the gunnery officer, demanding impatiently: 'Director, what's the delay in opening fire?' The gunnery officer pleaded: '"Daft, range?"', '"Can't make it out, sir"'. 'CPO Windsor, beside me, muttered: "Name and nature".'

The despairing gunnery officer hazarded a guess to the transmitting station below decks: 'Range 100' – signifying ten thousand yards – 'Commence, Commence, Commence.'

'The turrets hastily applied fuse settings, and successive "Ready" lights flicked on beside CPO Sharpe, the director-layer above the bridge. The warning "dong, dong" sounded just before he triggered the salvo, and *Laforey*'s six 4.7″s fired together, causing the whole ship to shudder. A second salvo had been fired before black puffs appeared behind the approaching bombers. On the fourth salvo, shells appeared to detonate among the attackers, and one fell away, trailing smoke, to plunge into the sea. No honest sailor could guess which ship's fire had inflicted the damage, but each one hoped for his own. The asdic officer, on *Laforey*'s bridge below, muttered unkindly "fluke", but Captain Hutton phoned, "Well done, sub".

The same gunnery officer wrote later: 'The main problem lay not in finding a target, but which one to choose. The cacophony of guns firing, shells bursting in the air and bombs exploding in the water rose and fell … [We saw through binoculars] swarms of torpedo-bombers circling out of range, low over the horizon … For about twenty minutes we slowly trained the director round following them, before they broke up into groups each of a dozen or more and turned towards us. Rather than individual gunfire a concerted H[igh] E[xplosive] barrage was ordered about two miles outside the destroyer screen, the likely dropping-area for torpedoes.'

In these two days, *Victorious* alone fired over a thousand rounds from her 4.5-inch guns, 1,553 from the Oerlikons, 5,673 from the pom-poms.

When the Italian 132nd Air Group took off from Pantelleria at 1730 on the 12th, for the first time in the war they enjoyed the reassurance of an escort of Macchi fighters of the 51st Group, led by a famous flier, Major Duilio Fanali. The German strike aircraft were likewise covered by their own single-seaters. At 1840, a few miles south of the fleet, their commander called the crews by radio-telephone, ordering his four flights to separate, each to attack a different ship. As they closed in, they felt awed by the size of the fleet below – and by the weight of fire from its warships, which caused some planes to break formation. One suddenly soared upwards into the path of Martino Aichner's aircraft, frightening his crew considerably and causing the pilot to break away. Another flight headed for a cruiser, while Aichner's section picked a destroyer. Their exuberant leader radioed his war cry before every attack: '*Dai alle panzone!*' – 'Sock it to the fat bellies!'

Aichner's crew were dismayed by plumes of water shooting into the air in their path which at first they assumed were being thrown up by shell explosions, before realizing that bombs were falling from other aircraft above them. They dropped their torpedoes less than a thousand yards from the targets and raced on between two destroyers and a cruiser. Aichner described his own dry mouth, 'shells exploding in front and behind, machine-gun fire stitching a patchwork of threads that seemed to tie up to the enemy gun barrels'. He had long since pulled his torpedo-release, but saw the pilot of a passing Macchi gesturing frantically. The plane's gunner appeared in the cockpit shouting: 'The torpedo, lieutenant! The torpedo! It's still underneath us.' Their armed weapon remained hung up beneath the aircraft, and nobody had the stomach for a second run. They turned disconsolately for home. Aichner's plane was last to land. On the ground, they learned that the release mechanism of another aircraft had likewise failed.

The pilot wrote: 'I felt very depressed to have made such a dangerous attack for nothing.' He was walking across to a friend's aircraft to discuss their shared experience when suddenly the airstrip erupted into spurts of dust. There was a roar of engines low overhead, long bursts of cannon fire: two Malta-based RAF Beaufighters were strafing them. As the planes faded into the distance, the pilot was devastated to discover that his friend Vittorio Moretti, who had survived the attack on the fleet, was dead. Moments after climbing down from his cockpit, 'he was hit by fire which blew him into the path of a propeller that was still turning … My heart leaps and my stomach churns at this new reminder of how narrow is the margin between life and death.' Several aircraft were burning fiercely. Aichner described himself and his comrades, following this and other such operations that week as 'exhausted, our nerves shredded'.

Meanwhile over the fleet Axis dive-bombers were achieving much more. In the Battle of Britain, Spitfire pilots regarded

Stukas almost contemptuously, as easy meat. Yet in the war at sea again and again these obsolescent dive-bombers proved deadly, partly because it was impossible for warships' guns to engage targets almost vertically overhead. At Midway, two months earlier, after the US Navy's torpedo-bombers had been almost wiped out in a vain assault on Japan's carriers, it was Helldivers that redeemed all, sinking four enemy flattops and almost unaided securing the most important naval victory of the Pacific war.

Now, on this August evening in the Mediterranean, mist which had impeded visibility for some hours cleared, just in time to give the attackers a clear sight of the fleet. Their escorts diverted many of the defending Hurricanes and Martlets. At 1837 one of a first wave of black-painted Italian Stukas near-missed *Rodney* by twenty yards on the port side, though one plane was downed by a Hurricane, a second by warship gunfire. The next wave, composed of twelve German aircraft, did even better, streaking down in succession upon *Indomitable*. For long, agonizing seconds the great ship became almost invisible amid splashes and plumes of water.

Roger Hill wrote: 'The carrier looks as if it had disturbed a hive of bees. The dive-bombers were zooming down on her and our own fighters, desperately protecting their home, were following the enemy planes right into the carrier's gunfire.' The Stuka pilots began their dives at around nine thousand feet, opening throttles, closing cooler flaps, opening dive-brakes. The planes' sirens, which back in 1940 terrorized so many Polish and French civilians as well as allied soldiers, had since been removed to save weight. The spectacle alone, however, was impressive enough, a procession of descending aircraft approaching speeds of almost 400mph in the dive, straining their gull wings almost to destruction before releasing bombs one after another, then pulling out at a thousand feet. Watchers on the ships felt a surge of unwilling admiration for the enemy aircrew who, in the words of NLO Lt. Edward Binfield aboard *Rochester Castle*, 'pressed home their attack with great determination ... I observed two crash by the bows of the carrier.'

The armament of the cruiser *Phoebe*, acting as the carrier's close escort, would not bear on the Stukas. An Oerlikon gunner, desperate to reach them, slipped on a patch of oil as he swung round in his harness, then clung to the trigger as he fell to the deck, so that the gun sprayed the cruiser's superstructure with 20mm cannon fire, mercifully without killing anyone. The rest of *Phoebe*'s crew were shocked spectators as the bombs struck *Indomitable* and flames spread. An officer wrote: 'I thought that she would roll over, like *Eagle*.' Denys Barton, naval liaison officer aboard *Ohio*, wrote in awe: 'It really was rather a fine sight, you could see the planes let go their bombs and tell whether they would hit ... I thought the ship was a goner.'

So did Hector Mackenzie, viewing the attack with less detachment from his action station on *Indomitable*: with radar reporting raiders approaching ahead, 'all eyes tended to look in that direction. Too late, several of us ... saw we had been taken in by the oldest trick in the book,' as the carrier's attackers appeared from another quarter. 'I recall a dreamlike hour from the moment I saw the Stukas dive ... I was scared out of my wits, convinced that the end was nigh. While the explosions rocked the ship there was considerable pain, in the ears and in the backside, from the whiplash effect aft of the bombs striking forward. For ten minutes one expected us to sink or blow up.' At 1835 three successive thousand-pound bombs struck *Indomitable*, while others exploded so close that deadly fragments raked the ship's sides. Both the carrier's lifts were wrecked, instantly rendering her incapable of operating aircraft. Her armoured flight-deck, such as other belligerents' counterparts, including the US Navy, lacked, alone saved her from destruction – four Japanese carriers had to be scuttled after sustaining fewer hits.

George Blundell on *Nelson* wrote: 'She completely disappeared, for all we saw of her for minutes were just columns of spray. Finally the maelstrom subsided and there was *Indom*, blazing both for'rard and aft of the monkey island, with great columns of smoke coming from her flight-deck.' Aboard the carrier, telegraphist Arthur Lawson described how 'suddenly the ship shuddered from stem to stern and one had the strangest feeling of great resistance being exerted by the whole vessel, to stop herself being pressed underwater. A hush prevailed immediately following the hits, followed by the bedlam of the damage control parties going about their grisly work.'

A year earlier, the imperturbable Tom Troubridge received signalled commiserations from his admiral after *Nelson*, then under his command, had been struck by an Italian torpedo. Troubridge responded: 'thanks very much, but a hit below the belt at my time of life does not hurt'. Now, however, the hits on *Indomitable* hurt very much indeed – the captain's own cabin was riddled with bomb splinters, and that was the least of the damage. The bridge broadcast microphone was accidentally switched on, and throughout the ship men heard their captain say to Rear-Admiral Boyd, beside him on the bridge: 'Christ, Denis, I believe they've buggered us.' An Albacore crewman wrote: 'fire seemed to be everywhere. Most particularly the Martlet parked by our Oerlikon position was fully fuelled and ammunitioned for action and blazing merrily above us.' The entire crews of two 4.5-inch gun turrets had been killed by blast: 'the stink of death was everywhere … one of the most enduring memories'. Almost fifty men were killed, a further fifty-nine wounded.

The huge ship, billowing smoke and with her steering jammed, swung in a wild circle. A flier watching from *Victorious* wrote: 'smoke and steam billowed up amid the wall of water; and for a quarter of a minute it seemed as if she could never reappear except as a smoking hulk. Then, slowly, as the mass of water heaved up by the near-misses subsided, she re-emerged, listing, on fire fore and aft, nearly stopped, but still afloat. From every ship men watched her anxiously; isolated from the disaster, yet sharing it, impotent to help, yet suffering as if it were to their own ship.'

For twenty minutes, *Indomitable* dragged herself through a turn, deck heeling, smoke pouring forth, then at 1704 an Aldis lamp on her bridge blinked a blessedly reassuring plain-language message: 'situation in hand'. The carrier once more steadied on a course.

Yet after such a blow it was impossible that *Indomitable*'s operational effectiveness, or crew confidence, could swiftly be restored. Twenty-year-old Hector Mackenzie wrote: 'It was not an easy time for morale … Damaged as we were on the enemy's doorstep [we expected] some concentrated raids to follow and finish us off.' Telegraphist Ted Venn said: 'everyone prayed for nightfall'. Men below in the hangar deck, deafened by the blasts and most vulnerable to the fires, came dangerously close to panic. The ship lost way, with one of her three propeller shafts out of action. For a time, she could make only five knots, though this progressively increased to fourteen, and later to even better speeds.

Overhead the air battle continued, with Hurricanes, Martlets and enemy aircraft of all kinds wheeling, twisting, exchanging fire, even amid the continuing roar and rattle of barrages from the ships. Four hundred feet above, pilot Blyth Richie chased a Stuka for a mile before making a beam attack from a range of a hundred yards: 'I ... saw part of the cowling fly off, then closed to astern at sixty yards and saw the gunner double up. It was now smoking and on fire on the starboard side. The wing dropped and it went into the sea from 200 feet.

'[Minutes later] I was flying through the barrage from the fleet when I saw five aircraft diving from 3,000 feet. I climbed to 2,000 feet and did a beam attack upon the quarter at 50 yards. I saw the pilot's cockpit shatter and the angle of dive changed from 65 degrees to 85 degrees and go straight into the sea. I followed him down but could not continue firing as I had no ammunition left.' Both the wounded German pilot and his gunner from the latter aircraft were rescued.

Meanwhile *Indomitable*'s wonderfully impressive damage-control parties were in action. Assisted by the pumps of the destroyer *Lookout*, which played hoses on to the flaming flight-deck, spaces in the carrier's starboard side were flooded, to correct her list to port: later, 760 tons of seawater had to be pumped over the side. Spirits rose when men heard engine revolutions increase as power was restored. But officers and ratings plying axes, sledgehammers and hoses amid twisted and blackened steel and shattered woodwork encountered horrors. Arthur Lawson said of a scene in the bomb-blasted wardroom: 'My most vivid memory was of seeing an officer finishing a drink he was having, with half his head blown away'; six stewards also died there. The engine-room staff, who overcame the terrors of their own predicament to restore the ship to high speeds and to prospective salvation, earned special admiration.

Though the carrier was not mortally hurt, her mauling caused Syfret to advance by twenty minutes the westward retreat of the fleet, now redesignated Force Z. They were less than three hundred miles from Malta, at the approach to the Skerki Banks and Sicilian Narrows. The admiral signalled Burrough at 1855: 'Force X proceed in execution of previous orders.' Meanwhile he redeployed his own ships for their return to Gibraltar, signalling the crippled *Indomitable*: 'You set the pace, but do buck up.' At 1913 the damaged carrier's speed increased to seventeen knots. The battleship *Rodney* was suffering boiler problems, accentuated by the violent manoeuvres of the past two days during which the fleet had made forty-eight emergency turns, together with many more drastic movements by individual vessels. Now, *Rodney* was ordered to join *Indomitable* and *Ithuriel*, all these 'walking wounded' making their best speed towards the Rock, by the most direct course, many of their crews – in the words of midshipman Mike MccGwire on the battleship – 'feeling somewhat ashamed for leaving [the remainder] in the lurch'. Meanwhile *Nelson*, *Victorious* and Syfret's other warships would linger overnight off the Tunisian coast to hold open the possibility of making an intervention next day, the 13th, if Italian heavy ships appeared intent upon attacking Burrough's charges. Thus, abruptly, the convoy and the Royal Navy's heavy units went separate ways, before the ships of Pedestal faced their grimmest hours.

2 THE PARTING OF COURSES

Syfret's decision, advancing the execution of the plan agreed back at the Admiralty weeks earlier, attracted criticism then, and has sustained controversy since, in the light of the painful events that followed. The fleet's retirement was certainly unheroic, yet it represented wisdom. Syfret's orders from Pound for Pedestal have not survived in the archives, but it is easy to guess where lay the limits to the discretion granted to the acting vice-admiral.

Every naval and military operation balances risk and advantage. To run a convoy to Malta, the directors of Britain's war effort agreed to hazard much. But it was never intended, nor was it remotely justifiable, to lose a substantial part of the Royal Navy's entire battlefleet in order to succour the island. Already one precious carrier had gone; a second was crippled. Only *Victorious* remained capable of flying off planes – the forty-seven aboard *Indomitable* were now useless. Despite the heroism of the Fleet Air Arm's pilots against far superior numbers of enemy aircraft, those August days emphasized that their fighters were embarrassingly outclassed. With the fleet's air cover reduced by two-thirds, its vulnerability increased proportionately.

The navy's after-action report on Pedestal was frank about the limitations that it had exposed in the ability of British warships to defend themselves against air attack. Since leaving Gibraltar Fleet Air Arm pilots claimed to have confirmed the destruction of thirty-nine enemy aircraft, plus another nine 'probables', for the loss in the air of thirteen of their own. Tom Troubridge, determined to see the glass half full, called the 12th 'a great day': confident that his carrier's fighters had shot down many attackers before the flight-deck was disabled. By contrast Rear-Admiral Lyster, with overall responsibility for the air groups, wrote with cruel honesty: 'it is a great disappointment to me that the fighters did not take a greater toll of the enemy'. In the post-mortem on the day's event, while Syfret gave credit to the Hurricanes for disrupting attacks, Lyster adopted a more cynical view, that enemy formations fragmented for their own tactical reasons. The final tally, according to Axis records, suggests that during the enemy's two hundred bomber sorties on 12 August eighteen attackers were shot down; several more were written off by ground strafing and operational accidents.

Neville Syfret, that Wednesday evening, bore a huge responsibility, which was all his own. Although he afterwards recorded nothing about his feelings, that hour during the early evening of Wednesday, when it appeared that *Indomitable* might be lost, cannot have been less than traumatic. Just as Tom Phillips would forever be remembered as the admiral who lost *Prince of Wales* and *Repulse* – off Malaya in December 1941 – so Syfret knew that he came close to becoming the commander who lost two of Britain's seven priceless aircraft-carriers in two days. If his fleet remained with the convoy for even a few more hours, subsequent events showed that it was not merely possible but probable that at least one more capital ship would be lost, to little purpose. Now that it seemed doubtful Mussolini's fleet would fight, the presence of *Nelson* and *Rodney* off the Skerki Banks – mid-Mediterranean reefs which limited the searoom for big ships – made them prospective banquet fare for the Luftwaffe and Axis submarines, rather than bulwarks.

Posterity can see the pathos of this parting of courses in mid-Mediterranean. The most conspicuously powerful elements of the fleet – as Dreadnought-era admirals measured naval strength – which had covered the Pedestal convoy from Britain, now reversed course, formed a single line ahead and drew away from the merchant vessels, cruisers and escorts commanded by Harold Burrough, at a combined rate of more than thirty miles an hour. For all the apparent might of Force Z, 'a brave concourse of ships', as the sentimental Commander of *Nelson* described it in his diary, its admiral dared not allow it to look the enemy in the face, figuratively, at the most peril-

ous phase of Pedestal. Richard Woodman, one of the wartime
Merchant Navy's most authoritative historians, has written: 'It
had ... much of the quality of a forlorn hope, for when Syfret
withdrew ... the material advantage effectively fell to the
enemy.'

Paul Kennedy wrote about the Royal Navy's experience of
the Second World War in his classic study *The Rise and Fall of
British Naval Mastery*: 'Air power had chased the battlefleet
from the ocean, and now posed a threat to every surface
warship.' Syfret's big ships dared not expose themselves any
closer to a land-based bomber threat. Only much later in the
war could the Americans, with their vast resources and air
capability – the US Navy eventually deployed almost a hundred
seagoing flight-platforms of all sizes – sail with impunity
alongside enemy land masses. Air forces set at naught the
potency of even the largest warships, and the planning of
Pedestal reflected rueful acknowledgement that this was so.

Though the story that follows is chiefly that of Harold
Burrough's men and ships, Syfret enjoyed not a moment's
peace of mind until Force Z was safe at Gibraltar. For many
miles across the sea, as its warships steamed their westerly
courses, they remained vulnerable to submarine and air attack.
That night David Scott, an eighteen-year-old midshipman at
the AA plotting table of *Rodney*, noted that he had been wearing the headphones linking him to the ship's director tower for
almost twenty-two hours. George Blundell of *Nelson* wrote:
'we were shocked and numbed by the fear, noise, heat, smell
and sheer hard work we had gone through'. Paint was peeling
from the barrels of guns overheated with incessant firing.
Captain David MacFarlane of the *Melbourne Star* wrote
emotionally of the parting from *Indomitable*, after witnessing
her ordeal at close quarters: 'it was a most impressive sight to

see her guns blazing furiously out through the smoke and flame, and later to see her steaming westward towards the setting sun, with her fires still burning fiercely'.

Destroyer stoker Chris Gould told his beloved fiancée Vera in south London: 'I'm just about all in, they gave us twelve hours of it today. I've had no dinner, tea or supper – I'm sure glad to be alive ... They tried a torpedo but the captain was too good for them and dodged it; he made a good show of dodging bombs. We were in a very bad position, also the light was with them. It was a beautiful day but hazy, so they flew in very low and we couldn't see them. The last attack was made with Stukka [sic] dive-bombers. Two came for us and about six went for *Indomitable* and got two [sic] on her ... She looked bad at first, but she is OK ... It seemed funny out of 50 ships they tried for us so many times!... I'm writing this at action stations, as we are remaining at them all night. What I would like is a nice bath and go to bed!'

Those of *Indomitable*'s Hurricanes which landed aboard *Victorious* had to be pushed over the side for lack of suitable lifts to strike them below. On both carriers, pilots slumped in exhaustion, overlaid with sadness for the dead. Beyond the fliers who crashed into the sea aboard their fighters, the bomb that tore open the wardroom anteroom of *Indomitable* killed five resting aircrew. The ship, before she was hit, had launched seventy-four fighter sorties, a record; hangar and flight-deck crews worked continuously for fourteen hours; the fighter-direction team closed up at their posts at 0530 and stood down only at 2130.

Hugh Popham wrote: 'the ship herself was strange to us, the run of her, that was as familiar as home, broken and interrupted by jagged rents in her plates, the charred and splintered woodwork, the great flapping tarpaulin anchored over the missing bulkheads in the anteroom. The strangeness of it all was enhanced by the faces that were missing.' Hector Mackenzie was distressed by the sight of a gaping hole in the wardroom's wooden panelling, through which had been blasted the head of twenty-five-year-old George Measures, his closest friend. Only Dickie Cork, the dedicated warrior, 'after the ceaseless strains and exertions of the day, remained still as cool and calm as if he had merely been doing half an hour's camera-gun exercises', in the words of a fellow pilot.

The *Indoms* held a brief ceremony to bury their dead over the side. Hector Mackenzie felt droll surprise that the ship carried sufficient white ensigns to cover each of forty-eight bodies: aircraft maintenance crews had suffered almost as heavily as gunners. It was a curiosity of naval life, naval war, that during interludes in mortal strife service rituals were resumed. The ship's bugler, just fifteen years old, played the Last Post before the corpses were tipped one by one over the side. Then the Royal Marine band struck up the hauntingly beautiful Evening Hymn. Mackenzie wrote: 'I do not know why, really, but we all felt better when that was over.' Following herculean efforts by her engine-room staff, *Indomitable* soon achieved twenty knots, and later twenty-eight. Only *Rodney*'s boiler-tube problems restricted the pace of the cripples' retirement to eighteen knots. That night, the familiar civilian catchphrase, about being obliged to conform to the speed of the slowest ship in the convoy, had special meaning for Syfret's charges.

Among those still steaming east, there were open expressions of dismay when the fleet turned back. American Gerhart Suppiger, on *Santa Elisa*, wrote: 'it was evident that most of the protection we had was from fighter planes based on carriers'.

Petty officer Les Rees, aboard the destroyer *Penn*, said: 'I am positive that "Uncle" Jim Somerville [former C-in-C of Force H] would not have left us before nightfall.' Senior officers were equally uneasy, even shamefaced. Before the fleet sailed from Britain, George Blundell of *Nelson* wrote: 'It makes me sweat reading the bit [in fleet orders] about the poor convoy getting through the last bit ... The last party that tried it got rather badly beaten up and, I suppose, this time we are doubling our stakes.' The carrier group had played a real part in the Pedestal battle for barely twenty hours, and quit the action when the heaviest air attacks still lay in the future.

Thirteen merchant vessels remained, with thus far only one absentee – *Deucalion*, left in the custody of *Bramham* since being hit by a bomb at 1330. After an hour hove to, the ship's Master Ramsey Brown restarted his engines and worked up speed to sixteen knots in an attempt to catch the convoy. The ship's rapid motion, however, forced oil out of her deck vents, driving the sea into a second hold. By setting three pumps to work and reducing speed to twelve knots, *Deucalion*'s No. 2 hold was eventually pumped dry, but there seemed no hope that the straggler could gain on her companion vessels. Eddie Baines, skipper of the shepherding destroyer, suggested that she might instead head south-east, hugging the North African coast. Ramsey Brown concurred, and later reported: 'we were getting on nicely, with every hope of getting the vessel to Malta', when at 1700 an enemy reconnaissance aircraft spotted the ship. At 1946 two Ju88s attacked with bombs, all of which missed. The ship seemed to have a charmed life.

Not for much longer, however. Just after sunset, as the ship passed the Cani Rocks six miles off the coast, two Heinkel 111 torpedo-bombers arrived overhead, which addressed their business in brutally effective fashion. One plane flew down the port side, to divide the attentions of the ships' guns, while the other braved a storm of fire to deliver a torpedo which exploded against the starboard side, killing an army gunner and igniting the cargo of high-octane spirit. Within minutes flames were soaring high into the darkening sky. *Deucalion*'s naval liaison officer reported later: 'In view of ammunition and the large quantity of spirit in the hold and consequent grave risk of explosion, abandon ship was ordered.' It was remarkable that five boats and several rafts were launched successfully, bearing away Brown and all his ninety-six crew and army gunners.

Because some of the crew had behaved ingloriously after the initial attack that afternoon, it was heartening that, in the final evacuation, there were heroes. In the rush for the boats, midshipman John Gregson – Alfred Holt's ships used the naval term – and a ship's apprentice named Peter Bracewell distinguished themselves. Gregson had been knocked unconscious by the explosion, then came round to find himself apparently alone on the blazing ship: 'I can still see the flames now, right up to the top of the mast,' he said years later. Then he found a man trapped beneath a fallen raft and freed him after desperate exertions, helped by Bracewell, who had also been left on board. Finding that the man had a broken thigh and fractured arm, they resorted to drastic measures, throwing him into the sea. The two teenagers then jumped in after him and towed the casualty four hundred yards to *Bramham*. As the captain said admiringly, it was 'a really gallant and plucky action by a boy of only 18'. Gregson later received the Albert Medal and Bracewell a Mention in Dispatches for their action.

At 2230 *Bramham*, crowded with survivors, hastened away at twenty-two knots in pursuit of the convoy, leaving *Deucalion*'s decks awash amid a chaos of exploding ammunition and fire in the superstructure. Brown and his men spent the next three days as passengers. 'In my opinion,' said the former Master, 'this was the worst part of the passage as we were crowded below decks, being subjected to constant air attacks and the noise of the guns above our heads and the explosions of the bombs was terrific.' One of the fourteen merchant-ships sent to succour Malta was gone. And far ahead of them in the darkness, as the destroyer sped onward to catch Burrough's ships, they glimpsed flashes and flames that signified fresh grief, new pain, for the men and ships of Pedestal.

The German and Italian pilots who landed back at their bases on Sardinia, Sicily and Pantelleria congratulated themselves on successful missions, courageously executed by most of the crews who attacked. An Italian pilot wrote about the return flight to Chinisia (Trapani), his Sicilian airfield: 'The great expanse of sea is uninviting at dusk, when one is unsure of the state of one's aircraft and one's heart leaps if the engine misses a beat.' Like many pilots operating from bases in their own country, Giacomo Metellini found that the drastic changes of scene from combat over the Mediterranean to nights with wine and girls in Sicilian towns imposed special stress: 'the abrupt, continuous toing and froing between facing probable death and embracing one more evening of life ... We knew that our circumstances were better than those of soldiers on the battlefield, but more than eighty per cent of the Italian aircrew who went to war in 1940 were dead by the end.'

Metellini, a thirty-year-old veteran from Trieste who had been flying in action since the Spanish Civil War, observed that most of the time he found combat over the Mediterranean 'very sterile and impersonal ... In a confined cockpit, numbed by the roar of the engine and crackle of machine-gun fire, buffeted by the violent manoeuvres we executed and sometimes surrounded by the black puffs of anti-aircraft fire, there was no time to think of wounds, pain, death. Even when we saw planes hit, friend or foe, we didn't think much about the fate of their pilots.'

This caused Metellini to be intensely shocked by the experience on returning from his latest sortie against Pedestal. After switching off his own engine, he ran to the damaged aircraft of a comrade, who had crash-landed in a field beside their airstrip. 'He was sitting motionless in the cockpit, a hand near his face.' Then Metellini saw that, in that hand, the pilot was clutching one of his own eyes, knocked out of his face by its impact upon his gunsight. Suddenly, the pilot's professed insouciance fell away. He succumbed to revulsion, at a glimpse of the reality of what might befall any one of them on any day of that bitter battle.

8

Force X

Following the departure of Syfret's capital ships, Harold Burrough's command, Force X, became a convoy under escort rather than a fleet. If there were still too many men – perhaps seven thousand – for his crews to be called a band of brothers, they were conscious of being no longer a mighty host. Of Burrough's cruiser squadron, *Nigeria*, *Kenya* and *Manchester* appeared formidable fighting machines, among the most graceful warships ever built, each five or six times the displacement of a destroyer. They were nonetheless neither as robust nor as useful as their size suggested. They had valuable communications facilities, but were relevant to the convoy's defence only if Italian cruisers sortied from Naples and Messina to engage in a gunfire duel. Against aircraft or submarines, the cruisers were prospective victims rather than effective protagonists. *Cairo* was a smaller and older vessel, designated as an anti-aircraft cruiser, with high-angle main armament, but such ships had little success in their role. There was still a relatively strong destroyer screen. With the Bofors and Oerlikons mounted on the merchant-ships, the convoy could put up a noisy anti-aircraft barrage – so long as it held together, retained coherence and discipline.

The air battles had already exposed a basic flaw in Pedestal's planning, which the prime minister should have anticipated. Because there was no British supreme commander in the Middle East, the navy prepared and executed the operation with no support from the British Army and little from the RAF. Every nation engaged in the Second World War suffered tensions between its armed services, but it is depressing to recall the experience of a Royal Navy officer who appealed to a British Army colonel on Gibraltar for help in finding quarters for survivors from sunken merchant-ships. The infantryman replied disdainfully: 'Surely this is hardly a military commitment?'

If Churchill had imposed his will upon the other services, stressing the critical importance to Britain's credibility of relieving Malta, its soldiers in Egypt could and should have been ordered to execute some diversionary activity against Rommel's forces, on a scale that would oblige the Luftwaffe to provide air support: to be sure, Eighth Army was a dispirited, oft-beaten force, whose C-in-C was sacked while Syfret's fleet was weathering its Atlantic passage. Something might have been contrived, however. Kesselring anticipated such an initiative in the Western Desert and was surprised that it did not materialize. Thus, with Auchinleck's forces passive behind their defensive line at Alamein, Axis commanders were for some days free to concentrate almost their entire Mediterranean air strength against Pedestal. Likewise, Sir Arthur Tedder dispatched only a token bomber force to attack enemy airfields in Sardinia and Sicily. German and Italian commanders, who expected such assaults, were astonished that they were able to launch air operations with negligible interference from the RAF.

On the early evening of 12 August, the Axis might have been forgiven some triumphalism after sinking one British carrier and crippling another. Instead, however, uncertainty and unease persisted in the enemy's camp, prompted by the scale of the British operation, which still seemed disproportionate for the mere relief of Malta. Luftwaffe chief Hermann Göring sent a personal message ordering that German air forces under Kesselring 'will operate with no other thought in mind than the destruction of the British convoy'. Axis aircraft were instructed not to waste effort on crippled vessels or stragglers, an order that would prove ill-judged. Rome was relatively slow to learn of the retirement of Force Z, of which word reached Supermarina only at 2030, ninety-five minutes after it took place – and remained apprehensive about a possible British amphibious landing at Tripoli, Rommel's supply base, on the morning of the 13th. To meet this contingency, all Axis convoy movements across the Mediterranean were suspended, and some strike aircraft and fighters were moved to North Africa from Sicily, together with supplies of fuel and ammunition. Rommel shifted artillery to the Sollum–Mersa Matruh area, in readiness to defend the coast. Yet ample German and Italian forces remained to intensify the agony of Pedestal.

Force X was still some 290 miles from Malta. That day, 12 August, had already been replete with drama and action, but among the merchant vessels only *Deucalion* had suffered much attention from the enemy. On *Ohio*, Dudley Mason wrote with almost inhuman equanimity about the tanker's experience, until that evening: 'the day passed fairly uneventfully. One or two isolated planes got through the outer screen and kept the gunners in action, and some bombs were dropped. Continuous salvoes of depth-charges and emergency turns to port and starboard every few minutes. Several vessels reported submarines, and I believe two were accounted for.'

Admiral Burrough, leading the port column in *Nigeria*, deployed around the merchant vessels an outer screen of ten escorts. He himself was a sunnier, more genial soul than Syfret, perfectly fitting the template for a senior British naval officer of that era: cheerful, enthusiastic, loyal, sociable, courageous, devoutly Christian though seldom contemplative, a passionate exponent of the spirit of 'can do'. He recalled with relish his teenage days as an impoverished 'middie' – midshipman – owner of only one pair of shoes. Ashore, even when he shed his uniformed splendour, he lived to a routine as strict as that afloat, punctuated by morning readings of *The Times*, especially its cricket reports; a glass of sherry before lunch; gin at six; fierce games of tennis and energetic vegetable gardening; renderings of Noël Coward songs when in celebratory mood; he loved military bands.

His two sons later joined the Royal Navy, and all his three daughters married into it. When Pauline became engaged to a young man whom the admiral deemed unsuitable, he took her aside and told her frankly: 'darling, if I had to choose six men out of any seven to take to sea with me, Michael wouldn't make it'. In those long-gone days, the sensible girl took her father's advice, instead married Andrew Cunningham's flag-lieutenant and lived happily ever after. She said of her father's role as a fighting sailor: 'he was born for it'. Like many British naval officers, Burrough knew Malta well, as a second home during pre-war years with the Mediterranean Fleet, providing an idyllic family life with plentiful servants, sun, swimming and sports.

From the moment that he assumed responsibility for Pedestal just before 1900 on 12 August, he enjoyed an hour of relative tranquillity. The admiral wrote later: 'In view of the magnitude of the enemy's air attack at 1830 to 1850, it seemed improbable that a further attack on Force X on any great scale would be forthcoming before dark, and it was hoped that the submarine menace was mostly over.' Men thankfully pulled off their helmets and stinking, sweat-stained anti-flash gear, lit cigarettes, savoured water or lemon juice. As always, there was ammunition to be replenished from the magazines in ships' bowels, opportunities snatched to wash and relieve nature. On some ships, there was hot food, though crews remained closed up, at action stations.

Burrough's brief interlude of calm would have ended sooner than it did had he known that at 1938 Renato Scandola of the Italian submarine *Dessiè* fired four torpedoes at three-second intervals, from a range of two thousand yards, aimed at two of Pedestal's freighters. As it was, however, no British ship even noticed their tracks, though Scandola later claimed to have heard explosions. His boat was one of six – the others were *Granito*, *Axum*, *Bronzo*, *Alagi* and *Ascianghi* – which had been cruising the area since early that morning, awaiting the convoy's coming. It says little for the performance of the escorts' asdic sets that none of these boats was detected before the first attacks came. Moreover, with *Indomitable* disabled and *Victorious*'s flight-deck over-crowded, it was deemed impossible to maintain even a token Albacore anti-submarine patrol as the carriers retreated towards the west.

At 1956, destroyers belatedly awoke to the proximity of *Dessiè* and began depth-charging. Scandola took his boat deep and held its crew and machinery mute amid repeated attacks that continued until 2127, the Italians claiming that 120 charges exploded in their vicinity. Among the most frightening aspects of such experiences was that a submarine's crew could often hear the resonating pings of the enemy's detectors on their own casing, tormenting them every few seconds with harsh warnings that their proximity was known. Carlo Pracchi, an engineer, described such an experience: 'The Asdic pings follow one after another until they get the echo … They're like whip-lashes against the hull, and when they get an instantaneous echo, they know they have hit the bullseye. The sounds are coming from starboard, midships. Thuds in the water, three at a time, and a few seconds later you hear three rumbles followed by the inevitable shaking. Sound source from 10 o'clock and then from 7 o'clock, and louder thuds. It sounds as if every gauge in the control room is shattering. All the valves are shut down.

'The glass of the depth gauge breaks. From the telephone comes the voice of someone shouting "Engines … signal leaks or other faults." We get the boards up to look at the water-level in the bilge tanks. I can see water running down the sides … I look up. In the meantime the bombs go off. I look up again to see where that water is coming from … I can see drips on the heads of the rivets in the frames. I touch one with my finger, and water runs down my arm. Then we get another depth-charge, and the drips start to run down by themselves. That earlier charge exploded underneath and brought us up; this one went off overhead, and pushed us down. Machine room crew informs: no damage … The main relays shut down. The lights have just gone. We'll have to use the emergency lighting. I feel my leg cramp up, then the tingling starts, then shaking … I try to stop just one knee from trembling, but it carries on and only makes my arms shake. In desperation I say: "Bloody hell I'm just panicking." [But] I can't get myself out of it … I really panic.' Another Italian submariner wrote: 'so many thoughts go through our minds in those moments that look like being our last'.

Dessiè survived. At 2212 Scandola rose warily to periscope depth and believed that he saw two merchant-ships in flames. Nearby was a thick, black cloud, shooting flames, 'which made me assume that another ship was on fire'. Scandola congratulated himself on a multiple triumph. In truth, he had hit nothing. There were indeed stricken British ships, but most were victims of another Italian submarine, which had delivered one of the most devastating underwater assaults of the war.

Renato Ferrini of *Axum* made his own attack twenty-five miles north-west of Tunisia's Cap Bon. He first spotted the British ships soon after 1800 and set his boat to close them at its best underwater speed, eight knots, even as air attacks were consuming most of Force X's attention. At 1927, Ferrini estimated the convoy's range at five miles, noting – inaccurately – fifteen merchant vessels, two cruisers and 'numerous' destroyers which seemed to be making a starboard zigzag in three columns. The submariner was conscious that, in the flat calm conditions, the 'feather' of his periscope wake must be exceptionally conspicuous. He thus repeatedly ducked and then rose again for a mere few seconds, before resuming his stalk. He was no ace – a month earlier, he had fired a salvo which missed the minelayer *Welshman* steaming at her full forty knots. Now, perhaps influenced by that failure, he determined to make sure.

When Ferrini peered through the periscope eyepieces at 1948, he glimpsed a cruiser, a supposed destroyer and 'large merchant ship', all of which represented superb targets. By chance, the Italian skipper struck just as the convoy responded to Burrough's order to close from four columns into two, necessitating a reduction in speed, to approach the minefields guarding the Skerki Channel. This sluggishness was almost certainly a tactical error, which made easier the task of attackers. At 1955, Ferrini fired four torpedoes from an estimated range of less than two thousand yards, the boat jolting at each release. The submarine's crew then entered a tunnel of suspense in which they remained for the duration of their weapons' run: 'After sixty-three seconds we heard the first bang, after ninety seconds another two explosions, very close. After four and a half minutes the escorts began depth-charging. I descended to [three hundred feet], shutting down all machinery.'

An Italian submariner wrote of 'silent routine': '[Our boat] lay on the bottom like a sleeping whale. Engines, pumps and all auxiliary equipment were shut down. Hard tack biscuits, jam, chocolate were distributed, in the knowledge that we could not surface before nightfall. The men took off their shoes as even the slightest noise might cause us to be pinpointed. Some stretched out on bunks, others sat on the deck, others again stood in clusters, making little jokes. Their laughter was subdued and every movement was sluggish, brief. At each ping of the Asdic on our hull my stomach tightened.'

The bombardment of *Axum* continued for two hours, the intervals between depth-charges slowly increasing as the attacking destroyer lost confidence in her task before finally abandoning it. Ferrini wrote: 'At 2250 I surfaced and saw dead ahead a big ship on fire, about [three thousand yards] away, another vessel in flames to starboard at an angle of 45 degrees. To port at an angle of 70 degrees, a third vessel is blazing.' Oddly enough, none of these hulks had anything to do with *Axum*, but its own achievement was quite deadly enough.

The first explosion heard by the submarine's crew was caused by one of its salvoes striking Harold Burrough's flagship, *Nigeria*. Twenty-seven seconds later two further detonations followed, close together. The second of *Axum*'s torpedoes struck

Cairo, the third the tanker *Ohio*, most precious merchant vessel in Burrough's charge. Inside a minute the convoy, which had escaped relatively lightly – even acknowledging the loss of *Eagle* – from the attacks of the previous forty-eight hours, reeled beneath a series of devastating blows. All three ships swung to port, falling out of formation to attend to their damage, while the rest of the merchant vessels steamed on, now in two columns led respectively by *Kenya* and *Manchester*, every man on the upper decks of passing vessels peering in dismay at the casualties. The convoy lost coherence, as ships manoeuvred in whichever way seemed best to escape collisions; *Port Chalmers* was obliged to go full astern while *Empire Hope* stopped her engines to avoid being rammed by *Brisbane Star*.

Nigeria was hit on the port side, abreast her forward funnel, precipitating a sheet of flame and causing the ship to circle, with her lower steering position wrecked. An officer saw a flash, heard a terrific explosion which shattered almost every light below decks. The ship immediately took on a sharp list and dipped by the head. Amid dense, acrid smoke below, all those men who could do so groped frantically through companionways, up ladders, to reach the upper deck. Anthony Kimmins said: 'the memory of *Eagle* was still fresh in our minds'. Amid brief panic, green rating Ron Stockwell asked a leading seaman 'do we go over?' – abandon ship. 'No,' said the veteran tersely. 'Help me to get some of this wood free' – for shoring bulkheads. Two men seized a personal initiative, cutting loose a Carley raft which slid into the sea, before themselves following it over the side: as the light faded, they were nearly killed by an Oerlikon gunner who mistook the float for a submarine and opened fire.

Against the setting sun, scores of shocked, frankly frightened sailors gazed up at the bridge, seeking a cue from their

senior officers: 'Admiral Burrough and the Captain were lean-
ing across the starboard side,' recorded Kimmins, 'looking
rather like yachtsmen at the tiller of a boat heeling well over to
a fresh breeze.' Burrough, knowing that reassurance was urgent
and critical, shouted down to the throng, 'Don't worry, she'll
hold!', with more confidence than he probably felt. He added
calmingly to all within earshot: 'Let's have a cigarette!'

For much of his life, the admiral cherished a private prayer
that began: 'Most merciful God, grant we pray thee that we
may never forget that as followers of Christ we are the observed
of all men, and that our failures may cause others to stumble,
that in a measure God places his honour in our hands ... Grant
to us the royal gift of courage, that we do each duty at once,
however disagreeable it may be.' In an increasingly irreligious
twenty-first century, it is easy to make light of such sentiments.
They mattered deeply, however, to such a man as Burrough at
such a crisis as his command now faced. Aboard *Nigeria* steering control was regained within ten minutes, but the cruiser
was listing, down by the head, and had lost forty-four dead,
among them two seventeen-year-old midshipmen. Some
ghastly scenes took place below, mercifully beyond sight of
those on the free side of locked hatches and watertight doors.
Damage control was a brutal science, its disciplines dictated by
the greatest good of the greatest number.

Telephone communications throughout the ship were cut
off, so that messages had to be passed to the lower steering
position by human chain. Burrough made an immediate decision to transfer his flag to the destroyer flotilla leader *Ashanti*,
which grappled the cruiser's side at 2015, a quarter of an hour
after she was hit. Her captain, Richard Onslow, wrote later: 'I
had a hell of a time getting alongside *Nigeria*, as her engines
were still going slow ahead and her rudder was jammed hard

aport, so she was steaming in a circle with me chasing her on the outside edge.' At last, however, the two ships ran against each other, allowing the admiral to crossdeck.

Burrough shouted to the teeming sailors on the cruiser's tilting deck: 'I hate leaving you like this, but my job is to go on and get this convoy to Malta and I'm going to do that whatever happens.' Their task, he told the *Nigerias*, was to make sure their own ship reached home. Some spectators found it moving that, as he was piped over the side, even at this moment of crisis he was given the courtesies of his rank: a Marine bugle call followed by three spontaneous cheers from men crowding the upper deck which, according to a spectator on a nearby destroyer, 'should have been heard in Sicily and Malta – they were certainly heard all over the fleet'. Anthony Kimmins said: 'it was a sad but great moment'.

However two men, still trapped below in one of *Nigeria*'s watertight compartments, when later freed complained bitterly about their own sensations: 'You blokes' – their word is likely to have been more expressive – 'were topside blowing bugles while we were down below dying!' Meanwhile a signaller who had either jumped or been blown into the water from *Nigeria* paddled a float to the nearby *Penn*, stuck up a cheeky thumb in the time-honoured fashion of a hitchhiker, and shouted, 'Anybody going my way?' The destroyer took him aboard, first of scores of survivors that by the next daybreak crowded her lower decks.

Burrough's transfer impacted importantly on the subsequent course of Pedestal. Cruisers possessed communications facilities with which those of a destroyer could not compare. Since the fleet left Britain, from *Nelson* Admiral Syfret had been able to exercise control of the motions of all his many ships. This sagged, indeed almost collapsed, when Burrough boarded

Ashanti, which lacked the wireless technology and personnel to operate effectively as an admiral's headquarters. In the words of *Brisbane Star*'s naval liaison officer, 'from then on the convoy was entirely disorganized ... We remained in a heterogeneous mass, [his own ship] proceeding at never more than 5 knots, and attempted to reform in our original two columns.' *Melbourne Star*'s NLO, Roger Frederick, described the convoy as 'scrummed up', with one column having reduced speed to eight knots. He believed that it was a grievous tactical mistake to restrict the cruisers to the relatively sluggish pace of the merchantmen. Frederick and other officers argued that the big, fast warships should instead have roamed, exploiting their speed to make themselves more elusive targets for submarines and aircraft.

Be that as it may, the flagship limped out of the battle. Burrough wrote: '*Nigeria* would obviously be unable to play any further part ... it was an unpleasant experience to have to leave [the cruiser] ... but I felt it was vitally important that I should regain contact with the convoy.' The transfer of command to the Tribal-class destroyer, less than a quarter the size of the cruiser, was perhaps the most controversial act of Burrough's command. The story of the hours and days that followed was that of the progressive fragmentation of Pedestal. Burrough knew less every hour about the circumstances of most of the merchant vessels, and they heard little from him. Captains, even those of the warships, often found themselves obliged to make their own decisions, to pursue their own courses, for good or ill. It was for this reason that, in the wardroom of the cruiser *Kenya*, a strong collective opinion was voiced that the admiral would have done better to shift his flag to their own ship or to *Manchester*, and it remains puzzling that he chose not to do so.

It quickly became clear that *Nigeria*, while badly damaged, should survive. *Cairo*, however, was in a much graver plight. At first, many of those aboard were slow to realize the scale of the havoc aft. Rating Bill Bartholomew, her captain's messenger, said: 'We felt a tremor go through the ship but that was all, the bridge being forward I hardly felt it.' The cruiser did, however: *Cairo* had lost most of her stern, with Y turret blown into the sea and machinery spaces flooded.

Bartholomew witnessed some temporary breakdown of order on the main deck: 'everybody sort of panicked, didn't know what to do'. At 2013, even as Burrough was leaving his flagship to board *Ashanti*, *Wilton* grappled the side of *Cairo* and hundreds of men leapt across on to her deck. A few petty officers even humped big green suitcases containing their possessions, which those long-sighted veterans had kept ready-packed. The destroyer took off twenty-nine officers and 383 ratings. Twelve men had been killed or mortally wounded, most of them stokers.

The port propeller of the old cruiser, commissioned in 1918, was gone altogether; the starboard engine was useless. When her captain learned this, and since no escort could be spared to tow the wreck, 'I decided to sink the ship.' *Pathfinder* was detailed to finish off *Cairo*, which proved no easy task: only one of four torpedoes fired by the destroyer hit its target, thanks to unnoticed damage to the launcher mounting; a pattern of depth-charges also had little effect. The embarrassed escort set off after the convoy, leaving *Derwent* to sink *Cairo* with yet another torpedo. Thereafter, at 2130, the latter joined two other Hunt-class destroyers escorting *Nigeria* on her unsteady course to Gibraltar, which the four warships reached safely three days later, carrying *Cairo*'s survivors. It was deemed essential to provide anti-submarine escorts for every big warship, intact or otherwise: Burrough's destroyer screen for Force X was severely weakened by loss of the detachment to shield *Nigeria* from further harm, together with other escorts sent to chivvy straggling merchant vessels.

Following the torpedo strikes on the cruisers, fragments of debris showered down upon the nearby *Ohio*. Then, in the words of Denys Barton, the ship's naval liaison officer, 'while we were still looking at *Cairo* there was a tremendous explosion just abaft our bridge. All this happened within a minute or two, though it seemed much longer.' *Axum*'s final victim was hit amidships on the port side, starting a big kerosene fire in the tanker's pump-room, which spread aft. On the bridge, Dudley Mason was blown to the deck by blast, then crawled dazed through thick smoke towards the chartroom, where he bumped heads with his third mate, who was crawling out. Steward Ray Morton, serving at a port-side Browning machine-gun, either jumped overboard or – as he himself asserted later – was blown into the sea by blast, fortunately wearing a life-jacket. Denys Barton said, 'my main recollection was an enormous flash'.

At first, both Barton and Mason thought the damage had been caused by a mine rather than by a torpedo. *Empire Hope* loomed up so close to the tanker that seamen threw out fenders to cushion against an expected collision. Most witnesses expected the worst. Allan Shaw said: 'we all thought this was it – when a tanker goes afire, you haven't got a great lot of chance'. The Americans on *Santa Elisa*, which passed *Ohio* on the starboard side, could see men dragging out hoses before, in the words of Lonnie Dales, 'the black smoke swallowed them up'.

At first, *Ohio*'s crew expected to abandon ship and there was a rush for the boats. Two army gunners, Eddie Smith and Bill Hands, together with fifteen-year-old galley boy Mario

Guidotti, one of the youngest crew members in the convoy, attempted to lower No. 5 boat, with those three its only occupants. In their panic they overturned it, tipping themselves into the sea, from which they were retrieved three hours later by the destroyer *Bicester*, as was Ray Morton. *Ohio*'s officers, with much else to think about, did not even notice their absence until next day. Meanwhile Doug Gray, the quiet-spoken Scots chief officer, led a party armed with Foamite extinguishers, addressing the blaze. With his ship still moving in sluggish circles, Dudley Mason stopped her engines lest she collide with the stricken *Nigeria* or *Cairo*, wallowing nearby.

At that stage Mason, understandably traumatized by the torpedoing, believed the ship doomed, efforts to quell the fires futile: 'I thought it was a forlorn hope.' Instead, amazingly, Gray's party suppressed the flames inside five minutes, with the help of seawater pouring through the gash in the ship's side. All the boilers were extinguished, however, and *Ohio* lay at a dead stop on the glassy sea, with a huge hole in the main deck, her steering motor and gyro-compass unserviceable. In the failing light, Burrough came alongside in *Ashanti* and spoke by loudhailer to Dudley Mason, saying: 'I've got to go on with the rest of the convoy. Make the short route if you can, and slip across to Malta. They need you badly.' Mason shouted back: 'Don't worry, sir. We'll do our best. Good luck!'

At 2030 the admiral ordered *Penn* to assist *Ashanti* in laying black and white smokescreens astern of the convoy to render the ships less conspicuous, silhouetted against the western horizon at last light. They had scarcely started doing so, however, when yet another calamity struck: the Luftwaffe returned. Its approach was undetected until the enemy formation fell upon the ships: according to the admiral's report, 'no RDF signals were received owing to the dislocation caused by the U-boat attack'. The German operation was conducted by thirty Ju88 bombers and seven Heinkel torpedo-bombers, painted in North African leopard-spot desert camouflage. This time, the attackers swept in with orders to attack the merchant-men, not the escorts. The bombers were also intended to monopolize the attentions of the ships' gunners, distracting them from the more vulnerable, low-flying torpedo-carrying aircraft.

Burrough recognized before leaving Clydeside that dusk would be an especially dangerous time. He requested that, even if the big ships turned round earlier for Gibraltar, air cover should be maintained over his merchantmen until dark-ness fell on the 12th. Yet *Victorious*, the only ship still capable of operating fighters, was overcrowded with aircraft, and flew off no further Hurricanes after the four which had been airborne landed back on at 2013: moreover, Force Z and Force X were already forty miles apart. Six Beaufighters from Malta appeared over Burrough's ships, but in the absence of VHF voice communication, it was impossible to direct them towards the enemy: they returned to the island without having engaged, save being fired upon by the guns of the convoy.

Thus the coordinated air assault that fell on Burrough's ships was unimpeded, save by blind shooting from the ships. These had lapsed into what a naval report called 'two rather vague and straggled columns', led respectively by *Manchester* and *Kenya*. The merchant vessels were almost bereft of warship close protection: most of the escorts were either hunting the Italian submarines or standing by the torpedoed cruisers. *Bramham* was absent with *Deucalion*. For anti-aircraft fire to protect the convoy, it was essential that ships should huddle, yet every clash and casualty prompted wider dispersal. *Dorset*'s naval liaison officer, Lt. Peter Bernard, afterwards deplored their nakedness: 'A really intense [anti-aircraft] barrage is effective, but it <u>must</u> be very concentrated.'

Pathfinder was circling *Ohio*, dropping charges in pursuit of *Axum*, when the German planes struck, some releasing bombs almost at masthead height. On the tanker's bridge Dudley Mason said: 'near-misses were many and frequent, throwing deluges of water over the vessel'. *Ashanti* found herself within a few feet of disaster: a bomb exploded alongside as the destroyer weaved at full speed. This caused a blowback in one of the boilers, precipitating a fire in the engine-room. Richard Onslow's chief engineer clambered to the bridge to report this, but received short shrift from his captain, who was preoccupied with dodging enemy aircraft. The visitor was spared only a glance and a laconic 'Thank you chief, put it out, please!' Half an hour later, the blackened, sweat-soaked engineering officer reappeared on the bridge, his overalls and gauntlets charred and soaked in oil, to report the fire extinguished. 'What fire?' demanded his captain blankly. Addressing the threats from submarines and aircraft, Onslow had forgotten the local difficulty in his own boiler-room.

One Ju88 dropped a stick of bombs that straddled the American *Santa Elisa*, while two more attacked *Empire Hope*, which had stopped her engines at 2035 to avoid ramming *Brisbane Star* just as the Luftwaffe swooped. *Empire Hope*'s persecutors near-missed her several times before scoring a direct hit. This bomb smashed through two hundred tons of coal stacked atop No. 4 hatch, then exploded below in ammunition and oil which ignited. Thousands of lumps of coal, hurled high into the sky from burst bags, rained down on every ship in the vicinity.

Ron Peck was at his station on *Empire Hope*'s deck when the bomb landed, throwing him off his feet and depositing him in the scuppers. 'The strange part was that I neither heard nor felt a thing, though it left me deaf in one ear.' As he picked himself up he sensed another bomb coming, and in a mad moment glanced down at his brand-new white canvas shoes: 'I didn't want to get them dirty so I nipped behind the deck housing, sure enough a bomb fell alongside causing an eruption of water which deluged the deck and threw the main engine off its bearings.'

A mate running out a hose grabbed steward Jim Parry and told him to turn on the water when he gave the word. Parry dutifully twisted the valve wheel, but the mate still yelled at him, furious that nothing was coming out of the nozzle. 'There's no bloody water!' shouted Parry, who later recorded: 'Everyone was hollering, but you don't really hear them, because it's so quick and sudden, but you can see blokes on the other side of the flames.' An officer ordered him to the boat deck, where he manned a foam extinguisher in a futile assault on the fire until told 'come on, Parry, into the boat'.

Captain Gwilym Williams ordered *Empire Hope* abandoned, and six lifeboats were lowered successfully. Two inexpert army gunners dropped a seventh boat into the sea without its drain plug, but Cadet Donald McCallum slid down a fall to secure it. The rest of the crew then hastened away from the flaming ship without noticing that Captain Williams and Henry Lafferty, the chief engineer, had been left aboard. A boat eventually pulled back to rescue them, and while they awaited it Lafferty returned to his cabin and changed into his precious best uniform. Once in a boat, they picked up two men who were clinging to a rope trailing from the stern, dangerously close to the screws. An hour later, *Penn* approached the huddle of boats and a voice called softly across the darkness, 'put out your lights!' – the red glows on every man's lifejacket – then added, 'come up the scrambling nets as quickly as you can, wounded first. Keep quiet.'

Once below decks on the destroyer, some men were pushed out of the way into a magazine, where the warship's stewards stood beside hundreds of cans of cordite, ready to pass up shells to the guns. Parry said: 'We made ourselves comfortable and just dropped off to sleep. I didn't know any more until morning. I might not have slept if I had known that we were locked in.' *Empire Hope* blazed for two hours before being dispatched by a destroyer torpedo at 2330.

A few minutes after Williams' ship was hit, at 2058 *Brisbane Star* was torpedoed by a Heinkel almost at the very point of the bow, causing an explosion which blew in the forward bulkhead and flooded a hold. Nonetheless the ship still had way, and for a time Captain Fred Riley attempted to keep up with the convoy. At her best speed of ten knots, however, it became plain that this was impossible. Consulting with his naval liaison officer, as full darkness descended, he determined to pursue his own course towards Malta, hugging the Tunisian shore.

At 2103 another Heinkel's torpedo struck *Clan Ferguson* amidships, the explosion flooding the engine-room and one hold, starting uncontrollable fires. As flames soared hundreds of feet into the air and debris fell among desperate men in the water, *Rochester Castle*, which had been steaming a few hundred yards astern, swerved to avoid the blazing hulk. George Nye, a British soldier watching from *Santa Elisa*, said: 'there were screams, quite a lot of screams, we assumed that people were getting burned'.

Tom Brunskill, one of *Glenorchy's* engineers, was another among hundreds of horrified spectators: 'the crew were jumping into the blazing sea and their cries of agony were something I never wish to hear or see again [though] … I cannot forget'. A more dispassionate naval report recorded: '*Clan Ferguson* was observed to blow up, apparently falling into two sections, each blazing fiercely for a considerable time.' The hulk was finally dispatched to the bottom around midnight by a gratuitous torpedo from the Italian submarine *Bronzo*. Beyond her main cargo, a million pounds in new-printed Maltese currency shared the ship's fate, though the notes were useless to any finder until serial numbers had been applied.

One boat and some rafts were lowered before the ship foundered, but few witnesses expected survivors. In reality, however, more than eighty men escaped. Only eleven perished, all save the surgeon being greasers or firemen. Arthur Cossar, *Clan Ferguson*'s Master, found himself drifting alone on a raft. When he saw what appeared to be the dim outline of a U-boat's conning-tower, he attempted in vain to attract its attention by flashing his torch. After his experiences, he deemed it comfort enough to be saved from the sea, no matter by which side. The submarine ignored him, but Cossar and most of *Clan Ferguson*'s complement eventually reached the North African shore, where they were interned by the Vichy French.

Rochester Castle's near-misses had fractured two frames in the engine-room, causing some water leakage. Cordite fumes, drawn into the engine-room from the ventilators on deck, threatened to make the machinery spaces uninhabitable. A sweating party of engineers sought to seal off the leaking plates with a cement-filled steel box, not entirely effectively. *Rochester Castle* nonetheless proved fit to steam on, for a time steering by the stars, until her gyro-compass could be reset. Meanwhile *Wairangi* survived an attack when her captain made an emergency turn to starboard after seeing torpedoes launched, one of which missed a few feet astern. *Ashanti* also dodged torpedo tracks.

The cruiser *Kenya*'s senior engineer, Cdr. Charlie Crill, visited all the ship's machinery spaces, to see and to be seen by the incoming watch. Unsurprisingly, after the recent experiences of *Nigeria*, *Cairo* and *Ohio*, and now the other merchant vessels, these supremely vulnerable men were not much at ease. Crill said: 'one of my chief stokers had a terrible twitch, poor man, but he stuck it like everyone else'. Only a few minutes later, however, at 2112 *Kenya* was hit in the forefoot by a torpedo fired from the Italian submarine *Alagi*. A US Navy cruiser officer described the experience of taking such a blow as 'about the same as driving a car at high speed when you hit a pile of logs. You'd be knocked up in the air, probably sideways, and you'd come down on the concrete on the other side with all wheels flat.'

The tracks of the enemy salvo had been spotted by an alert lookout on *Port Chalmers*, and after an instant general alarm *Kenya* put her helm hard astarboard. This caused one torpedo to pass under the ship, two more to miss astern, before the fourth struck home. A bridge watchkeeper recorded: 'there was a crash as we were hit, and the bows dipped right down, so far that I feared she would never come up. The sea came over the bridge, washing us all off our feet and drenching us to the skin, but she came up at last, and we staggered on.' It is hard to overstate the shock such an experience imposed upon hundreds of men including the bridge crew, many of them briefly submerged. Stoker petty officer Buster Crabb, at his station for'ard, was one of those honest enough to admit that he lost control of his bowels, a common occurrence that week. Before he could resume his duties with a damage-control party, his first thought was to jettison the embarrassing evidence, white boiler suit and underclothes, through a convenient scuttle, from which it providentially fell upon the Master at Arms, the least popular man on every warship. After some desperately tense, terrified moments, order was restored. *Kenya* was hurt, but nobody died, and the cruiser proved able to maintain convoy speed. By the standards of that terrible day and night, she escaped lightly.

Nine minutes later, at 2121, *Kenya's* captain Alf Russell announced by emergency signal to all ships that in view of the flagship's disablement he was assuming command of Pedestal. Burrough was infuriated by this message, and countermanded it as swiftly as he could. It also incurred the anger of Syfret, who wrote in his subsequent report: 'this, as it proved incorrect, statement did not help to improve an already confused situation'. But it was a reflection of the prevalent chaos, and of persistent communications difficulties, that some ships thereafter supposed Russell to be in charge; they believed their admiral had become a mere survivor passenger aboard *Ashanti*.

In reality, Burrough was hastening south-eastwards in pursuit of his ships, impatient to reassert control. The chaos that followed *Axum's* attack, followed so swiftly by the dusk air attacks, caused most of the surviving merchantmen to reduce speed to four or five knots, and to mill in disarray, rather than to maintain distance and columns. Syfret admitted frankly in his own report: 'the effect of this series of disasters was to cause the convoy to become scattered'. Moreover, during the darkness that had now descended, such carnage – the sea erupting into flames, explosions, screams, floating wreckage – caused some men to despair.

One of those who lost his head was the convoy commodore, RNR officer Cdr. Arthur Venables, aboard *Port Chalmers*. Venables wrote later that following the bombing attack at last light 'flames from one of these [stricken] ships made it like daylight. A submarine was now sighted on port beam. Course was altered to port and I determined to try and save the ship by leaving the convoy from the rear. Ship proceeded full speed to westward.' An unidentified naval hand subsequently underlined the last word on this report and added a derisive exclamation mark. Venables' change of course took place around 2130. *Melbourne Star* signalled the commodore for orders and was told 'TURN BACK'. Her captain, David MacFarlane, wrote accusingly in his ship's log: 'He forced us to turn around by crossing our bows.'

Melbourne Star, together with *Dorset*, briefly followed *Port Chalmers*, making the reasonable assumption that Venables must know what he was supposed to be doing. *Melbourne Star*'s NLO wrote: 'The state of affairs at this time was chaotic, some ships on fire, some sinking and destroyers going to the rescue.' But after he and MacFarlane had conferred, they determined to refuse Venables' lead, as did Jack Tuckett of *Dorset*. MacFarlane, wrote the NLO, 'from the time of this disaster, was determined that the only thing to do under the circumstances was to make for Malta, in which I and all the officers were in full agreement'. They hastened onwards at sixteen knots on course 120 degrees, passing Zembra Island at 2330 and leaving Venables in *Port Chalmers* to suit himself.

Essentially, the commodore sought refuge in flight. 'Two other rear ships were informed of my intention and turned 16 points to follow me. They were not seen again, as presumably a destroyer found them and ordered them to rejoin convoy. This destroyer overtook me and gave instructions to proceed to Malta, which was my intention as soon as circumstances appeared favourable.' The escort was *Bramham*, laden with *Deucalion* survivors. She was catching up the convoy from astern when she met *Port Chalmers*, heading in the opposite direction. Eddie Baines closed alongside and sternly instructed Venables to resume course for Grand Harbour.

Contrary to Venables' claim that his change of course was only temporary, he had signalled Alf Russell in *Kenya*, whom he supposed to be in charge, and urged that, after so many ships had been lost, the survivors should make for Gibraltar. Russell messaged brusquely back: 'the convoy will carry on to Malta'. An officer on the cruiser later wrote of the captain: 'his firm handling was invaluable and saved the whole operation from possible catastrophe'. *Almeria Lykes* also intended to turn back, but Roger Hill came alongside the American vessel in *Ledbury* and 'told them they hadn't a hope in hell without an escort', adding: 'all the English ships are heading for Malta'.

The Pedestal vessels were now entering an area notorious for Axis mining. At 2118 the sweeper *Intrepid* received a signal from *Ashanti*, instructing her to 'leave the convoy'. This baffled her captain, who questioned it and found that instead he was being asked to 'lead the convoy', along with *Intrepid*'s sweep-equipped sisters *Fury* and *Icarus*. The sweeping destroyers duly took up station ahead, while all ships streamed anti-mine paravanes. As for the misread signal, a destroyer officer made the point that many telegraphists, 'Hostilities Only' ratings, were less than proficient, and now also desperately tired after seventeen hours of almost continuous action. Moreover *Ashanti*'s transmitters and signallers, even boosted by Burrough's staff, who had boarded the destroyer with him, struggled to send and receive the scores of messages which the admiral needed to pass if he was to restore any sort of grip. The captain of *Derwent* afterwards complained that because destroyers had no cipher staff, and the ship's doctor who usually helped to encrypt and decrypt signals was fully engaged tending the wounded, he himself was often reading messages twelve hours after their dispatch. He was obliged to ask officers at their action stations to decrypt some traffic.

After *Ohio* was hit at 2000, it seemed likely that it would take several hours to get her under way again. In reality, her engine-room crew raised steam amazingly quickly, their labours surely lent urgency by awareness of the perils that threatened her every moment that she drifted. At 2130 the tanker's engines began to turn. The ship steamed erratically onwards, her helmsman struggling to correct a chronic veer to port caused by a burst plate protruding from her side beneath the waterline and acting as a drogue. To escort her, the admiral sent back *Ledbury*, which eventually located the tanker in full darkness. Through a loudhailer Roger Hill spoke as cheerfully as he could contrive to Dudley Mason and Denys Barton, the ship's NLO, telling them to steer a course of 110: 'The admiral is waiting for you with the cruisers and destroyers. We'll get the Malta Spitfires tomorrow and be in Malta for lunch.'

Mason explained to Hill that his gyro-compasses were still out. The doughty destroyer captain responded that he would show a blue stern light, which *Ohio* could follow. This indeed she did through the remaining, painfully eventful hours of the night. As *Ledbury* and *Ohio*, in their pursuit of the convoy, passed the blazing wreck of *Clan Ferguson*, Roger Hill wrote: 'we felt naked and exposed, silhouetted against the fire and it seemed an agonizing age before we got clear of it'. It was a warm, clear night. As they headed towards Cape Bon, for a time the tanker managed nearly sixteen knots. *Ledbury*'s entire crew remained at action stations, where they ate soup and stew. Then gunners sought brief, broken sleep, lying on deck beside their weapons. Hill dozed in his high chair on the bridge, lulled by the relentless pinging of the asdic and terse reports of the watchkeepers.

Ahead of *Ohio* and *Ledbury*, the overwhelming sensation on many merchant-ship bridges was loneliness. Already that evening, they had seen other vessels like their own founder or become abandoned in flames. They had little or no notion of the whereabouts of the admiral, or even of other vessels. *Brisbane Star* was pursuing her own course, close to the North African coast. A young officer aboard her, Cadet John Waine, wrote later: 'the action … in the Narrows was a shambles'. *Port Chalmers* was still trailing behind, after her lunge towards the west. The other seven merchant-ships were obliged to plough slowly onward, now on a south-easterly heading, passing the coast some four miles off Cap Bon, waiting for the enemy to deliver his next blow. A rating on *Nigeria* ended a letter written to his parents with rueful words: 'PS the "glorious Twelfth" wasn't so glorious!'

Admiral Syfret responded to the devastation that had befallen Pedestal by detaching the anti-aircraft cruiser *Charybdis*, escorted by two destroyers, to quit the fleet and turn back east, to reinforce Burrough's depleted squadron. It seems plausible that not all the men on these three warships, who had been heading towards the haven of Gibraltar after two days' harrowing experiences, were enthused by the change of course. On their night passage to join Burrough, they enjoyed some luck, of which they were oblivious: two Italian submarines, *Bronzo* and *Alagi*, sighted them without attacking, because they mistook the British ships for part of Da Zara's squadron. The cruiser and her escorts duly reached the convoy at 0245 next morning, in time to share the last phases of its agony. Meanwhile *Axum*'s captain, Renato Ferrini, gazing from his bridge across the darkness at the flames leaping from the hulks of *Empire Hope* and *Clan Ferguson*, which he wrongly supposed to be his own victims, made no attempt to deliver another attack. He said later that it had become essential to recharge his boat's batteries. In truth, he had already done more than anough to serve the cause of the Axis.

The Luftwaffe attackers who had delivered the final air assault of the day landed back on Sicily having suffered the loss of just one Ju88, a small price for drastically diminishing Pedestal's cargoes for Malta. As for the exhausted men on Burrough's ships, it was bitter fruit to recall that, for so many hours, they had successfully fought off aircraft and submarines. Of five Italian boats that attempted attacks, one was sunk, one damaged, the others driven off. Although enemy aircraft had crippled *Indomitable* and *Foresight* and sunk *Deucalion*, most of the fleet had survived its ordeal. Then, suddenly, in a heartless hour a single submarine got lucky, and the dusk lunge by daring and skilful Luftwaffe pilots caught the convoy off-guard. In consequence, the entire balance of the battle had been transformed, not to the advantage of Burrough's ships.

9

Scuttling Charges

As *Nelson* ploughed west that night, George Blundell read signals about the tribulations of Force X and wrote in his diary: 'shortly after we left them they got beaten up badly – it's a pity we couldn't have stayed on a little longer'. Burrough's captains might have said: quite so, though it is unlikely that the continued presence of the fleet would have averted the calamities. Just before midnight on 12 August, the three minesweeping destroyers leading the convoy passed Cap Bon lighthouse, its powerful beam catching the blacked-out silhouettes of the foremost ships – others were trailing up to thirty miles behind.

Sailors aboard *Ashanti*, accustomed to destroyer informality, found it hard to get used to the awe-inspiring presence of the admiral and his staff on her overcrowded bridge. A sailor said: 'Gor blimey, you've never seen so much brass in your life.' *Ashanti*'s first lieutenant, deciding that war correspondent Anthony Kimmins was the least essential presence around Burrough, removed him to crew an Oerlikon gun, which the writer seemed to find diverting.

By 0100 on 13 August the leading ships were two hundred and twenty miles from Malta on their designated southern, then easterly, dogleg course, designed to skirt the Axis minefields south of Pantelleria. In the darkness the admiral could exercise effective command only of the warships close at hand. He had no grip upon those merchantmen that had become invisible, scattered across hundreds of square miles of sea. Burrough acknowledged later: 'It is almost impossible to give proper protection to a scattered convoy with such a small escorting force.' Horatio Nelson, 140 years earlier in waters not very distant, fumed at lack of frigates as the curse of his campaign against the French. Now, lack of destroyers, their latter-day counterparts, menaced the very existence of Pedestal. It was almost certainly a mistake, though a readily understandable one, to have sacrificed three escorts to accompany the crippled *Nigeria* to Gibraltar when Force X faced a new menace.

Deployed in clusters close to the Tunisian coast and a few miles offshore, nineteen Axis fast torpedo-boats waited in ambush. Tedder's Middle East Air Force, conscious of the threat the boats represented, bombed their North African harbours on successive nights, 5 and 6 August, but without success. The light craft, constructed of wood or aluminium, were too fragile to achieve much in rough weather or on an open ocean, and their high-performance engines were notoriously prone to breakdown. To have a realistic chance of success, they needed to close to almost point-blank range, less than a thousand yards, before unleashing their twin torpedoes – a *Doppelschuss*, in German parlance. The calm prevailing in the early hours of 13 August off Cap Bon nonetheless offered ideal conditions for them to achieve their worst.

Like submariners, boat crews saw themselves as an elite, liberated from the rigid discipline of big warships, revelling in the thrill of racing into action aboard the fastest fighting craft afloat. The boats drifted on the glassy sea with power switched off, because only thus could crews hear anything save the roar of their own engines; and as soon as they moved at speed, their churning white wakes became visible, even in darkness. Most were Supermarina's sleek and deadly MS and MAS boats – *motosiluranto, motoscafo armato silurante* and *moto-torpediniera*, outstanding examples of Italian design skill.

There were also German hundred-ton E-boats – a British designation, for the Kriegsmarine called them *Schnellboote* – each powered by three 1,630rpm Daimler-Benz engines and carrying a crew of around twenty. They had become familiar scourges in the English Channel and North Sea, attacking convoys and duelling with their British MTB and MGB counterparts, always at night, because in daylight when they were visible a single shell or burst of 20mm fire was likely to be fatal. Late in 1941, Grand Admiral Erich Raeder, Germany's naval commander-in-chief, approved the transfer of the 3rd E-Boat Flotilla to the calmer and shallower Mediterranean. The boats moved southwards by way of the French river and canal system, a pleasing experience for their crews. In February 1942, operating from bases on Crete and at Sicily's Porto Empedocle, they began laying mines off Malta, exploiting their speed to dash to and fro under cover of darkness.

In the torpedo-attack role, they achieved a notable success in June, off the North African coast, by sinking the cruiser *Newcastle* and destroyer *Hasty*. They also mauled several British ships that were evacuating fugitives from Tobruk. By early August two of the 3rd Flotilla's boats were at Suda Bay, Crete, two were at Mersa Matruh, three at Augusta in Sicily, unfit for operations due to technical and manning problems; other boats were under repair in Palermo dockyard. Consequently, just four participated in the attacks on Burrough's ships, guided towards their prey by a Luftwaffe Heinkel 111 reconnaissance aircraft, which transmitted regular position reports.

For their counterparts of the Italian Navy, Pedestal provided their finest hour of the war. Gunfire and searchlights might have enabled the British warships to protect the merchant vessels had the convoy been tightly closed up. As it was, however, forty minutes after midnight, when the first Axis boats started their engines and sprang across the water to engage, most of the escorts were separated from the scattered merchantmen.

The three destroyer/sweepers were well ahead, followed by *Kenya* and *Manchester*, and astern of them the cargo liners *Glenorchy*, *Almeria Lykes* and *Wairangi*. *Rochester Castle* and *Waimarama* steamed somewhat behind these leaders. The destroyer *Bramham* was more distant, hurrying to rejoin after retrieving survivors from *Deucalion* and recalling the convoy commodore to his duty. *Ledbury* was leading *Ohio*, perhaps twenty miles behind the van, and slowly closing; *Ashanti* was also catching up. *Dorset* and *Brisbane Star* were sailing independently.

Melbourne Star, on course 086 at 2350, steering for Cap Bon, was overtaken by the destroyer *Pathfinder* which challenged, 'what ship?' The vessel gave her identity and asked, 'will you lead on?', which the warship answered affirmatively. David MacFarlane of *Melbourne Star* then passed *Santa Elisa* and shouted through a loudhailer: 'I'm going on to Malta! Will you follow?' 'Yes!' responded an American voice from the other ship's bridge, and for a brief time *Santa Elisa* did so. The relative positions of all these vessels shifted frequently during the hours of darkness, and it is impossible to track and rehearse those of each one. Here it suffices to say that only the three lead destroyers maintained their appointed courses and speed, while behind them the cargo vessels steered increasingly erratically, shifting their helms as the perils of the moment seemed to demand.

The new chapter of sound and fury opened when a paravane streaming off the bow of *Manchester*, the largest warship under Burrough's command, detonated a mine as the cruiser rounded Zembra Island, a 1,500-foot-high rock formation, richly endowed with wildlife, nine miles off the Tunisian coast. The explosion inflicted no damage, save to the paravane. Then *Kenya*'s main armament fired starshell illuminants, which cast a milky light over the sea and revealed the racing approach of four Italian MAS boats. The British opened fire and turned to starboard to comb the tracks of the enemy's torpedoes. Their shooting was too wild, the Italians too nimble, to inflict damage – the MAS boats activated smoke dischargers to screen their retreat after launching torpedoes. All missed, however. This first attack was a failure.

Many other mosquito craft were coming, however, skippered by officers who were game to engage much more closely. British crews were exasperated, indeed enraged, by the bright beam of Cap Bon's Kélibia lighthouse, on Vichy French soil, lighting up the sea for ten miles – together with their own ships. Several captains requested permission to extinguish the light by gunfire, but were refused: it had been the Royal Navy's choice to sail so close to neutral Tunisia. The planners adopted a route that set at naught Pedestal's advantage in steaming through the most dangerous miles of the dash for Malta at a moonless period: they argued that hugging the coast was less lethal than braving the minefields to the north-east or taking the direct route within a few minutes of Sicilian airfields. And who can say whether they were right? There was never a safe course to Malta, only a choice of variously hazardous ones.

At 0105 *Icarus* narrowly avoided a torpedo, and fifteen minutes later two Italian boats, Giorgio Manuti's *MS 16* and

Franco Mezzadra's *MS 22*, raced out from behind the hulk of a long-beached British destroyer to launch an attack on *Manchester*. Manuti, a thirty-three-year-old native of Bari, was a regular officer who became one of the Italian Navy's most decorated war heroes. Mezzadra, twenty-three, had commanded his boat for only four months. Neither recorded the sensations of attack that night, but another mosquito-craft skipper did so in similar circumstances at the same period: 'The six engines burst into life in quick succession, the boats swung around to port and steadied. The throttles were lifted swiftly; the roar of the engines reverberated as we settled to thirty knots. I for one had a queer feeling in the pit of the stomach, but what man, determined upon performing an unpleasant duty, has not?… I was standing on the canopy now, straining into the darkness with the glasses. I did not want to open fire until we had got in close, but [then] the enemy's guns spoke, a single brilliant stream of light tearing towards us, then another, then another. A tornado of fire was blasting at us now as we tore into the enemy line, the sky seemed alight with hurtling meteors and comets, but we were a low, difficult target, half covered by our great white pressure waves thrown high round the boat, throbbing at thirty-eight knots.'

On this August night in the Mediterranean the big British cruiser's bridge crew spotted the attackers heading pell-mell towards her. The main armament fired several deafening salvoes, its flashes lighting up the ship and the sea. The exploding shells failed to check the charging Italians: the processes of fire-controlling, loading and triggering the six-inch guns were far too sluggish to keep pace with boats weaving at a land speed of almost fifty miles an hour. Only automatic weapons – Bofors, Oerlikons, pom-poms – had a realistic chance of hitting the attackers. No warship had enough of these, and guard rails hampered their depression to fire below the horizontal.

The two MS boats launched their first torpedoes at ranges of eight hundred and six hundred yards respectively – and missed. One passed just ahead of *Manchester*, the other vanished unseen. The boats' second shots, however, were made at almost point-blank range. At least one torpedo, possibly both, struck the cruiser's starboard side, blasting open its aft machinery spaces. 'The first explosion plunged the engine-room into darkness,' said leading stoker Albert Slater, 'and the second flung me from the platform on which I was standing. Only an uprush of water prevented me breaking my back in a fall. The rising water carried me up towards the ceiling. I gave up all hope, waiting for the moment when the water would finally fill the entire engine-room. But suddenly [it] ceased rising, only a few feet from the ceiling. Then steam spread in a scalding cloud.'

Slater struggled to a hatch, through which friendly hands dragged him upwards to safety.

The attackers escaped unscathed back into the darkness, while *Manchester* reeled beneath grievous damage. Even with her engines stopped, for ten minutes she continued to circle with jammed rudders, slowly losing way. Below decks, those of her engine-room staff who were unable to find a path to safety, as Albert Slater had been fortunate enough to do, suffered terribly as swirling seawater and scalding steam overwhelmed their spaces. The ladder leading to the hatch at the top of one boiler-room collapsed, trapping thirteen men who were steamed alive: 'the screaming went on and on', recalled one of those who heard it, petty officer Rob Cunningham.

The ship assumed an eleven-degree list, with her four-inch magazine and main radio-room flooded, as well as some machinery spaces. The cruiser dispatched a terse message to Burrough, which was read by every escort: 'Engine-room flooded, on fire'. At 0200, half an hour after the torpedo struck, Captain Harold Drew decided that his ship, two miles off Kélibia, was unsavable. He ordered Confidential Books – codes and ciphers – thrown overboard in the usual weighted bag, and transferred 172 officers and men to *Pathfinder*, which slid alongside. Curiously, Drew authorized this partial evacuation before he received detailed damage reports from below. These were delayed by the lack of electrical power: engineering crews were obliged to explore the steely, echoing, unnaturally silent darkness below with the guidance only of torches.

Several compartments were flooded, including one engine-room, but some of the specialist officers concluded that the ship might be made to move again on one shaft of her four; steering could be restored soon afterwards. By 0245, transferring oil from her starboard fuel tanks to port and jettisoning her torpedoes had reduced the list, but the ship was ill-fitted to defend herself, after most of her guns' crews had been allowed to abandon their posts and vanish into the darkness aboard *Pathfinder*. Dawn would break at 0530, rendering *Manchester* mortally vulnerable to air attack.

Drew had already commanded his ship through a similar crisis a year earlier, in the same region of the Mediterranean: on 23 July 1941 twenty-six *Kenyas* died after she had been torpedoed from the air, hit in her machinery spaces. Now, the captain told his senior engineer, according to Commander William Robb's subsequent evidence: 'I will explain the position to you as I see it. I have to make the decision whether to try and save this valuable ship, against the risk of further heavy loss of very valuable lives. We are on the edge of a minefield, with E-Boats all around us. I am afraid we may be put in the position where we could not move, and the ship may be captured by the enemy.'

Robb claims to have responded: 'We stand a chance, Sir, and are in the same condition as last year as regarded machinery.' At that moment a voice from *Pathfinder* shouted across the darkness: 'All aboard who are coming.' Drew called from his bridge through a loudhailer: '*Pathfinder* shove off when ready, and thank you very much.' The destroyer offered to take a few more men, but the cruiser's captain declined: 'No, thanks very much. We will look after ourselves now. Please send something back to see what has happened to us, even if it is only a plane.'

The engineer officer later claimed to have renewed his altercation with the captain, urging that they should persist with efforts to save the cruiser. Drew demurred: 'we are of no further use to the convoy, and it would be unfair to detach an escort for us ... Here we are close to the shore, and if we act quickly we can get the ship's company ashore as a body, and they may be of further service to the country.' Robb returned to the engine-room, inspected the progress of his repair parties, then again reported to the captain: one engine was ready to turn at any time, which should enable them to steam at ten knots, and they were almost ready to connect emergency leads to the steering motor. Drew responded dismissively: 'we don't stand a chance, old boy'. He ordered the cruiser to be abandoned and sunk – sea-cocks opened, scuttling charges detonated.

Junior engineer officer Phil Rambaut had joined the navy in 1939 in hopes of a better death than he might meet ashore: 'if one was to go, better to die by being drowned or blown to smithereens, wearing a clean uniform, than dying of gangrene in the mud. The thought of having to shoot someone I could see was utterly abhorrent.' Now Rambaut was twenty-seven, and as he laboured to run power leads to the steering, he was no longer in the least clean. He reflected upon the custom of Gieves, the famous naval tailor, of cancelling an officer's outstanding account if he was killed in action.

Rambaut said later that the scuttling decision inspired 'mixed feelings. On the one hand I was out of the war, which was a relief. On the other, perhaps we should have had a go at getting to Gibraltar.' He and his mates recalled that the cruiser, after being hit the previous year, had made it back to the Rock on one shaft. The engineer noted that Drew was 'obviously a very disturbed man – [he] endeavoured to put it about that there had been no steam available to work the engines', which was untrue.

The ship's crew, of whom more than six hundred remained after the departure of those who had crossdecked to *Pathfinder*, took to boats and rafts. Those in the former began to row for the shore just as day was breaking. Bill Terry, the Master at Arms, groped his way through the mess decks to check that they had been fully evacuated. With boats overcrowded, some men set out to swim the two miles to land – and succeeded. Others continued to drift aboard Carley rafts.

Then an Italian aircraft dropped a single torpedo, which missed the ship's bow – ironically failing to accomplish what the British were already attempting. Drew called the attention of officers sharing his boat to this attack, which he said explained and justified his decision. Only thirteen members of the ship's company died. Beyond those transferred to *Pathfinder* and raft survivors later rescued by destroyers, the remainder reached the Tunisian shore, where they languished in Vichy internment for the ensuing three months, until freed following the allied invasion of North Africa. One such man recorded with unashamed gratitude: 'we were all delighted to be alive, and sleep was all we needed'.

Drew's decision to abandon *Manchester* prompted a controversial court-martial, discussed below. The evacuation reflected
the sense of despair that overtook some officers, of both the
Royal Navy and Merchant Navy, after the procession of disasters that had befallen their ships since Syfret's fleet parted
company, a mere eight hours earlier. Logic was on the captain's
side. It is highly likely that, had he kept his crew on board and
attempted to creep back towards Gibraltar, Axis planes would
have sunk the cruiser next day, perhaps with heavy loss of life.
The enemy's air forces in the region were far stronger than they
had been in the previous year when the ship made her successful escape after being torpedoed.

But the Royal Navy's fighting captains were expected to
consider other issues at least as important as logic: the tradition of the service, including Nelson's legendary 1805 signal
that began 'England expects ...', and the need to set an example
of defiance, determination and sacrifice, not only to other
ships of Pedestal, but also to the world. In the early hours of 13
August, Harold Drew behaved with impeccable good sense
after his ship had been crippled, but Churchillian crusades
were not to be won by good sense, nor even by displays of
humanity. Much of the rest of the navy was as unconvinced by
Drew's conduct as were some of his own officers.

And even while *Manchester* was enduring her hours of agony,
a swarm of new attackers fell upon the convoy. Until *Pathfinder*
turned aside to assist the stricken cruiser, she had been shepherding *Melbourne Star*, and for a time *Glenorchy* and *Almeria
Lykes*. When the destroyer abandoned those vessels – probably
mistakenly, on a cold judgement of the priorities of Pedestal
– she signalled them to continue on their southward course,
but thereafter each of the three Masters considered their vessels

to be sailing alone. By the time the destroyer, now laden with the cruiser's survivors, again overtook *Melbourne Star*, the faster *Glenorchy* and *Almeria Lykes* had vanished ahead. The American ship briefly pursued her own, direct easterly course as her captain attempted an independent dash towards Malta. After a time, however, he was dissuaded by his naval liaison officer, who said that such a venture would be suicidal. The ship once more turned south, and suddenly loomed out of the darkness just ahead of *Glenorchy*. To avoid collision the latter was obliged to ring down her engines to full astern, precipitating a shower of sparks from her funnel – deadly telltales for predatory torpedo-craft.

These began to dart in and out of the confusion of British ships, firing machine-guns and torpedoes; receiving spurts of return fire; creating a pyrotechnic display which gripped the fascinated gaze of all those who witnessed it. What took place that night off the Tunisian coast resembled a deadly game of blind man's buff, with thirty-odd warships and merchant vessels groping their way through a darkness repeatedly torn open by fighting lights, gunfire, flames and the Kélibia beam. Several merchant-ship captains attempted lunges aside from the most threatened places of the moment. Drastic helm orders were issued, to avert both collisions and torpedoes. In those hours, Burrough could do little or nothing either to reassert his own control of the convoy or to repel the attackers. Chance, spasmodically influenced by individual captains' skill or courage, determined which ships survived – for a few more hours anyway – and which went to the bottom.

At 0200 *Kenya* momentarily flicked on a searchlight to show *Almeria Lykes* and *Glenorchy* her own position, but this provided a beacon for the enemy. Eight minutes later two torpedoes fired from close range by twenty-seven-year-old

Antonio Calvani's *MS 31* struck the latter freighter on the port side. It was an important difference between warships and merchant vessels that for obvious reasons the latter's builders did not duplicate vital machinery or compartmentalize them to survive battle damage, as was the practice of naval constructors. With the engine-room flooded, all power lost and the ship going down rapidly by the head, Captain George Leslie ordered *Glenorchy* abandoned. He himself, however, stayed on his bridge, to become one of seven men aboard who perished. Survivors said later that when Leslie heard that the chief engineer, a close friend, had died below, he lost interest in his own fate, rejecting urgings to make for a boat. He told his Third Officer: 'You are a young man, Mr. Simon, and life is just beginning for you.' Simon found this remark grotesquely fatalistic. He himself made for a boat after jettisoning the ship's Confidential Books.

Others of the crew endured struggles to reach the upper deck: Tom Brunskill, Second Engineer, had served on the Alfred Holt & Co. ship since first she put to sea, in 1939. Now he found himself trapped beneath a grating by the buoyancy of his own lifejacket while a torrent of oil and water swirled around him: 'I made up my mind my time had come.' He said a mute farewell to his wife, and sought forgiveness of sins from his Maker. Then, suddenly and inexplicably, he found himself free of the grating, breathing fresh air, though unable to see. He groped along a companionway until he reached the upper deck, where he shouted: 'Is anyone there?' Another desperate voice pleaded for help in getting free of debris. Brunskill explained his own blindness, but managed to feel his way towards the trapped man – the Fourth Engineer – and drag him clear.

He himself was led out on to the deck, where he regained some vision by wiping diesel oil from his face, though still shocked and concussed. Unknown hands summarily pitched him down into a boat, where he landed with a force that stunned him into unconsciousness. When he came round, he was astonished to find himself alive. They rowed to the Tunisian shore, five miles distant, where local villagers plied them with watermelons. Tom Brunskill said: 'I do not think I have ever tasted anything more delicious.' In return, the British survivors gave local people chocolate from their lockers.

Glenorchy's radio officer, Greenwood, later told Italian interrogators that he himself was among those left behind when the two heavily laden boats pulled away from the doomed ship. He assisted three other men, including the chief officer, to unlash and launch a raft, which they hoped that Captain Leslie would board. Instead, as they drifted, they pulled three more survivors out of the water. Eight men, in all, clung to the raft until they were rescued at first light by the very MAS boat that had sunk their ship.

Antonio Calvani's boat had been suffering problems with its steering gear, but he and his crew took pains to behave generously that morning. They retrieved another two swimmers, who proved to be officers from *Manchester*. A droll altercation followed between rescuers and rescued. The Royal Navy's men protested furiously that they had been picked up inside French territorial waters, and that therefore their captors had no right to hold them. Calvani chivalrously agreed and delivered them to the Vichy French ashore, to join their crewmates in detention. He refused the same privilege to the *Glenorchy* men, however, on the grounds they had been taken prisoner in international waters. The triumphant Italian captain eventually took them ashore in Sicily, where they watched him being garlanded with flowers by admiring locals, and were invited to share the wine and food lavished on the MAS boat's crew.

Through the hours of darkness the battle ebbed and surged, sometimes erupting into a storm of gunflashes, tracer and flame, then dying away for a time, until the Italian and German attackers – all of which the British branded as E-boats – identified a new victim and dashed forward to launch their torpedoes. The warships responded to radar sightings of an enemy craft by firing starshell or illuminating searchlights that briefly bathed it in whiteness, from which speed enabled it to twist away back into blackness. One of the crew of *Pathfinder* painted a miniature: 'Our engine-room telegraph clanged the order for full speed and our helm was put over to ram. At the same moment, we caught the E-Boat squarely in the beam of our big 44-inch searchlight, and every gun that would bear, 4-inch, pom-poms and Oerlikons, opened a rapid fire. Enveloped in shell splashes, the E-Boat shot ahead, and swung around laying a smokescreen. She disappeared behind the smoke, and we followed her through it. Once again the searchlight caught and held her, and all the guns opened up; and once again she disappeared in a large cloud of smoke.' *Pathfinder*'s quarry was damaged, but escaped, as did every Axis boat that night.

The Americans aboard *Santa Elisa* joined other crews cursing the lighthouse beam that repeatedly swept their ship, which was trailing the convoy's leaders. They could see flashes beyond the cape, which at first they supposed to be caused by shore batteries firing on the convoy, but in truth indicated warships attempting to engage E-boats. Captain Thompson and his NLO Lt. Cmdr. Baines descended from the bridge to the chartroom, where they could discuss their options with a light on the map. They decided to head east towards Pantelleria, altering course easterly first to 170, then to 145, risking a known Italian minefield rather than continue towards the flame-flashing horizon. Purser Jack Follansbee said: 'As far as the eye could see the Mediterranean was filled with fire. Fire shooting from the torches of doomed ships, fire from burning heaps of wreckage, fire from burning oil on the surface of the sea.' One man refused to return to his post in *Santa Elisa's* engine-room after experiencing the earlier attacks there.

Ensign Suppiger mustered his gunners and ordered them to bring up more ammunition and clean the weapons: one Oerlikon was out of action, almost all the Bofors barrels were worn out and only four hundred 40mm rounds remained. It was obvious, however, that more air attacks would come at dawn, for which they needed to be as ready as they could make themselves. *Santa Elisa's* naval signallers broke radio silence to seek counsel from the admiral, but received no response from *Ashanti's* overburdened telegraphists. Few merchant vessels recorded successful wireless contacts with Burrough's temporary flagship during that night, or the following day – direct voice communication, bridge to bridge, was the principal influence upon their movements – when it could be contrived. *Santa Elisa* sailed onwards, a mile or two north-east of the course of most of the other vessels.

At 0222, the minesweeping destroyers and *Kenya* led the convoy out of the Sicilian Channel, followed closely by the other American vessel *Almeria Lykes*. Alf Russell, the cruiser's captain, ordered a change of course away from the coast, and half an hour later, around 0245, his bridge lookouts sighted first *Santa Elisa*, out to port on her own, then the cruiser *Charybdis* and two destroyers. Soon afterwards *Wairangi* joined them, at 0300, the very moment when enemy boats again charged in. Escorts fired starshell, which did more to illuminate targets for the Italians than to assist the British to sink attackers. *Ledbury* briefly switched on her fighting lights, which caught glimpses of enemy craft bathed in a distant glow from the flames of *Glenorchy*. Roger Hill also changed course north-eastwards, to lead *Ohio* away from the attackers while giving a wide berth to the Italian island of Pantelleria, where flak and spasmodic flashes testified to a night attack on its airfields by three Malta-based Beaufighters. Italian aircrew returning from attacks on Pedestal ate dinner that night by candlelight because a British bomb had hit the electricity transformer.

Ledbury became temporarily lost, 'for we had been fighting and manoeuvring the ship the whole time', in the words of her first lieutenant. Likewise *Bramham*'s captain asked his navigating officer for their position, to which the hapless Sub-Lt. Mitchell replied that he 'just didn't know'. This outburst of frankness somewhat alarmed others on the bridge. Soon afterwards they encountered a large, dark shape which they feared might be an enemy cruiser, but proved to be *Dorset*, 'as lost as we were … So we said "jolly good, and nice to see you"', and the two ships proceeded on eastwards. John Waters of *Dorset* wrote: 'what should have been the darkest night to give us the chance to get through the narrow straits of Cape Bon turned out to be the most lit up … We picked our way among this mob of destruction, and wrecked blazing ships, helpless to do anything but plod on.'

Meanwhile *Port Chalmers* took station behind the destroyer *Penn*, which was also risking the minefields to avoid the E-boat battle. Fred Jewett, at his action station in the port waist of *Ashanti*, was stricken with paralysis by a sudden glimpse of twin torpedo-tracks, 'these two fingers pointed towards us'. For several seconds he found himself unable to speak or move. 'I knew exactly what they were, but anyway I couldn't have done anything.' The torpedoes, set deep to sink a cruiser, passed harmlessly beneath the hull of the shallow-draught destroyer.

Aboard *Santa Elisa*, men had been struggling to catch snatches of rest, exhaustion overcoming the grim fascination of the flames on the sea a few miles away. Mate Fred Larsen dozed in the wheelhouse. Eighteen-year-old Southerner Lonnie Dales was manning one of the aft Oerlikons. Purser Jack Follansbee, responsible for another gun on the port side, was huddled with a blanket around his shoulder when he was shaken awake by a British gunner who said: 'sorry to wake you again, sir, but some bloke says 'e 'ears what sounds like 'undreds of aircraft'. 'Some bloke' was a US Navy man, one of the crew of the four-inch gun on the stern. Then British and American sailors alike caught the sound – a low roar of engines.

The British soldier appeared beside Follansbee again, saying: 'Everybody's 'eard it now! They think it must be motor boats. Maybe motor torpedo boats!' Aboard *Penn*, Les Rees recorded 'At one moment it would be absolutely quiet and pitch-black. Then out of nowhere would come the unmistakable roar of an E-Boat's engines. Suddenly it would be like Blackpool illuminations, with searchlights and destroyers' red and green fighting lights coming on all over the place. What a night! If anybody got any sleep he was certainly not on *Penn*.'

Nor *Wairangi*. At 0324 a torpedo, probably fired by Siegfried Wuppermann's E-boat *S58*, struck the Shaw Savill cargo-liner. She was a veteran of the Australia–New Zealand run, now carrying eleven thousand tons of military stores including ammunition. A massive explosion followed, causing a waterspout to soar. As lifeboats were lowered, the chief engineer told the captain there was no chance of restoring power. The ship signalled the escorts, suggesting that *Wairangi* should be sunk with gunfire, to put her out of her misery, though no destroyers were available to fulfil this task. The crew took to the boats: daylight found them still rowing around the hulk. Even as they languished, three Luftwaffe Heinkels made successive attempts to torpedo *Wairangi*, without success; the bombs of an Italian aircraft likewise missed. Later that morning, destroyers retrieved seventy-nine survivors from the lifeboats. A few hours later, the ship disappeared beneath the surface of the Mediterranean.

Five minutes after *Wairangi* was hit, around 0330 lookouts on *Rochester Castle* sighted yet another enemy boat – it is impossible to know whether it was German or Italian – drifting two hundred yards away with its engines stopped. Gunner Joe Tunney poured tracer towards the craft from a bridge Oerlikon, and the Bofors gun on the fo'c'sle also opened fire, but moments later a second enemy boat loosed a torpedo which struck her No. 3 cargo hold, filled with flour, blowing a huge hole twenty-five feet by eighteen, wrecking two lifeboats and putting the ship down by the head. She shuddered and briefly listed, reeling under the shock.

A shout went up from an unidentified voice, 'abandon ship!', and NLO Edward Binfield threw the ship's Confidential Books over the side. He was premature: calmer voices regained control. Binfield wrote: 'as ship appeared to be steering well and not in immediate danger of sinking, course was maintained'. When next a sailor standing among a knot of fearful men on the boat deck demanded to be told when they could lower away, the First Mate shouted defiantly back: 'she's still floating, isn't she?' The ship's engines were undamaged. *Rochester Castle*'s tough Master, Richard Wren, held his course for Malta at thirteen knots, 'zigzagging and steering by the stars until compasses were set back on gimbals'.

Almeria Lykes received a signal, ordering her to take station behind the cruiser *Charybdis*, when a few minutes later, at 0414, she became yet another victim. The US merchantman had been an unhappy ship since leaving the Clyde, manned by a motley crew of fifty-one men, most of them older than the average. British liaison officer Lt. Cmdr. Hugh Marshall wrote: 'They were a mixture just labelled American. To me they did not seem to realise properly that there was a war on – let alone that they were in it.' From Gourock to Cap Bon, they complained bitterly and often about the quality of their rations, and much else.

When *Manchester* was hit, *Almeria Lykes*' Master, Willie Henderson, urged a fast, solo dash for Malta, but after a bitter argument Marshall persuaded him to hold his course with the convoy. *Ledbury* and *Ohio* caught up the American ship as they passed Kélibia Light, and Roger Hill ordered her to follow him: 'Instead she went off on her own. I could not leave the tanker, so I had to let her go.' The merchant vessel was zigzagging at thirteen knots when the Italian *MAS 554* approached unseen on the port side and launched two torpedoes from 550 yards, one of which hit the forepeak, precipitating a terrific explosion. In most cases, ships hit forward proved able to survive, if their machinery spaces were unharmed, which was the case with *Almeria Lykes*. Henderson nonetheless gave an immediate order to abandon ship.

What followed was one of the most controversial, at times black comic, episodes of Pedestal. It seems fair to suggest that, by the early morning of 13 August, the confidence of the ship's crew in 'Limeys' in general, and the Royal Navy in particular, had fallen low. The British gunners and passengers aboard the ship had been allocated their own lifeboat, which was now lowered. When it splashed away from the side, however, it was discovered that none of the soldiers manning it had ever before pulled an oar: chaos ensued. The Americans' lifeboats, propelled by hand-screw, had no trouble reaching a destroyer, *Pathfinder*, whose captain promptly demanded that they return to their own ship, which 'appeared to be perfectly all right. I persuaded her company to return to her and endeavour to steam her to Malta.'

In truth, the naval officer's urgings were less than successful. Captain Henderson agreed to reboard the vessel, but found few men willing to accompany him. Eventually he himself and three officers returned, while the others sat passive in the lifeboats. This group examined the situation below, then clambered back down into their boat and rowed over to the ship's NLO, Commander Hugh Marshall, drifting in the British boat nearby. They told him that, while *Almeria Lykes'* engines were intact, the damage forward would make it impossible to maintain any speed. Moreover, the crew flatly refused to resume manning the ship. The British officer then agreed to accept the Americans' verdict that she must be sunk. A small party yet again reboarded, laid and detonated scuttling charges. Disobligingly, however, *Almeria Lykes* remained afloat.

At 0930, when destroyers reluctantly took aboard the unheroic occupants of the ship's boats, the freighter's deck was awash, but she still refused to vanish beneath the surface of the Mediterranean. This was the only one of the night's merchant-vessel casualties about which real doubt persisted, whether she might have been saved. In Admiral Syfret's written orders, issued to every captain before the fleet entered the Mediterranean, one passage was underlined: 'no ship is to be scuttled if she is capable of steaming and there is no immediate risk of capture'. Syfret's subsequent report declared uncompromisingly 'The tale [of *Almeria Lykes*], as recorded in the report by her Naval Liaison Officer [Marshall], of the abandonment of this ship is one of shame.' The probability is that a more determined crew could have saved her, might have got the vessel to Malta. Following the experiences of the past days and nights, however, the solidarity of the Anglo-American alliance, and of its immediate representatives off Tunisia that morning, had been stretched beyond breaking-point. It was an act of folly that Britain's Ministry of War Transport had engaged an American ship, with such a crew, for this most hazardous of ventures.

Dorset lagged well behind the convoy, and lost touch with her own temporary guide, the destroyer *Penn*. The merchant vessel's bridge watch saw the flashes and flames in the darkness which pinpointed ships in utmost trouble, and received a radio warning of E-boats lurking. Jack Tuckett, her Master, decided against continuing to follow the other British ships and instead adopted a course much further offshore, which should enable him to intercept the others' dogleg route at daybreak. This took him through the known Axis minefield, and at 0330, twenty-eight miles off Kélibia Point, the ship's port paravane indeed detonated two mines. Tuckett said that the effect resembled that of depth-charges exploding alongside. The ship appeared undamaged, however, though he suspected that her plates were strained.

Soon afterwards a German E-boat closed on his port side. It was momentarily engaged by one of *Dorset's* Oerlikons, which then jammed. The exasperated gunner, John Waters, vented his frustration by hurling a torch, the only missile to hand, into the midst of the enemy crew, though he confessed that this absurd gesture did not seem to do them much harm. An apprentice, Desmond Dickens, who also glimpsed 'a black shape on the port beam' just twenty yards away, said bitterly: 'you could even hear the voices of the swine on board'. Lookouts believed they saw a torpedo track at this time, then watched another ship hit – possibly *Almeria Lykes*. Despite all this mayhem and the proximity of the enemy, *Dorset* came through the night not much hurt.

Aboard *Santa Elisa*, in the early half-light of dawn just after 0500, British NLO Lt. Cmdr. Barnes glimpsed two hundred yards astern the Italian *MAS 557*. Its skipper, twenty-four-year-old midshipman Giovanni Cafiero, had chosen his own ambush point forty miles south-west of Pantelleria, some distance from his assigned rendezvous. He and his crew had been drifting for several hours, their engines shut down. Now, glimpsing a target, Cafiero ordered the engines started, and raced in pursuit of the freighter, twisting and weaving his sprightly craft to and fro across the ship's wake, his machine-gunners exchanging fire with those of the American vessel. Lonnie Dales had just strapped himself on to his own gun when a burst of incoming automatic fire, green tracer from the sea, hit his loader in the neck. The man collapsed dying in a pool of blood, though Dales was so fiercely concentrated on his shooting that he noticed only when he found the Oerlikon drum empty, and nobody heaved up a replacement. Two more British soldiers were killed on the bridge, a fourth at the Bofors on the main deck. The port 40mm fired so close inboard that one of its rounds exploded in their own rigging. Captain Thompson threw himself flat on the deck with his bridge crew, as wheelhouse windows shattered: 'the bullets made a steady clattering sound a few inches in front of our eyes'.

Cafiero fired a torpedo, which missed astern as Thompson swung the ship hard to starboard. In the poor light his gunners struggled in vain to get a good aim. Dales afterwards persuaded himself that four boats had been pursuing them. There were probably only two, of which the second – Giuseppe Iafrate's *MAS 564* – appears to have fired from point-blank range the torpedo which exploded in the ship's forward hold, laden with aviation fuel. There was 'a terrific flash and explosion forward' followed by a waterspout that drenched most of the crew on the upper deck, while several gunners were blown overboard by blast. The ship immediately began to list to starboard and desperate voices called 'Let's get off before she explodes.' The heat intensified as flames ignited sacked coal, stacked on a forward hatchcover. Thompson ordered the engines full astern, to check the ship in her course, then called for *Santa Elisa* to be abandoned: 'it was a hell of an order to have to give. It was my first ship. But I had the crew to think about.'

Fred Larsen was dispatched to Thompson's quarters in search for the captain's dog, which had vanished, never to be seen again, poor mutt. The subsequent lowering of boats was shambolic. Larsen, having clambered down a scrambling net to board one, thanked his stars that he was wearing a steel helmet when two British signallers jumped on top of him, wearing boots. Only much later did he discover that their impact had fractured his spine. He took charge of the hand-screw-propelled 22-man boat as it headed away, carrying twenty-eight of *Santa Elisa*'s crew.

Ensign Suppiger dropped his .45 automatic pistol. Larsen, conscious of the distress, trending towards hysteria, among some of the boat's occupants, promptly appropriated the weapon, ignoring Suppiger's protests. Men in Captain Thompson's boat pulled mates out of the water, some badly burned as blazing fuel spread across the sea. 'Tommy' administered whisky, hidden aboard each boat by purser Follansbee. The captain said: 'As the men rowed I heard a strange sort of whisper among them. I had never heard anything like it before. I guess it was what you'd call panic ... I stood up in the bow of the boat and said "Listen you men, if anybody opens his yap, I'll clout him." I guess the men were surprised to hear that kind of talk from me. Anyhow, after that there was silence.'

No. 3 boat got away successfully, commanded by Chief Mate Englund, a Swede. This also contained Dales, Follansbee, Randall and some burned men, whose morale slumped when Englund said, somewhat hysterically: 'Row like hell! She could blow up at any minute!' For a few moments, panic threatened to overturn the boat, until Dales and other men calmed the company. The sun had just begun to rise when they saw a destroyer closing from the north, escorting *Port Chalmers* and Commodore Venables. When *Penn* brought-to, lowering her scrambling nets as *Santa Elisa*'s boats came alongside one by one, a naval officer shouted: 'Make it snappy down there! We can't sit here any longer!' Ninety-six survivors from *Santa Elisa* were taken aboard, of whom three bad burn cases died on board. A decision about whether or how to scuttle the abandoned vessel was resolved at 0600 by a lone Luftwaffe Ju88, which dived and dropped a bomb on her foredeck, causing the ship to blow up. The destroyer was half a mile away, but wreckage rained down on her upperworks. *Penn* then sped away, in pursuit of the remainder of the convoy.

The German 3rd Flotilla's four boats played a less effective role in the battle than did the more numerous Italian MS and MAS craft, but the log of Siegfried Wuppermann of *S58*, the local Kriegsmarine operational commander, conveys from an Axis perspective the drama – and, for the young men of the crews, the thrill – of the night. Its factual inaccuracies, for instance about the type of British ships the E-boats engaged, are unsurprising. Moreover, in the confusion of darkness it remains impossible to be certain which Italian or German boat achieved what hits, especially later in the action: there is no doubt about the identity of the Italian captains who accounted for *Manchester* and *Glenorchy*. For the later successes, however, both MAS boats and E-boats made rival claims. Among the disputed torpedo hits, Giovanni Cafiero's 557 claimed *Santa Elisa*, Iafrate's *564 Rochester Castle*, Marco Calcagno's 554 *Almeria Lykes* and Carlo Paolizza's 553 *Wairangi*. All received Italian decorations. Since the confusion of the night afflicted both sides, it is likely that all the German and Italian captains sincerely believed they had been responsible for the sinkings. The 3rd Flotilla commander logged:

0040 – Sighted muzzle flashes

0250 – [Saw] starshell and flares

0308 – Sighted destroyers and tankers

0319 – E-Boats in pursuit

0322 – Torpedo detonation on tanker, boats reported under heavy fire from destroyers

0327 – Starshell illuminates big tanker. Cut to one engine, speed twelve knots, fired at seven hundred metres, 'ship [possibly *Wairangi*] hit amidships, fire broke out, very heavy shocks to own boat'

0339 – Convoy now in disarray.

0341 – Enemy ceased fire

0405 – [S58] turned back to reach harbour before daylight

Wuppermann's *S58* was badly damaged in the action, with four men wounded. The flotilla commander, Friedrich Kemnade, reported to Rome that 'Wuppermann has done well to get it home.' His achievement was thought especially praiseworthy because he had a scratch crew who had never before served afloat with him. Albert Müller's *S59* claimed to have twice hit a freighter, though she was not seen to sink. Horst Weber of *S30*, for whom this action was his first since returning from a long spell in hospital recovering from wounds, reported twice hitting a big freighter at 0508: 'the vessel immediately burst into blames, as it was carrying fuel'.

By any reckoning, this had been a disastrous night for Pedestal: in the four hours since midnight, cockleshell Italian and German boats had damaged *Rochester Castle*, sunk or fatally injured *Manchester*, *Glenorchy*, *Wairangi*, *Almeria Lykes* and *Santa Elisa*. Although Wuppermann's boat was damaged, none of the enemy craft had been sunk, which reflected an embarrassing failure by the escorts. 'On no occasion did destroyers ahead appear to locate an E-boat, or U-boat on the surface,' Alf Russell of *Kenya* commented harshly in his report about the three destroyer/sweepers leading the convoy. One of their captains, Colin Campbell of *Fury*, agreed with Russell that the warships would have been better employed giving close cover to the merchantmen, rather than clearing a narrow lane in the minefield through which almost the entire surviving convoy was funnelled, to the satisfaction of enemy boat captains.

The courage of the Italian and German crews deserves admiration. Their attacks were of a kind that often failed, in the hands of all navies, because crews failed to get close enough to give themselves a realistic prospect of hits. On the night of 24 October 1944, for instance, during the great naval battle of Leyte Gulf, thirty American PT boats fired torpedoes at a Japanese force of two old battleships and a heavy cruiser, escorted by just four destroyers. All missed. Later in the same battle, only one of twenty-seven torpedoes expended by a destroyer flotilla achieved a hit, though other American destroyers later fired more accurately. Italian and German skill and guts, closing targets to fire at point-blank range while facing heavy albeit ineffective defensive fire, were the decisive factors in their success in the early hours of 13 August.

An important aspect of Pedestal was that each successive disaster fed the next. Loss of *Eagle* weakened air cover through the day that followed – not so much by the diminution of absolute fighter numbers as by depriving the fleet of one-third of its take-off and landing platforms, a critical constraint on the number of planes that could be kept airborne. The crippling of *Indomitable*, following *Eagle*'s loss, persuaded Syfret to advance the withdrawal of Force Z, so that the Luftwaffe was able to deliver unimpeded its dusk assault on Force X. Burrough's transfer from his crippled flagship to little *Ashanti* weakened his grasp upon the entire convoy. *Nigeria* and *Cairo* were the only vessels equipped with VHF fighter-direction communications, such as might have enabled Burrough's officers to vector Malta-based fighters towards Axis aircraft.

The admiral bore no responsibility for the limitations of air defence and anti-submarine technology, which allowed wave after wave of attackers to strike at the convoy while paying an amazingly small price for their successes. Moreover, it was the Admiralty planners, and not Burrough, who selected the convoy's course, close to the Tunisian coast. As for the merchant-ships, their crews and gunners had thus far displayed remarkable staunchness in the most testing circumstances, with the exceptions of Commodore Venables in *Port Chalmers* and the company of *Almeria Lykes*. The E-boats' successes reflected an absence of higher direction and tactical skill among Pedestal's escorts, which made the losses in those hours perhaps the most culpable of the whole saga, insofar as one can consider anything blameworthy that takes place in a battle against such odds.

War correspondent Anthony Kimmins, with Burrough aboard *Ashanti*, acknowledged: 'It was an ugly, uncomfortable night, and I must confess to a sigh of relief as the dawn broke.' The destroyer's men were intrigued by the appearance on deck of a bedraggled canary, obviously a survivor from a sunken ship, liberated from some cage as she foundered. Yet that Thursday morning – though most men had long since lost track of days of the week – despite the amazing experiences through which the warships and merchant vessels had passed, and the physical and mental exhaustion of their crews, some of the most dramatic episodes were still to come. At 0800 on 13 August, with the leading ships 140 miles from Malta – less than ten hours' fast steaming – and now within the radius of operations of the island's RAF Beaufighters, both sides still had everything to play for.

The Admiralty's dispositions assumed that more than a few ships of the convoy would be lost. The survival and successful passage of even a few would provide the beleaguered island with sufficient provisions and ammunition to hold out for months more – if only the tanker, with her priceless oil, could be among them. Six other merchant vessels were still steaming, though little was known about *Brisbane Star*, striking an independent course close to the Tunisian shore. *Ohio*, thanks to her huge buckled sideplate, weaved erratically behind *Ledbury*. Nonetheless, by 0540 in early-morning light the destroyer and the tanker were close behind the convoy. Forty minutes later, the destroyer led Dudley Mason's ship to take up a position at the rear of a column, and reported to the admiral.

A fresh signal from the Admiralty to Moscow informed the prime minister of events in the Mediterranean, in bald and not wholly accurate terms: '1. Pedestal according to plan with following exceptions. INDOMITABLE damaged by air attack and reduced to 20 knots. 2. Cruisers NIGERIA, KENYA and CAIRO mined or torpedoed. Condition not known. 3. Destroyer FORESIGHT torpedoed but proceeding to Gibraltar in tow. 4. Five ships of convoy sunk after dark tonight probably by E-boats.

'There may be a battle with enemy cruisers in the morning … One [Italian] capital ship at sea but appears to have her eyes to East.'

This signal to Winston Churchill, who was as shaken by its tidings as might have been expected, emphasized the uncertainties that clouded the Admiralty's picture of events, especially the assertion that Italian cruisers might engage at dawn, when in reality they had turned back several hours earlier. Only *Waimarama*, *Dorset*, *Melbourne Star*, *Port Chalmers* and *Rochester Castle* were within distant sight of Burrough on *Ashanti*, though even these ships were too widely scattered for the escorts to be able to deliver a concentrated barrage in defence of any single vessel. 'As it grew light on that Thursday morning we realized we were alone,' said a *Dorset* gunner John Waters. 'On the horizon we could see the smoke of the rest of the convoy … Jerry started to concentrate on us … we were soaked with the near-misses.' The Luftwaffe was back, to renew its assaults on the remnants of Pedestal.

10

Retribution

During the hours of darkness on 12/13 August, even as Harold Burrough's ships were enduring their latest ordeal, far out of their sight and knowledge events were unfolding that yielded not merely a gleam of hope for their survival but a significant riposte. When Syfret's fleet turned west the previous evening, some two hundred miles north of Palermo in Sicily the four light cruisers and escorting destroyers of Admiral da Zara's squadron were steaming southwards, still expecting to join other Italian warships to launch an attack on Pedestal at first light next day. By 2305, the ships had closed the distance to within thirty miles of the island of Ustica, just north of Palermo – and correspondingly closer to Burrough's ships – when a Wellington operating out of Malta was ordered to drop illuminants above the enemy and to continue to make plain-language sighting reports every half-hour. A second Wellington bombed the cruisers, albeit ineffectually. They also messaged a non-existent RAF Liberator force.

This was an inspired initiative by Air Vice-Marshal Keith Park. Knowing the Italians' morbid terror of air attack on their fleet, this New Zealand hero of the Battle of Britain strove to convince them that such a threat existed. In truth, only the RAF's twenty-four Beauforts were capable of launching any such assault, as Syfret's Albacores receded into the distance. It is unlikely that the British would have achieved much against fast and well-armed warships, however courageously the pilots pressed their attacks. Nonetheless Supermarina, in Rome, continued to insist that it was acceptable for Da Zara's ships to advance to an engagement with the Royal Navy only if they received air cover.

Since the Italians had relatively few fighters, the Luftwaffe would have to provide this. Yet almost its entire regional strength was committed to maintaining surveillance over, and to attacking, Burrough's ships. Given the Germans' poor opinion of both the will and capabilities of the Italian Navy, it is easy to understand why they decided that their Ju88 bombers and Bf109 escorts were better employed sinking British ships by bombing than supporting some half-hearted Italian surface thrust. Yet here Kesselring was almost certainly wrong. If he had provided the Italians with the fighters they wanted, they would have had no further excuse not to commit their cruisers, and Da Zara was capable of fighting them well. As it was, however, the wrangle between generals and admirals about air cover, which Germany's Admiral Weichold described as 'heated', reached deadlock.

A further significant factor has only recently been identified. At 2250 on 12 August, a Cant night-reconnaissance aircraft of the 146a Squadron, patrolling over the central Mediterranean, signalled a sighting of three large British warships heading east, which its observer Oscar Ferrara identified as a battleship and two cruisers. In darkness, his mistake was unsurprising, but in reality these were only the anti-aircraft cruiser *Charybdis* and two destroyers, *Somali* and *Eskimo*, sent back by Admiral Syfret to reinforce Burrough's shrunken escort force.

This report nonetheless made a dramatic impact on senior officers burning midnight oil at Supermarina. They were provided with a fresh excuse for a display of pusillanimity such as they favoured anyway. Admiral Arturo Riccardi's title as under-secretary for the navy made him the senior officer of both Italy's maritime and air forces, under the Duce's supreme authority. The unfailingly cautious Riccardi now claimed to consider that the risk of an engagement with British capital ships made it too dangerous for the navy to commit the cruiser forces then at sea, 'posing a risk that could not yield a corresponding return'.

In the Operations Room of Supermarina, deputy chief of naval staff Admiral Luigi Sansonetti adopted the same view. These top brass were opposed by only one officer, and he the most junior present. Thirty-two-year-old Commander Giuseppe Roselli Lorenzini, assistant to the navy's chief of staff, urged interception at all costs. A veteran submariner, who had spent two years commanding boats in both the Atlantic and Mediterranean, he was disgusted that his chiefs were unwilling to seize a historic opportunity. It was decided to refer the issue direct to Mussolini: the Duce was invited to make a personal determination, whether his beloved fleet should risk all – as the Italians saw it – by attacking Pedestal, or return to port.

Characteristically, he made the latter choice. When the signal recalling the cruisers was dispatched from Rome at 2345 on 12 August, Commander Roselli Lorenzini stormed out of Supermarina's operations room, slamming the door behind him. Admiral Weichold said after the war: 'In this fashion, a splendid opportunity for a crushing victory by the Italian ships was thrown away, even as they were already at sea and heading for the battle area. It was a strategic failure of the first order.' Gianni Rocca, one of the foremost historians of the Italian

Navy, has written: 'If at dawn even a few elements of the Italian fleet had been at sea off Pantelleria, no English ship could have escaped destruction.' Harold Burrough agreed, saying later: 'I was always grateful to Mussolini. There is no doubt in my mind that had the Italian cruisers arrived that morning, there would have been a massacre. We would have been wiped out.' Patrick Gibbs, who would have commanded an RAF torpedo-bomber strike against the cruisers had they engaged, said that the map track of what the fliers expected to prove a suicide operation remained indelibly etched in their memories: they believed that they owed their own lives to having been spared from it.

At 0156 on the 13th, the Italian 3rd Cruiser Squadron set a new course towards Naples, while the 7th Cruiser Squadron headed for Messina. An unromantic historian must concede that the relevant orders were given before Keith Park's airborne stratagem had time to influence the Italian change of heart. Burrough's judgement was almost certainly correct: the Italians thus forsook one of their best opportunities of the war to inflict a crushing defeat on a British force at its last gasp.

Instead, a junior officer of the Royal Navy seized an opportunity to strike back, to formidable effect. The submarine *Unbroken*'s captain, Lt. Alastair Mars, was a forceful twenty-seven-year-old who had joined the service as a cadet, and since served six years in submarines. *Unbroken* was his first command, in which he sailed from Barrow-in-Furness eight months earlier, leaving behind a young wife and baby daughter. The elaborate kapok jacket that he affected on the bridge caused him to be nicknamed 'Pansy' by his sailors. A highly strung, passionately enthusiastic and ambitious young warrior, he told the thirty-two officers and men of his ship's company when first they put to sea: 'we have two jobs – to be successful and to survive'.

In August he had everything to prove to them, because he had yet to achieve any significant success against the enemy. He was also conscious of the odds against longevity facing operational submariners, whether British, American, German or Italian. Before leaving home, he attempted to get his life insured for a thousand pounds, for the benefit of his wife and unborn baby. The company that he contacted quoted a premium of £500, more than his annual pay: 'I decided I could not afford to get killed even if a suitable opportunity presented itself.'

Unbroken was one of the smallest submarines in the navy, its dark-blue paint meticulously maintained, because in the clear waters of the Med rust patches became conspicuous from the air. Mars and his tiny crew put to sea knowing nothing of Pedestal. Indeed throughout its patrol the crew were oblivious of the bloody events taking place a few hundred miles south-west of them. But their orders, and those of the seven other British submarines operating across the region, were drawn up with the obvious purpose of intercepting Italian warships, perhaps Mussolini's entire fleet, if it put to sea to confront Syfret. Unless or until this happened, their business was to make as much trouble as they could, against targets within reach of the enemy's naval bases.

Unbroken spent the night of 8 August surfaced near Longobardi, a small town high on Italy's instep, watching the railway that ran along the shoreline some fifteen hundred yards distant. The darkness was starless, but the bridge look-outs could see the red and green track signals, also pinpoints of candlelight in the windows of peasant homes. The crew of the submarine's gun, locked and loaded, stood hunched on the casing, their feet only a few inches above the water. At 2210, they were thrilled to sight a locomotive, its headlamp scything through the darkness, and soon afterwards a northbound passenger train. This was not the target they sought, however: their mission was to attack goods wagons, heading south towards Rommel's armies in North Africa. Half an hour later, the exuberant Alastair Mars spotted just such a train in the distance.

The submarine slid rapidly inshore, closing the range to nine hundred yards. At 2246 *Unbroken*'s little three-inch gun fired, the shell's explosion on the track precipitating a vivid blue flash and causing the overhead power cables to collapse. Round after almost point-blank round battered engine and trucks until fires were raging along the track. The submarine broke off its miniature bombardment after just two minutes: then it slipped back into the darkness, towards the most important rendezvous of its patrol.

Unbroken was one element in what had been planned in Valletta as a cordon of British boats awaiting the Italian cruiser squadrons, in positions which the staff hoped would cover all eventualities. Most of the submarines were posted much further south, but *Safari* and *Unbroken* were deployed as backstops, on the Italians' likely route home. The early hours of 10 August found the latter's crew two miles off Cape Milazzo lighthouse, eighteen miles west of the great naval base at Messina.

At the moment when Syfret's fleet was passing through the Straits of Gibraltar, Lt. Mars was peering through his periscope and seeing ... nothing. His asdic was out of action, but suddenly his hydrophone-operator reported 'Mushy H.E.' – hydrophone effect, usually caused by ships' propeller noise – 'bearing one-eight-five'. Mars himself took up the earphones and listened, hearing unfamiliar sounds. Puzzled, he wondered if the submarine had been detected by some new, unfamiliar enemy listening device. He remained baffled when daylight broke: the noise persisted, but no ship was visible. At 0900, however, at the periscope he spotted an approaching patrol tug. The uneasy skipper took his submarine down to eighty feet, and after fifteen minutes heard the crump of an exploding depth-charge, followed by four more, which caused the hull to shudder.

He guessed the explosions were four hundred yards away, and ordered the boat into silent routine, with all non-essential machinery shut down, only the gyro-compass and electric motors still using power, sapping precious battery life: 'I glanced around the control-room. Apart from the licking of lips and wrinkling of foreheads, the others gave no indication of the tense anxiety they must certainly have been feeling. I experienced the same combination of nerves, claustrophobia and fury that I had suffered when sneaking through [a] mine-field for, as on that occasion, we were essentially on the defensive, powerless to hit back, unable to run away – simply wallowing there, waiting for the enemy's accuracy of fire to improve.'

No more charges followed the first pattern, however. Mars edged his submarine cautiously up towards the surface. He wrote: 'There is always a fascination about peering through a periscope, even for the thousandth time. At first you see nothing but the clear blue water thirty feet above your head. As you rise the light broadens until, with a crystal flash, the periscope bursts through the surface and you are gazing at the world above the sea.' Access to the eyepieces made a submarine commander the most privileged man on board: underwater, he alone could enjoy a window on the outside world. Now, Mars glimpsed his attacker a mile away, and soon afterwards spotted an Italian minelayer approaching, escorted by a flying-boat. *Unbroken* blew her vents once more, descending to a hundred and twenty feet, just in time to receive another pattern of depth-charges. These never came close: 'we were disturbed by nothing more than heart-jumping *krrumps* and occasional shudders'.

The attacks persisted, however. In the Mediterranean at high summer, temperatures in a submerged submarine could soar as high as fifty degrees Celsius, 122 degrees F, causing panting men to use up precious oxygen even more swiftly than usual. Mars wrote: 'My clothes were soaked in sweat, the air in the boat was thick and oily, and my nerves were in a wretched state. If only we had been able to hit back! If only there had been some movement or action to take our minds from the agony of the situation! But no. We could only wait and hope and pray, brooding and exaggerating, picturing a torn, smashed hull and a bubbling, choking, lung-bursting death ...'

At 1500 the attacks faded, tension eased and Mars ordered tea and sandwiches all round. He resisted the temptation to sneak back to periscope depth, instead cruising ever so cautiously, holding his depth at a hundred and twenty feet. Then suddenly the hydrophone-operator reported, 'H.E. to starboard!', and a new pattern of charges fell – this time, they were close. Fortunately for *Unbroken*, her tormentors thereafter moved away.

Mars studied his charts and considered options. It seemed obvious that, since the Italians knew a British submarine was lurking, they would not permit their cruisers to pass nearby, through his designated patrol area. He determined to take *Unbroken* elsewhere, but did not dare to dispatch a radio signal informing Malta because of the certainty that the enemy would eavesdrop on the transmission even if they could not decrypt its meaning. Signals dispatched to a patrolling British submarine from its flotilla headquarters were repeated at appointed times, in the expectation that sooner or later it would be able to surface and receive them. Boats never acknowledged, however, to escape enemy radio direction-finding: the Admiralty was vividly conscious of how much vital intelligence it gained about Dönitz's U-boats by 'DF-ing' their transmissions, even when Bletchley Park could not crack texts.

Mars finally decided to take his boat thirty miles north-west, to watch the sea between the islands of Stromboli and Salina, where those cruisers must pass which used Messina as a port. As evening fell, the Italian warships abandoned their search for his submarine, having dropped a total of seventy depth-charges. After the tense, indeed terrifying day, Mars ordered the coxswain to issue tinned tongue for supper – a taste treat. At 2230, *Unbroken* slid warily up to periscope depth, where her captain was relieved to see nothing save a single overhead star. They surfaced, glimpsing the faint glow from Stromboli's volcano. Mars ordered: 'Half buoyancy. Group up. Half ahead together. Start the engines.' Then they began to steer westward, knowing that the Italians lacked radar to pick up their low silhouette in the darkness.

The young officer knew that if his personal initiative went awry he would receive a rough reception on returning to Malta. Submariners, however, were expected to think for themselves, to show dash and buccaneering spirit. This was why such men as Mars, like his Italian, German and American counterparts, volunteered for the service. That night of 10/11 August, the young lieutenant was sufficiently exhausted to sleep for four hours, twice his accustomed quota during an operational patrol, before they dived again at dawn. Soon after midday, the boat reached the position its skipper had chosen, where Mars and his crew settled to wait. Through the balance of that day, even as *Eagle* was suffering disaster at Helmut Rosenbaum's hands, no vessel crossed *Unbroken*'s line of sight. Its skipper felt oppressed by the need to achieve a success, not only because he had struck out on his own, but also because his previous attempted attacks on enemy shipping had been frustrated. He feared the effects on the crew's confidence if he failed to deliver: 'They were working hard and risking their lives, and when it came to the great moment' – he referred to an earlier aborted attack under his command – 'it appeared as though I had let them down.'

The captain, and the captain alone, gave the order to fire. To contrive a sinking required both luck and the finest possible judgement by the officer hunched before the periscope lenses. Mars wrote of his assessment of his own lack of achievement as a skipper: 'the situation was such that one of three things had to follow – success and a new lease of life, a watery grave, or my being relieved by a new C.O. As far as I was concerned, only one of those three alternatives was at all satisfactory.'

After lingering almost inert through the afternoon of 11 August, however, the whole of Wednesday the 12th passed in the same torrid fashion for the crew of *Unbroken*, while a few hundred miles south-westwards the ships of Pedestal were suffering and sinking under air and submarine attack. The Mediterranean heat, sufficiently oppressive below decks in a surface warship, became almost intolerable within the hull of a submarine. During the darkness of 12 August, *Unbroken*'s crew recorded a midnight temperature of eighty-eight degrees. In the early hours of the 13th, Mars was dozing in a canvas chair on the bridge when he was tapped on the shoulder: 'Captain, sir! The P.O.Tel [petty officer telegraphist] has a signal in the control-room. It's emergency.' Mars slid down the steel

ladder to be handed a pink slip: 'From Vice-Admiral
Commanding at Malta: enemy cruisers coming your way'.

The young officer was momentarily horrified, because the
message assumed that *Unbroken* occupied its designated patrol
area, thirty miles away. Then he received an explanatory signal
from submarine headquarters. At 0300 a British aircraft – the
Malta Wellington – had pinpointed enemy cruisers off Sicily's
Cape de Gallo, steering east at twenty knots. At that moment
Safari, another British submarine, was north-west of Palermo
and within a few thousand yards of the cruisers, but in the
darkness failed to sight them and achieve an attacking position. The other six British boats at sea in the central
Mediterranean were many miles away, south of Pantelleria.

The Italians were due to reach *Unbroken*'s designated position at 0730 – two and a half hours from that moment. The
little boat was all that lay between Da Zara's warships and the
safety of their anchorages. Mars at first considered making a
dash back to Cape Milazzo – he might get halfway there, taking
the huge risk of racing at full speed in daylight, on the surface.
Then *Unbroken*'s skipper reflected – intelligently and imagina-
tively – that after being spotted by a flare-dropping British
aircraft the Italian admiral would surely change course. Only
land lay to starboard of him, and thus Da Zara would have to
turn to port ... where the British boat lay in waiting: 'I could
see no other alternatives and made a decision which, as far as
Lieutenant Mars was concerned, was momentous. We would
stay where we were.'

He ordered an early breakfast for the crew, and was himself
eating bacon and eggs when the hydrophone-operator
summoned the officer of the watch: 'H.E. ahead, sir!' Mars
sprang up, sending his half-full plate crashing to the deck,
dashed the few steps to the control-room and seized the peri-

scope handles as he ordered 'Diving stations!' Then the hydrophone-operator called: 'H.E. bearing two-three-oh ... Heavy
units ... Fast'. Mars swung the periscope, willing his eyes to
penetrate the early-morning haze, yet seeing nothing. He
himself took up the headphones, heard a confused jumble of
noise from ships that were obviously big but still distant. He
returned to the periscope: 'My heart thumped like a trip-ham-
mer, but I noticed with satisfaction that my hands were steady
and my eyes were clear. Then I saw them as their masts broke
through the haze on the horizon. In my joy I could have danced
a jig.'

The ecstatic Mars counted two ... three ... four Italian cruisers approaching in line ahead, already within ten miles. He
dropped his big periscope, raised the smaller, less conspicuous
attack periscope. His first lieutenant broadcast a calm situation
report to the crew, who were listening intently at their posts in
machine and torpedo spaces. Mars' twenty-year-old arma-
ments officer, Ted Archdale, was manipulating the 'fruit
machine', calculating angles and ranges for the torpedoes'
discharge: 'course for ninety-degree track, sir, one-four-oh'.
Unbroken glided forward, holding a depth of fifty feet. Mars
sought to close to point-blank range, with a mere three
hundred yards between his boat and the targets, before turning
to present his bow, and his tubes, to fire. The short, chunky
young officer wrote: 'As [the first lieutenant] relayed the information to the rest of the boat, I felt a moment of queasiness. It
was a feeling I knew of old – a boyhood memory of a twelve-
stone brute flashing down the touchline of a rugger field, the
full back at the other side of the field, and only myself to bring
him down ...'

Around him in the control-room the planesmen sat with
their backs to the periscope, attention fixed on the depth-

gauges, while an engine-room artificer watched the pressure indicators. Another rating hunched over his headset, listening to the now-repaired asdic. Until the very instant of firing, the British attack could be frustrated by an alert Italian asdic-operator aboard one of the eight destroyer escorts flanking the cruisers, or a sighting by one of the enemy aircraft overhead. The squadron could escape the submarine by making a drastic emergency turn away, such as Syfret's and Burrough's ships had done so often in recent days, and *Unbroken* was incapable of chasing after them.

Mars' luck held, just as Ferrini's had done against *Nigeria* and *Cairo* on the previous evening. The Italian warships, in defiance of their orders, were not zigzagging, and had slowed to eighteen knots. The men aboard *Unbroken* had no inkling of the devastation *U-73* and *Axum* had wrought in recent days. But they did understand that such a target as was now bearing down upon them seldom presented itself to a British submarine. 'This was a "now or never" opportunity,' wrote Mars. 'There would be no second chance. We had to deliver a single, swift knock-out punch.' This would also be the Royal Navy's only window of opportunity for exacting some portion of retribution for its shocking losses two hundred miles southwards.

With the lens of the attack periscope raised mere inches above the surface of the sea, Mars squatted to peer through its eyepieces. The Italian squadron altered course to starboard, thus presenting one of the two eight-inch heavy cruisers as the most accessible target for *Unbroken*, while the two six-inch light cruisers steamed beyond, with five of the eight destroyers behind them. He ordered 'down periscope', told the 'fruit-machine' operator to assume an enemy speed of twenty-five knots. The man responded: 'Director angle, green three-six-and-a-half', meaning that they needed to lay off that far – to ease the boat a trifle to starboard – for a deflection shot. Two of the four torpedoes in the bow were set to run at a depth of fourteen feet, two at sixteen. It was 0800. Once more, the attack periscope rose a fraction above the almost motionless sea. From the 'fruit machine', a voice said: 'Three-and-a-half minutes, sir.'

Mars eased his boat just inside the track of the approaching destroyers of the Italian screen, then cursed when the leading vessel altered course to head straight for the submarine. The Italian should pass a few feet above *Unbroken*'s hull. If the enemy boat hit it, however, it would slice through the conning-tower as surely as *Wolverine* had torn open *Dagabur* less than thirty-six hours earlier. Mars, his whole being fixed upon the attack, determined to accept the risk, held his boat steady, rejected the 'safety first' option of a crash-dive. The destroyer speeding towards him was now a thousand yards away, while the cruisers were three times that distance, still a full minute, an apparently interminable minute, beyond the reach of the little cluster of young men pouring sweat in the submarine's control-room. *Unbroken*'s skipper found his mind filled with a nightmare vision of the destroyer striking his boat – perhaps on the gun, on the periscope or against the bridge. Then 'she was too close for me to see more, but I was a hairsbreadth off her port bow … With a deafening roar she rushed past, and I caught a momentary glimpse of a scruffy-looking sailor smoking a 'bine [Woodbine – cheap cigarette] as he leaned against a depth-charge thrower.' Then the target filled Mars' periscope sight and he called, 'Fire One!' The boat jumped, he lowered the periscope, then after eight seconds ordered, 'Fire Two!', then three, then four.

The tell-tale bubbles and surging wakes of Mars' torpedoes were spotted running by lookouts aboard the destroyer

Fuciliere and cruisers *Bolzano* and *Gorizia*, all of which began to turn to comb the tracks. *Bolzano* was nonetheless struck amidships, precipitating a thunderous explosion that flooded engine-rooms and a magazine, starting a major fire which spread from an oil tank. The six-inch cruiser *Muzio Attendolo* saw her neighbour's plight and also began to take evasive action – again, too late. The detonation of another of *Unbroken's* torpedoes tore an eighty-two-foot gash in the light cruiser's bow.

Bolzano had already suffered at the hands of a British submarine: a year earlier and not many miles east, *Triumph* torpedoed her, consigning the heavy cruiser to a dockyard for three months. Now, as dense black smoke once more surged through her hull and superstructure, some of the crew leapt into the sea, from which they were rescued by small boats which put out from the nearby shore, crewed by old men, women and even children. Vincenzo Costantino, who gloried in the rank of Second Trumpeter on board, had been bleakly anticipating this disaster, as had many of his shipmates, since the RAF's Wellington O-Orange illuminated them with flares during their night passage. Costantino, and even his captain, knew nothing of the fact that it was Alastair Mars' luck and intuition, rather than the enemy aircraft sighting, which had contrived their undoing.

Costantino had just come off watch, and was sharing coffee and hardtack biscuits in the cruiser's No. 6 mess when the explosion rocked the ship. Almost immediately the lower decks filled with choking smoke, through which he and scores of others groped towards the open air, blindly and frenziedly twisting multiple butterfly nuts to unlock the big hatch sealing their compartment. The first sight that met his eyes on getting his head out was that of the nearby *Muzio Attendolo*, also stricken: 'I shouted for help, calling the name of Beato our senior trumpeter. On the deck it was hell, with dead and wounded men being carried aft, their burns covered in black cream. Then Beato came, and we headed for the stern.'

Renato Ditre was likewise having breakfast in the petty officers' mess when the torpedo and boiler explosions rocked the ship. There was a dash for the nearest hatchway, which opened forward, but those who opened it were greeted by a lash of flame which inflicted severe burns on several men. Slamming the hatch closed, a crowd of fugitives surged towards the canteen's rear hatch, amidships close to the funnel. Ditre said: 'Once out on the lower deck I could only make for the prow.'

He saw desperate men, one with his clothes on fire, his face disfigured by burns, who collapsed even as he sought to flee. A crowd gathered at the foot of a ladder, fighting for access. 'I was not among this scrum and so managed to attract the attention of others on the upper deck, who opened another hatch.' Ditre, like most of the ship's company, convinced himself that *Bolzano* was doomed. Then, however, the wind changed and the ship drifted around, opening a gap in the flames and smoke which allowed men like himself, trapped forward, to dash unscathed towards the stern.

Vincenzo Costantino described Luigi Merzagora, their captain, as remaining calm, summoning the destroyer *Santa Barbara* alongside not only to take off wounded, but also to remove shells, in danger of detonation, from gun positions menaced by fire. Another destroyer three times passed a towing cable to the cruiser, but each attempt to take the strain caused the link to break. Then Merzagora resorted to braggadocio such as it is hard to imagine being emulated by any of the British captains who, during the past three days, had found themselves in similar circumstances. He called to the crowd of sailors gathered on the ship's stern to escape the fires forward: 'Men! Whose is the *Bolzano*?' They shouted back: 'Ours!' Then the captain cried out: 'I salute the King! I salute the Duce! Abandon ship!' Hundreds of men leapt from the cruiser on to the adjoining deck of the destroyer.

The submarine that had wrought the havoc meanwhile dived steeply to eighty feet, heading south-west at its maximum underwater speed of nine knots. Every man aboard *Unbroken* knew that it was vital to exploit the interval between firing torpedoes and their detonation to put distance between their boat and the cruisers' avenging escorts. Mars wrote: 'The scene in the control-room might have been transplanted from a militant Madame Tussaud's – the tense, still figures, some standing, some sitting, others crouching, all rigidly silent, unblinking and tight-lipped, straining to catch the sound of a torpedo striking home. For two minutes and fifteen seconds we were like that, until a great clattering explosion brought a back-slapping roar of triumph ...' Fifteen seconds later, there was a second explosion. 'Were we capable of lyric poetry', added the euphoric Mars, 'we'd have composed a Psalm of Thanks, for we felt as boastful and as proud as David must have felt that afternoon in the valley of Elah,' where he slew Goliath.

Four minutes later, a shower of depth-charges fell, some of them close enough for the submarine crew to hear 'rain' – gushers of water thrown up by the explosions – falling back into the sea. Mars also heard the metallic clicks of primers, which preceded detonations by half a second. Through the next ten hours, they endured repeated attacks, over a hundred charges in all, mercifully well above *Unbroken*, which lurked at a hundred and twenty feet. Then the Italians sailed away, and with the coming of darkness the triumphant submariners surfaced, revelling in the hot but blessedly fresh air that seeped down the conning-tower into the hull when the hatch was at last opened.

Muzio Attendolo's bow fell off under tow. The cruiser reached Messina at 1854, but never thereafter put to sea to fight. *Bolzano*, which burned all day, had to be temporarily beached ashore, then was towed to La Spezia, where she was later scuttled. Mars had written off two major Italian warships, delivering a punishing riposte for the losses the Royal Navy had endured while he was at sea. Count Ciano wrote in his diary on the 13th: 'It appears that things are developing well for us, but we have paid a high price, with the loss of the *Bolzano* and damage suffered by the *Attendolo*.'

At noon on 18 August, *Unbroken* and its crew sailed into the Number One billet of the submarine base at Malta's Lazaretto, to receive a royal welcome. Mars was awarded a DSO for his achievement. He knew how fortunate he was to have survived. Later that year, he calculated that of the seven boats of *Unbroken*'s class launched by Vickers, four were already lost, their crews dead, while a fifth skipper had been relieved of his command. It would be extravagant to claim that the submarine's successes amounted to full compensation for the losses endured by Pedestal, but they showed that the Royal Navy in the central Mediterranean could still strike bold and hard.

11

Blenheim Day

First light on Thursday 13 August found most of the surviving British ships barely 150 miles from Malta. Winston Churchill wrote of this date as 'to me always Blenheim Day', anniversary of his great ancestor the Duke of Marlborough's 1704 battlefield triumph over the French. In 1942, however, that date wore a different hue. *Penn*'s cook observed with gloomy relish to a huddle of survivors, below decks on the destroyer: 'we don't know if there's any convoy left after the night'. Lookouts spotted a hulk and the warship headed towards it in search of lifeboats or bobbing figures in the water. Around the wreck, for hundreds of yards blazing oil covered the sea, the black smoke mocking the brilliant blue sky above.

That day Hitler's headquarters issued a bulletin, headed 'GIANT CONVOY DESTROYED IN THE MEDITERRANEAN'. It then listed the British naval losses, more or less accurately, save that the Germans were convinced that they had badly damaged the USS *Wasp*: 'three cruisers as well as 6 large merchant-ships and tankers, with a combined gross tonnage of 51,000 BRT, were sunk or badly damaged. The convoy was blown to pieces. The bulk of the escorting forces are heading back west. The remaining ships are trying to reach Malta. The fight continues, and further great successes can be expected.'

Some British seamen aboard the remaining merchant vessels concurred with the German view of their prospects. At least as many, however, had become passionately committed to their task. On *Melbourne Star*, David MacFarlane the Master and James Fleming, his helmsman, maintained a superb calm through successive attacks, giving a masterclass in ship-handling. *Port Chalmers'* chief engineer Stan Bentley was described by another officer as 'a tower of strength to his men. His cheerful and cool demeanour when on deck was a fine example to all.'

On *Rochester Castle* Royal Navy signaller Jack Harvey wrote: 'There came a turning point in that voyage when all I wanted to do was to get that ship to Malta. It may have been sheer bloody-mindedness.' Since being torpedoed during the night, the vessel had again caught up the escorts, taking station behind *Kenya* and *Charybdis*. Amid the huge drama of the convoy, the pettinesses of daily life were not forgotten. Harvey, a twenty-five-year-old veteran, bickered with a much younger mate about who should wash up the plates after the meals they snatched at action stations. An astonishing number of sailors were unable to swim. This issue now loomed understandably large in their consciousness, including that of an Irish comrade beside Harvey in the ship's wheelhouse, who made no secret of his terror.

Nineteen-year-old Fred Jewett, on *Ashanti*, was impressed by Admiral Burrough's sangfroid when a telegraphist read off to him a list of ships hit or sunk during the previous night. The message interrupted a rugby football conversation between Burrough and Richard Onslow, the ship's captain. In response, Burrough nodded, 'Very good,' then resumed the sporting discussion. This was, of course, a display of conspicuous calm by a commander mindful of scores of flapping ears around him on that crowded little warship. It was, nonetheless, impressive bravado, which may have come easier to an officer steeled by memories that included witnessing the slaughter of most of his own guns' crews at Jutland.

In the wake of the convoy's passage lay lakes of fuel oil; the scattered debris of half-a-dozen ships; knots of humanity, figuratively and sometimes literally clinging to the wreckage. Partly as a consequence of the faltering communications, sixty-four survivors of *Empire Hope* had spent the night drifting on the sea – the flaming sea – close to where their ship had foundered. That they were forsaken reflected disbelief that many men could have lived through the explosion which tore apart the vessel's hull, together with the confusion that now prevailed, but the castaways nursed understandably bitter sentiments about their abandonment.

As they floated in oil-soaked lifejackets, at intervals petrol cans broke loose from the wreck on the seabed below, burst to the surface and added new fuel to the fires. Eighteen-year-old apprentice Albert Allison was making his first trip to sea. He won admiration and gratitude from his shipmates by using a small makeshift raft to haul man after man from the water, paddling eighteen in succession to the relative safety of a larger float.

Dawn revealed the survivors divided between four Carleys. Around noon, there was a sudden eruption of the sea beside them as the Italian submarine *Bronzo*, identified later by the paintings on her conning-tower of the Disney cartoon character Pinocchio, surfaced and examined them. An English-speaking member of its crew shouted down to ask if they were in distress. The question was arguably superfluous, but on receiving an affirmative *Bronzo* signalled Italian air–sea rescue headquarters on Sicily, giving their position, before disappearing towards Tunisia. Submariner Livio Villa wrote sententiously, but with some justice, 'It is not in the rule-book to save shipwrecked mariners, but the good seaman of our race is humane even to his enemies, who are both sailors and men.' When two years earlier the Italian submarine *Alessandro Malaspina* sank the tanker *British Fame*, it towed the survivors in their lifeboats within reach of shore, before cutting them loose and submerging.

At 1320 on that 13 August of 1942 a big German Dornier flying-boat landed, picked up more than thirty cruelly sun- and salt-baked *Empire Hopes*, and removed them to become PoWs. The same aircraft later returned for a second load, but found the sea so strewn with floating and burning debris that the pilot decided it was unsafe to land. The men who remained were obliged to spend that night and most of the next day on the rafts. A small Italian Red Cross seaplane landed and removed a further seven seamen to Sicily. Later still, an Italian MAS torpedo-boat, from the flotilla which had attacked Pedestal, retrieved as trophies the ship's captain and several more men, who were likewise taken home to Italy. Around noon on 16 August, Second Officer Arthur Black's raft drifted ashore on Cap Bon, where its parched occupants were greeted by local people who treated them kindly, providing food, wine, cigarettes. They were thereafter interned by the Vichy French, and liberated in November, save for an eighteen-year-old fire-man who was apparently killed while trying to escape.

Seven hundred miles further east, on 13 August the first ships of Admiral Syfret's fleet approached Gibraltar unscathed. Its cripples, headed by *Indomitable*, headed for the dockyard.' There was a minor incident in the air during the last phase of Force Z's passage: that afternoon, Hurricane pilot Sub-Lt. Peter Hutton landed back on *Victorious* after a patrol to report that he had shot down Marceau Méresse's H-246 Air France flying-boat. This was a chance encounter, with the very same civilian airliner that three days earlier had provided the Axis with its first detailed sighting report on Pedestal's course and composition. In reality, the aircraft survived, permitting Méresse to give an outraged account of the incident. Once more on his Marseilles-to-Algiers shuttle run, he had begun the descent to North Africa when his co-pilot tapped his shoulder and pointed out four British fighters on his port side.

'What had we to fear? For several days they had shown no hostility to our flying-boat with its very conspicuous neutral *tricolore*. The planes followed us for several minutes' – awaiting radio orders from the fleet – 'before suddenly making an approach from starboard. I waggled my wings in greeting, throttled back the engines and accelerated our descent. At almost the same moment I heard the dry clack of bullets hitting us and, curiously, I also [smelt] cordite.' Four passengers were killed and several others badly wounded, including the cabin steward and wireless-operator, but although the British airmen persuaded themselves that the flying-boat was doomed, it contrived a landing off the North African coast. Méresse's theatrical indignation found few sympathizers, save among fellow adherents of Vichy: the pilot had torpedoed his claims to immunity by transmitting intelligence which aided the Axis.

Further east that morning, some hours' steaming behind Syfret's fleet the destroyer *Tartar* was towing *Foresight* towards Gibraltar at a speed of 7½ knots, with a little assistance from the cripple's engines despite worsening flooding below. Then, fifty miles north of Bône at 0848, the two warships' crews were electrified by the sight of four tracks speeding through the sea towards them. The torpedoes had been fired by Helmut Rosenbaum's *U-73*, nemesis of *Eagle*. The German skipper reported a hit on a destroyer. In truth, his salvo missed: the sinking of the British carrier proved to have represented the high point of Rosenbaum's career as a submarine captain. *Tartar* nonetheless slipped her tow and mounted an ineffectual depth-charge attack.

Robert Fell, captain of *Foresight*, signalled *Tartar* urging that it seemed best to take off his crew and abandon efforts to salvage the damaged destroyer – 'the risk of the loss of your vessel is so great that it is not worthwhile. Especially in view of loss up to date.' An hour later, at 0955, *Tartar*'s commander St John Tyrwhitt recognized that the crippled vessel, her main deck almost awash, could not prudently be dragged any further. *Foresight*'s men, less five killed during the initial attack on the ship, transferred to *Tartar*, which then sank the destroyer with depth-charges and a torpedo, before racing at twenty-seven knots to catch up *Nigeria* and her escorts. She joined the cruiser's screen at 1735, having been repeatedly fired upon by enemy shadowing aircraft.

Burrough's remaining ships were now, theoretically at least, within reach of cover by RAF Beaufighters taking off from Malta. It was debatable how much protection these sluggish twin-engined aircraft could provide against faster and fitter Ju88s, but they might at least have disrupted enemy bombing. Hereafter, however, the cost of the loss of *Cairo*'s and *Nigeria*'s VHF fighter-direction voice radio became apparent. Again and again, Park's aircraft ranged over the sea in search of the convoy, only to experience utmost difficulty in finding it, or doing any good when they did so. Aboard *Ashanti*, the admiral's flag-lieutenant – his personal signals officer – strove desperately to manipulate a wireless set to achieve contact with RAF aircraft that were in plain sight. For an hour he insistently called 'Bacon–Apples', 'Bacon–Apples' until he was thrilled to receive a crackling response. His hopes were dashed, however, when the owner of the distant voice proved to be sitting not in a cockpit above, but instead in the wireless office of the nearby cruiser *Charybdis*. Burrough wrote: 'Poor "Flags" assumed an expression of despair and flung down the receiver.' A later Admiralty post-mortem concluded that, during the pre-planning for Pedestal, a liaison officer should have been flown home from Malta accompanied by a signals specialist to arrange procedures, wavelengths and suchlike, which would have prevented the chaos that prevailed on the ether.

The ships never achieved communication with their would-be saviours in the sky. On the morning of 13 August, three Beaufighters took off from Malta, one of which returned to base having failed to find the convoy, while the other two began to patrol blind overhead, as long as their fuel endurance allowed. Within minutes, however, they were bounced by two German single-engined Bf109s, which riddled one British aircraft, killing the pilot and causing his observer, Jock McFarlane, to bale out into the sea. Shortly afterwards an Italian Stuka crashed in the same area, after being shot down by a Malta-based Spitfire. Its crew, Guido Savini and Nicola Patella, escaped from the cockpit before the wreck sank.

The latter managed to clamber aboard the plane's dinghy, but Savini found himself drifting away in his Mae West, tossing unhappily on a gentle swell. A passing Italian aircraft failed to see the two red flares that he fired, but they were spotted by Patella, who paddled to Savini's position and dragged him aboard. After some hours, in fading light they heard a whistle being blown and propelled themselves clumsily towards the sound. They glimpsed a red-headed swimmer whom they pulled into the dinghy, and found to be their fellow distressed airman Jock McFarlane. The three, unable to communicate in any common language, drifted together for the ensuing sixty hours, until they were picked up by a passing Dornier flying-boat, which McFarlane vainly attempted to resist boarding. The Scottish airman spent the rest of the war as a PoW. Here was yet another extraordinary miniature within the huge canvas of Pedestal.

Jack Tuckett of *Dorset*, on finding his ship alone during the night, with flashes and flames erupting in the direction of the distant convoy, made the bold decision to strike his own course, further offshore. Though he and his crew spent some fearful hours, every moment anticipating the explosion of a mine or assault by a predatory E-boat, Tuckett's decision proved inspired. At daybreak he and his ship were unharmed, and only eighty-four miles from Valletta, at worst six hours' steaming. But they had run thirty miles ahead of the convoy, with not another vessel in sight, having steamed at full speed, without zigzagging.

Now that they were no longer shielded by darkness, it seemed a near-certainty that either enemy aircraft or submarines would find them, with consequences that suggested little prospect of securing escape or rescue. Tuckett's wirelessed requests to Malta for air cover generated no response. Thus the captain, whose luck and judgement had hitherto served his crew well, made a fateful call: he ordered the helmsman to turn about, to rejoin the rest of the convoy, pursuing safety in numbers. At 0730 *Dorset* reversed course towards the west, steaming some twenty miles back the way she had come, to the bewilderment and consternation of Tuckett's crew.

Meanwhile *Melbourne Star* had also run ahead of her companion merchant-ships during the night, and found herself alongside the leading destroyers. In the words of David MacFarlane her Master, 'none of them seemed to want us there, and we began to think that we were nobody's baby'. At 0505 *Ashanti* ranged alongside, and through a loudhailer Harold Burrough ordered her to about-turn, to join the only other two merchant vessels in sight, two miles astern. MacFarlane replied complacently that he was perfectly happy where he was, which prompted a brusque rebuke from Burrough: 'I am the admiral!' *Melbourne Star* reluctantly reversed course. Burrough was seeking to restore order after the convoy's overnight disintegration into chaos. Every British naval commander was haunted by the two-month-old memory of PQ17, when the doctrine of *sauve qui peut!* – every ship for herself – precipitated a disaster at which more than a few Pedestal naval officers and ratings had been bystanders.

They were tantalizingly close to their destination. Seven merchant vessels survived, half the number that had set forth from Gourock. Yet the weather, though now poor over Sicily, was good enough to favour new air attackers over the sea to the south. Thus the strain on the admiral, after the horrors of the night, was if anything greater than before. Unshaven for days and lacking a moment since quitting *Nigeria* to change clothes or cleanse himself of oil, Burrough wrote: 'We were all feeling the lack of sleep, but our troubles were not over … I have to confess that those six hours' – between the first and last air attacks on the morning of 13 August – 'were the longest I have ever experienced.' It was a ghastly experience for a commander, to preside over carnage on this scale when there was little or nothing that he could do to avert it.

Anthony Kimmins wrote sympathetically of 'Admiral Burrough's stubble growing longer and longer as the days and nights went on. His face growing redder and redder with the sun. His eyebrows rising higher and higher in the effort to keep his eyes open.' Even if the admiral was not an old man, at fifty-four he was no longer a young one, blessed with the resilience of his more youthful captains. And almost every man of the convoy, young and old alike, was now sleepwalking. A rating on *Ashanti* said: 'Most of us were bloody knackered, absolutely exhausted by the effort, the constant concentration.'

The destroyer *Ledbury* had guided *Ohio* through the hours of darkness, showing a stern light to Dudley Mason on the tanker's bridge. He, in turn, used the ship's telephone to transmit instructions to the improvised rear steering position, manned by chief officer Gray, to whom *Ledbury* was invisible. At 0630 Roger Hill signalled *Ashanti*, reporting the position of *Ohio*, five miles astern of Burrough, and of other ships behind, some straggling but still steaming, others mere abandoned hulks. *Port Chalmers* had veered ten miles to the north-west, after Commodore Venables' attempted retreat, but now closed up behind the other vessels. The convoy slowed to allow *Ohio* to catch up, which the tanker achieved half an hour later.

Rochester Castle and *Dorset* formed a port column, behind *Kenya*. The starboard column, following the cruiser *Charybdis*, comprised *Waimarama* and *Melbourne Star*, the whole formation still led by the destroyer/sweepers, with *Ohio* bringing up the rear. Many captains were uneasy about the unhurried pace set by the admiral. But the events of the night convinced him that dispersal was lethal: thus, when in daylight he regained some control of his ships, he emphasized togetherness at the expense of haste. Roger Frederick, *Melbourne Star*'s NLO, wrote in his report about this new phase of Pedestal: 'there was a strong feeling both before, during and after the disaster [of the morning's air attacks], that not enough advantage was made of speed of the ships'.

Burrough, unaware that *Manchester* was abandoned and sinking, dispatched *Eskimo* and *Somali*, the two Tribals that had joined during the night with *Charybdis*, to return to her last-known position. They arrived in time to retrieve 149 of her crew from Carley floats, and to witness through binoculars the bulk of the *Manchesters*, including Captain Drew, being formed up on the beach by Vichy troops to be marched into internment. The two fleet destroyers were then ordered to hasten back to Gibraltar, perhaps because Syfret decreed thus – there is no record of the signal traffic – though they would have well served the convoy by staying to strengthen its close escort.

Now began the last phase of the ordeal of Burrough's ships. At 0803, thirty miles south-east of Pantelleria, three Italian bombers appeared, lingering on the horizon beyond range of the ships' guns, obviously reporting the new British position. Seven minutes later, twelve German Ju88s launched shallow dive-bombing attacks, undeterred by the barrage from the ships' gunners, many of whom cursed their own lack of sunglasses, as planes flew at them out of the morning glare. Three chose to descend upon *Waimarama*, which was following the cruiser *Charybdis*. A crew member observed of the vessel, at 12,843 tons the largest in the convoy, that with her cargo of cased high-octane aviation spirit, alongside vast quantities of ammunition, 'the whole ship smelt like a refinery'. The cost of bearing such a burden now became explicit.

The first aircraft, facing a storm of flak, bombed wide. The second, however, hit the hapless cargo-liner four times, and also near-missed *Melbourne Star*, a cable's length behind. A stupendous explosion followed, as *Waimarama*'s eleven thousand tons of munitions and fuel blew up with a force that destroyed the next attacking aircraft, killed eighty-three men on the ship and sent debris soaring hundreds of feet into the air and across the sea – fragments caused casualties and superficial damage on *Rochester Castle*, half a mile away. Edward Binfield, that ship's NLO, recorded bleakly of *Waimarama*: 'ship observed to disintegrate and disappear in an enormous pyre of flame and smoke'. Another witness said: 'the whole ship exploded in a cloud of grey vapour, not the black and red depicted in films'. Within seconds, much of the vessel vanished towards the bottom, leaving a pall of smoke hanging over the debris and flaming oil, which continued to blaze for many hours.

Even in that week of slaughter, every witness was stunned by the violence of the end of *Waimarama*. Lt. Denys Barton, on the bridge of *Ohio*, called it 'one of the grimmest things I have ever seen'. Steward Jim Parry, an *Empire Hope* survivor aboard *Penn*, noted with a numb curiosity that metal debris from the explosion appeared to drift down, almost as if the fragments were paper, 'very slowly … it seemed as if it was floating, almost'.

Onlookers were astounded that some of the ship's crew lived through the explosion. *Ledbury*, the escort closest to the scene, received a succinct order from Burrough: '[rescue] survivors – but don't go into the flames'. The intrepid Roger Hill ignored the warning, took his little ship straight in regardless, an action that won the admiration of all who witnessed it, Royal Navy and Merchant Navy alike. As hundreds of men aboard *Penn*, both crew and survivors, watched *Ledbury* race into the burning sea, all chatter fell away, so that the only sounds on the upper deck came from the hissing funnel and sighing ventilators. Jim Parry said: 'everybody was so emotionally uptight that nobody talked. We just watched this blinking flame.'

A few minutes after *Waimarama* was hit, at 0815 the same Luftwaffe squadron struck at *Rochester Castle*, which was steaming almost parallel with *Melbourne Star*, behind the exploded vessel. Both the merchant-ship and cruiser *Charybdis* put up a terrific barrage, including rocket-propelled aerial cables and mines – PACs and FAMs – which seemed to shake the German pilots, but *Rochester Castle* was bracketed by near-misses that threw up waterspouts which, for some seconds, hid the ship from the rest of the convoy. Witnesses saw her emerge painfully slowly, with fires raging on her decks and a mass of splinter holes in her sides.

Seventeen-year-old cadet David Lochhead had been manning a telephone on the open deck behind the wheelhouse without flinching, passing orders to the aft Bofors crew, but now he was carried below after suffering severe lacerations from bomb splinters. For the ensuing forty minutes, coura-geous fire parties led by chief officer Arthur Culpin fought the inferno, hurling burning deck cargo over the side and playing hoses on the flames. The chief engineer felt obliged to flood the small-arms magazine, lest fire overtake it, but the ship had escaped serious structural damage. Smoke and flames on deck made her plight look worse than it was: *Rochester Castle*, a handsome Union-Castle refrigerated fruit and mail ship diverted from the South Africa run, maintained full speed.

Even before *Ledbury* reached the scene of *Waimarama*'s passing, *Melbourne Star* was perforce sailing into it. The ship was just four hundred yards astern when the explosion came. A great shower of debris, some of it flaming and including shards of steel plating five feet long, rained down, hammering and clattering on decking and superstructure. The base of a ventilator smashed into a machine-gun position, wrecking the weapon and shocking its gunner. Next day David MacFarlane discovered an unexploded six-inch shell, blown from the lost ship, embedded in the bulkhead above his own cabin. The vessel became 'surrounded by flames and it appeared as though [we] must blow up', in the words of NLO Roger Frederick. As *Melbourne Star* bore down on the stretch of fiery sea, from his post high on the monkey island the Master ordered hard aport, but the ship could not change course swiftly enough to escape. MacFarlane then embraced his only other expedient – to make the vessel's ordeal by fire as brief as possible. He called to Second Officer Bill Richards in the wheelhouse for full speed. Richards snatched at the brass handle of the telegraph and rang down to the engine-room.

MacFarlane recounted later: 'The sea was one sheet of fire, and as we were so close, we had to steam through it.' He himself leapt down from his lofty vantage point to seek temporary refuge in the wheelhouse from the blaze that swept over the superstructure. 'It seemed as though we had been enveloped in smoke and flame for years, though it was only minutes. The heat was terrific.' Wooden lifeboats, hanging out from both sides of the ship, were scorched and charred; much of *Melbourne Star*'s paint was blistered away. MacFarlane, fearing that his ship might suffer the same fate as *Waimarama*, shouted to every man within earshot to head for'ard: the fo'c'sle was the point furthest from the likely seat of a cargo explosion.

Then yet another German aircraft descended, releasing a stick of bombs which near-missed *Melbourne Star* and precipitated panic. Thirty-three men, most of them gunners manning the aft Bofors, Oerlikons and six-inch gun, leapt overboard into the sea. Roger Frederick saw yet more hands racing around the deck in futile quest for a way of escape. Then, to the amazement of them all, *Melbourne Star* emerged beyond the burning oil, blackened but unbowed. As the crew steadied after the outbreak of hysteria which Frederick estimated to have lasted three minutes, those still aboard returned sheepishly to the fo'c'sle guns and other action stations. The naval officer afterwards applauded the performance of *Melbourne Star*'s entire company, refusing to censure even those who had, literally, jumped ship. 'The fighting spirit was magnificent. Every single man on board made all effort throughout to fight off the enemy … No gun ever lacked for a volunteer, not even after so many men had gone overboard … No blame whatsoever can be attached to the men who jumped overboard. It appeared an obvious thing to do at the time.'

The men who abandoned *Melbourne Star* found themselves mingling in the sea with survivors from *Waimarama*. Seventeen-year-old Freddie Treves, an ex-Pangbourne Nautical College cadet, was one of the youngest men on the convoy. *Waimarama*'s captain had entrusted this novice to the care of a veteran, fifty-six-year-old steward Bob Bowdrey, 'a lovely old man', in Treves' words. 'His two sons were fighting and he had rejoined the Merchant Navy against his wife's wishes, saying "if my sons are going to this war, I'm going too".' Treves had run to his quarters beneath the fo'c'sle as the alarm went, moments before the explosion, which blew him through the door on to some bags of lime. Bowdrey threw himself protectively on top of the young man. Treves, stunned, 'thought I was going to die … There was black smoke everywhere, flames were burning aft of the bridge.'

Bowdrey, a non-swimmer, clambered off the teenager's prostrate form. As the listing wreck began to sink, both men jumped some sixty feet, from the side furthest from the sea. The cadet was wearing a kapok buoyancy suit and steel helmet – the latter saved him from being injured by the mass of fragments which rattled down upon them from the sky. On all sides hysterical voices were shouting and screaming, as flames crept towards them. Treves heard repeated cries of 'I can't swim! I'm drowning!' Himself a good swimmer, he strove to calm the frantic men around him, somewhat absurdly – as he himself wryly observed later – blowing his whistle. He took hold of wireless-operator John Jackson, a non-swimmer, and towed him for five minutes to the relative safety of a drifting timber, clear of the flames. Jackson said: 'The heat was terrific … I'm quite sure that I definitely owe my life to this cadet.'

Then Treves spotted Bowdrey, whom he had come to love, and was appalled to see him standing screaming on a raft which was drifting into a flaming stretch of sea: 'it was a picture I'll never be able to forget'. Knowing that he could never tow the heavy raft, 'I turned over and swam away. This has haunted me all my life. I was a coward. He had done everything for me and I didn't do anything for him.' The world thought differently, however, accepting that Treves could have done nothing to save Bowdrey, while he had already saved Jackson and other struggling swimmers: the cadet was later awarded the British Empire Medal.

Ledbury spent two hours picking up survivors from *Waimarama* and *Melbourne Star* – one of the latter's crew decribed the destroyer's performance as 'magnificent'. Roger Hill comprehensively flouted the admiral's injunction against risking his own ship's safety: ratings played a hose from her foredeck to force flaming oil away from swimmers. Hill ordered Charlie Musham, the ship's gunner, to lower the whaler and take four rowers in search of survivors. Musham embarked on a grim race to reach men in the water before burning oil did; some won, others lost. Most of those whom they picked up proved to be suffering from burns.

Amid the smoke and fire, Hill glimpsed a figure holding an arm aloft, sitting on floating debris with flame all around. This was Alan Bennett, a seventeen-year-old scullion – the wonderfully anachronistic Merchant Navy term for a kitchen boy. The destroyer captain took his ship as close as he dared, shouting down to the first lieutenant, directing rescues on the main deck, 'for Christ's sake, be quick!' Behind Hill, his yeoman of signals muttered, 'Jesus, it's just like a film!' Then it got more so. Seaman Reg Sida jumped over the side attached to a line, seized a survivor on a raft close to the flames and the two men were hauled aboard the destroyer, clutched in a desperate embrace. Twenty-seven-year-old petty officer cook Charlie Walker emerged through a hatch from below, on to the main deck. He took one look at the scene, tore off his apron, kicked away his boots and leapt overboard. Walker, captain of the ship's water-polo team, had already made another desperate swim a fortnight earlier, diving into the sea with his captain in an attempt to rescue the crew of a downed RAF Sunderland.

Now he began heaving men out of the water, into Ledbury's boat. The cook said later of one man he picked up: 'Oh, he was really bad, he was burned to hell, he was. He was really cooked in diesel oil. His nose was like a little pear drop, and he was cooked in diesel, and he was going "water … water …"' Then Walker swam to the raft to which Bennett clung, pulled him into the water and towed him to the side of Ledbury, which had just begun to manoeuvre to escape the spreading flames. Hill shouted down through his Ardente loudhailer: 'Hold on like hell, I am going astern'. Terrified that the cook and the survivor would be washed away, he backed Ledbury off. Walker clung doggedly to the scrambling net and was hauled aboard a few moments later, still attached to his human burden. For his impulsive, extraordinary act of mercy on 13 August, he received the Albert Medal.

On the destroyer's deck, Lt. Tony Hollings said, 'the spirit of the men in the water was wonderful. There was one' – long afterwards identified as bosun John Cook – 'who shouted out to us "don't forget the diver, Sir!" as we went past.' This was a catchphrase borrowed from Tommy Handley's famous BBC radio comedy show *ITMA*. Hollings shouted down that they had to pick up some others in even worse case than themselves, and the man cried back: 'That's alright, sir.' By the time the destroyer had collected forty-two survivors, eighteen from *Waimarama* and twenty-four from *Melbourne Star*, the convoy was thirty miles ahead. Fourteen of those who leapt overboard from the latter vessel, including the Royal Artillery lieutenant in command of the Bofors, paid with their lives for misjudging peril. In the course of a subsequent fast passage to catch up, at 1050 *Ledbury* suffered another attack by seven Ju88s, which resulted only in 'the usual near-misses'. The terrifying had become the commonplace.

Others were meanwhile less fortunate. The second big air attack of the day developed an hour after the first, and followed a pattern that became familiar. Nine Ju88s staged a diversionary raid on Malta, to draw off the island's Spitfires, while larger formations headed for the convoy. One group of enemy aircraft appeared on the port side, drawing fire from the ships, while a more serious threat closed from the starboard quarter. An officer who served in the Mediterranean wrote: 'one knew that if the ship receives a direct hit, one will either be killed or thrown into the water to struggle gasping for breath and probably drowning. The bomb comes down, you try to find cover and cannot, the ship heels over trying to dodge the path, [the bomb] misses and a great spout of water is thrown up only yards from the ship … If anyone tells you that in those circumstances he was not frightened I would say he was either a monumental liar or a very, very brave man.'

When twenty-six German Ju88s launched low-level bombing assaults on *Ohio*, one of the first was hit by flak. It struck the sea off the tanker's bow, then bounced on to the tanker's foredeck, where the wreck came noisily to rest, by some fluke without catching fire. A second attacker's bombs near-missed so closely that the ship's forepeak tank was flooded, her bow plating buckled. Half an hour after the Ju88 attack, eight Italian Stukas – 'Woodpeckers', in Regia Aeronautica slang – followed, which again repeatedly near-missed.

One of these aircraft, flown by Sergeant Oscar Raimondo, was hit by the tanker's gunners seconds after releasing its bomb. The Stuka crashed on to *Ohio*'s poop, with another deafening impact, separated by only two inches of steel from eight thousand tons of fuel. From the stern steering position, Doug Gray excitedly telephoned Dudley Mason on the bridge to report this unwelcome new arrival. The Master replied laconically: 'Oh, that's nothing. We've had a Ju88 on the foredeck for nearly half an hour.' At 0941 Ju88s near-missed *Kenya*, then at 1000 more aircraft attacked the tanker. One dropped six bombs which bracketed her so close that in Mason's words his own vessel 'appeared to be lifted right out of the water, and shook violently from stem to stern'. Within the half-hour, the effects of this attack became painfully apparent: two electric fuel pumps died, causing the main engines to stop. By 1030, the tanker was wallowing, dead in the water. With lighting knocked out, engineers worked below in the echoing darkness of the machinery spaces illuminated only by torches.

The story of *Ohio* during the ensuing forty-eight hours is best unfolded as a single saga in the chapter that follows. All that need be said here is that both British and Axis commanders made poor decisions on 13 August, based on a common failure to recognize that getting the tanker into Malta, or preventing her from doing so, was an objective that should transcend all others. Luftwaffe chief Gen. Deichmann, writing in his post-war memoir, deplored orders imposed on II Fliegerkorps that day to commit aircraft to attack the retiring Force Z, against which they achieved no successes, rather than to concentrate exclusively on destruction of the remaining Pedestal merchant vessels and above all *Ohio*. German and Italian strike aircraft divided their attentions. And Admiral Burrough, exhausted after the events of the preceding sixteen hours since he assumed command of Pedestal, treated the tanker as merely one among his many responsibilities.

The attack that stopped *Ohio* also addressed *Dorset* just as she rejoined the convoy, escorted by the destroyer *Bramham*. Here was another tragic earnest of the absence of a warship fitted for fighter direction: even as the Italians approached, Spitfires arrived overhead from Malta and flew unwavering into the path of *Dorset*'s guns. A pilot, John Mejor, wrote in his logbook: 'we waited up sun but the swine came in down sun. Bags of flak. We dived through and fired short bursts of one-second each at three Ju87s. Observed strikes on tail of one. I saw two Ju88s making for the largest ship. Peeled off through flak to make head-on attack. I was hopelessly out of range but fired to scare them – it did. They dropped bombs far wide of targets and turned away. I followed. Rear-gunners poured tracer at me and I took took poor view of it. Ran out of ammunition.'

One Spitfire was shot down by *Dorset*, its Australian pilot Bob Buntine killed. The ship's NLO Lt. Peter Bernard repeatedly shouted to the gunners to cease fire, in vain. He wrote in mitigation: 'It must be remembered that, after the attacks we had been subjected to, the gunners were very excited and not in the mood to run risks.' This sort of mishap might have been avoided had the British ships possessed phone links between their gunners and the bridge. They did not, however. Thus, facing the tension and indeed terror of the moment, and with nobody directing their fire, gun crews cut loose at anything airborne within their range. Since the previous day, they had been given little cause to suppose that any portion of the sky above their heads was tenanted by friends. Ron Linton, one of *Dorset*'s Oerlikon gunners, shrugged: 'we are told, if you see a plane it's the enemy, so shoot it. We believed we had no air cover.'

And yet more attackers were coming. The Italians' Savoia Sparrowhawk torpedo-bomber squadron was led on the 13th by their most experienced crews, again accompanied by a fighter escort. Martino Aichner wrote: 'That day there is a greater commotion than on any other, heightened by the presence of thick black rain clouds over the sea.' The crews were further rattled by the fact that they were now attacking across sky patrolled by some British aircraft. At 1120, they found the weather clearing as they approached the British convoy. Spitfires from Malta engaged their fighter escort, but most of the strike aircraft got through.

One Sparrowhawk dropped a torpedo from extreme range, but the others sought to close the distance. Ted Gibbs, commanding *Pathfinder*, wrote wryly: 'One can get tired of being attacked by aeroplanes; we had to alter course to avoid the torpedoes, so we decided to do the attacking ourselves. We increased speed to thirty knots and turned straight out to seaward to meet the Savoias.' The Italians approached at masthead height, to be met by every gun on the destroyer, firing from both sides.

Gibbs continued, 'the noise was tremendous, and I personally remember this as the most exhilarating and enjoyable moment of the war. The nearest Savoia was almost within biscuit toss, and the whole formation was surrounded by shell bursts and streams of tracer bullets, through which they dived, twisted and climbed, dropping their torpedoes in almost all directions except that of the convoy.' One Sparrowhawk caught fire, others turned back towards Pantelleria. *Pathfinder* reversed course to rejoin the convoy, triumphant about having broken up the attack. *Ashanti*'s gunnery officer noted that, while some Italian pilots closed in bravely, they dived at so steep an angle that their torpedoes had no time to steady at their intended depth-settings before reaching the warships.

One enemy aircraft had aimed straight, however. *Port Chalmers*' Captain Henry Pinkney ordered the helm put over to comb two approaching tracks. The first torpedo slid beneath his vessel, while another passed down her starboard side. Then the anti-mine paravane trailing beside the bow began to vibrate wildly. Guessing the cause, Pinkney stopped his engines. The paravane, hoisted clear of the water, proved to have a torpedo caught in its wires by the tailfin. This tangle of technology, explosive and cable was hastily unshackled by First Officer Bill Craig and set free to plummet towards the bottom, where the torpedo detonated with sufficient force to cause the whole ship to lift in the water.

Twenty minutes after this attack, at 1140 fourteen German Stukas dived on the ships, concentrating on *Dorset*, which had rejoined the convoy only five minutes earlier after reversing course and engaging the Malta Spitfires. She was now strad-dled by three bombs exploding close alongside. Seaman John Waters said: 'We made no attempt to move … we were too exhausted to think of means of escape …' He and his mates stood watching, waiting for an explosion, which did not come. But one near-miss opened plates in the ship's starboard side, causing water to pour into the engine-room. Another blew a further hole in her side and disabled key machinery – a piston cooler and refrigerating discharge pipes. A steward who had been carrying a teapot to a gun's crew found himself clutching only its handle, the remainder having been blasted away. Fire broke out close to high-octane fuel, and the chief engineer reported the main engines and pumps out of action. One of the ship's mates said bitterly: 'if I could find the bastard that gave the order to come back' – at first light that morning – 'I'd shoot him'.

Captain Tuckett ordered his ship abandoned. The crew were taken from their boats on to *Bramham*. Yet, once aboard, Tuckett reflected that *Dorset* showed no sign of sinking. His first suggestion, to Eddie Baines, was that the destroyer should take the merchant vessel in tow, which the naval officer was willing to attempt. Baines was obviously unimpressed by the hasty manner in which the crew had quit *Dorset* and noted sardonically in his report that they 'had no desire to go back to her'. John Waters was one of twelve men, most of them more or less shamefaced about quitting their charge, who now volunteered to row back and reboard the ship with Tuckett and the NLO. Waters wrote: 'Looking back, I can find no excuse for what some of us did, unless we were perhaps slightly insane through exhaustion.' The men's first action, once back on board, was to hurry to their quarters to retrieve abandoned possessions. In the galley, they found saucepans boiling over. In a spasm of domestic pride, they removed these from the hot stove, before making their way to the fo'c'sle to address the tow.

They were about to connect a cable to a guide rope passed between the two ships when German aircraft returned and dived to attack. As bombing and gunfire resumed, the captain and his party abandoned their labours and returned to

Bramham, where the Master reported to Baines that his ship's main engines were now under water, the shelter deck ablaze. It was a fundamental difference between warships and merchant vessels that the former were constructed with multiple power sources while most of the latter had only a single engine-room. During the hours that followed, while *Dorset* wallowed, a stream of divergent or contradictory orders were issued by higher commanders. Yet again these reflected the problems of control besetting the convoy.

At 1305, 'acting on a "carte blanche" from C.S.10' – meaning that Burrough told Baines to act as he thought best – the destroyer captain decided to sink *Dorset*. Thirteen minutes later, however, this order was cancelled and *Bramham* was told to try to save the ship. Baines took the destroyer close in, to try to tow from alongside while using his own ship's pumps to check the merchant vessel's flooding. Then *Penn* appeared, and instructed *Bramham* – presumably with Burrough's concurrence – to leave *Dorset* and proceed to the assistance of *Ohio*, some five miles astern, which Baines duly did. At 1430, the Senior Officer Malta Minesweepers arrived on the scene in *Speedy* and assumed formal responsibility for salvage operations around both *Dorset* and *Ohio*. Before anything was done for the former, however, two hours later the ship was once more attacked by German Ju88s, which achieved a direct hit that set her foredeck ablaze. Other bombs near-missed in the sea alongside. *Dorset* settled by the stern, though remaining stubbornly afloat for several hours longer.

The ship's fate became one of the most controversial episodes of Pedestal. It is hard not to conclude that Tuckett's morning decision, to reverse the ship's course, was a misjudgement. *Dorset*, already within easy range of Spitfire cover, would have had a much better chance of survival had she dashed alone for

Malta. Once again, the communications nightmare reared its head: there were no means of summoning close air cover for the ship through what should have been her final miles to Grand Harbour. It remains an open issue why, even in the absence of voice control, the ground-based fighters were not more successful in preventing the last enemy attacks within sixty miles of Malta in the course of 13 August.

Questions were also asked whether Tuckett and his crew abandoned their ship prematurely. There can be no right answer to this challenge. How much risk could crews be expected to bear? How many air attacks, bombs, torpedoes were the men of the Merchant Navy who had sailed *Dorset* through the past days and nights to be asked to endure? That very morning they had witnessed the colossal explosion and conflagration after *Waimarama* was hit: their own cargo could generate a matching calamity.

Most war veterans believe that courage is capital, not income, that all men possess variable but expendable reserves. It seems fair to suggest that a grit which might have been expected of *Dorset*'s crew in the first days after they passed the Straits of Gibraltar was no longer to be had in the hours before they approached Malta. They had borne much to get their ship thus far. Now, as they saw flames soaring around munitions and high-octane fuel, they lacked the almost manic strength of will that would have been needed to linger aboard to try to fight the ship through the final miles.

And whatever *Dorset*'s crew did or did not do, the vessel remained afloat for hours, within ten minutes' flying time of Malta, while no attempt was made to tow her in. This partly reflected the fact that Burrough felt obliged to regard the welfare of his surviving cruisers as a responsibility as onerous as getting the merchantmen through to Grand Harbour. A thin case can be made that the cruisers' presence contributed to deterring an Italian fleet sortie. In the absence of this, however, these big, expensive warships had proved a liability to Pedestal rather than a contribution to its fighting strength beyond what their guns added to anti-aircraft barrages.

In other circumstances – not least the December 1939 Battle of the River Plate, following which the pocket battleship *Graf Spee* was scuttled – British cruisers showed their worth, but not in the Mediterranean in August 1942. Anti-submarine and anti-aircraft protection for the merchant vessels was weakened by the need to provide escorts for the cruisers – witness the three detached to cover *Nigeria's* retreat to Gibraltar, three more sent to succour *Manchester* and thereafter her survivors.

The last assault by Axis aircraft on the afternoon of the 13th was executed by Ju88s which took off from Sicily at 1445. While searching for the convoy west of Bizerte, they spotted and bombed two surfaced submarines. The first escaped unscathed, but the second was significantly damaged, its captain and seven of his crew injured. Both were Italian – *Alagi* and *Dessiè*. Meanwhile, the Italian Sparrowhawk squadron landed back at their field, disconsolate after losing Lt. Guido Barani, shot down by the British. Martino Aichner described how a meeting of pilots was summoned to discuss the results – or rather, lack of them – from the first strike, and to consider whether a second sortie should be dispatched. 'The clouds are getting thicker and blacker and it is becoming stormy. It would be pointless, suicidal, to send out aircraft in these conditions.' A new attack was postponed until the 14th. The grateful Aichner and his comrades were stood down.

* * *

Edward Binfield, the NLO on *Rochester Castle*, wrote with studied understatement of that last Italian torpedo-attack: 'Avoiding turns taken and attack proved harmless. The remainder of the voyage was without incident.' At 1630 on 13 August, just west of Malta, Rear-Admiral Burrough passed the three merchant vessels of Pedestal that were still steaming in his wake into the custody of the island's minesweeping flotilla. These little warships led the survivors through the defensive minefields, under fighter cover from the island, while the admiral, with the cruisers *Kenya* and *Charybdis* and four of his seven remaining destroyers, turned west for Gibraltar.

It was a measure of the ferocity of the three days and night of battle that since 10 August *Kenya* alone had expended 425 shells from her six-inch main armament; 1,780 rounds of four-inch; 5,050 pom-pom; 1,178 Oerlikon; 3,220 Vickers machine-gun. Other ships had consumed ammunition in like proportion. Many gun barrels were so worn that they posed a severe danger of premature explosion – such accidents killed several men on the passages both to and from Malta, one of them aboard *Ledbury*.

Burrough, at last relieved of his gravest burden – he still had to get his ships back to Gibraltar – for the first time in three days and nights felt able to leave the bridge of *Ashanti*, to strip off his bedraggled whites, still coated in black oil from *Nigeria*. He took a salt-water shower on deck, before the eyes of what he described drily as 'an interested ship's company', unaccustomed to naked admirals. The genial Burrough thought the spectacle embarrassed them more than him. His Force X experienced further U-boat, E-boat and almost incessant air attacks off Cap Bon during its withdrawal, but suffered no further ship losses, though both *Kenya* and *Ashanti* were damaged by near-misses.

Syfret wrote in his post-Pedestal report: 'the magnitude of the task with which Force X and the [merchant] ships were confronted might well have daunted the stoutest hearts. The display of courage, fortitude and determination in fighting a way through all the difficulties and in spite of severe losses ... reflects the greatest credit on captains, masters, officers and men and in particular on their resolute leader Rear-Admiral Burrough.' All this was true. His leadership of Force X had been an object lesson in steadfastness under the most desperate circumstances.

Burrough can be criticized for some local decisions, mentioned above, but it was absolutely no fault of his that the Axis deployed formidable air and submarine forces with access to bases across the entire central Mediterranean; that British anti-submarine technology and tactics were weak; that naval carrier fighters were embarrassingly inferior to the enemy's aircraft. Through twenty-four hours on 12/13 August 1942, his Force X faced overwhelmingly hostile odds. Burrough lacked any ready means of improving its outcome.

At 1825 minesweepers led *Melbourne Star*, *Rochester Castle* and *Port Chalmers*, carrying between them 23,000 tons of general cargo – mostly food – and 5,500 tons of military stores, into Grand Harbour. In the words of a submarine captain, this once magnificent anchorage of the Mediterranean Fleet, beneath the battlements of the old castle of the Knights of Malta, was now 'a heartbreaking scrapyard of bomb- and mine-shattered hulks'. But to the crews of the three ships on that momentous evening it appeared an almost mystical haven. Men aboard *Rochester Castle* noted ruefully that the ship's swung-out lifeboats were so burned and peppered with shrapnel and bullet holes that they would have been almost useless in a sinking. Of the entire convoy, *Port Chalmers* alone sailed in unscarred. History does not record whether Commodore Venables showed a flicker of embarrassment as on her bridge he shared this moment of triumph.

The ships were greeted by bands and ecstatic crowds, foremost among them the governor, Lord Gort. A local woman wrote: 'what a glorious sight that was! The bastions around the harbour were lined with people. We waved and cheered until we could cheer no more.' An RAF pilot was also watching: 'As the three ships come through the harbour entrance, just about maintaining steerage way, the cheering of the Maltese ... slowly subsides until there is absolute silence. Some of the men, mostly elderly, take off their hats and the womenfolk in their black hoods and cloaks cross themselves. At the word, a bugle sounds the "Still", and not a soul moves.'

After disastrous earlier experiences, when newly arrived merchant vessels succumbed to air attack inside Grand Harbour, Operation Ceres was immediately set in motion – a meticulously planned, round-the-clock offload, engaging soldiers, troops and dockers, to transfer cargoes into bomb-proof caves ashore. The *Times of Malta* reported the ships' arrival under the headline: 'MERCHANT NAVY DEFIES AXIS BLOCKADE ... SHIPS THAT CAME THROUGH LIVING HELL'. The accompanying editorial declared in ringing tones: 'Through the mercy of Providence and the courage of our seafarers, Malta has been given succour in an hour of need borne by people and garrison alike with the fortitude and an abiding faith in the justice of our cause.'

Even as the convoy completed the last miles of its passage to Malta, the destroyer *Ledbury* was retracing a course west. At 1115 that morning Hill's ship returned from her heroic rescue operation, weaving a course between fragments of the wreckage of *Waimarama*. She reached *Ohio*, adrift, while the nearby *Penn* was depth-charging an asdic contact; ten miles ahead, Roger Hill and his bridge crew could see the convoy enduring an air attack. *Ledbury* then received a signal from the admiral, ordering her return to the Gulf of Hammamet to search for *Manchester* – this reflected Pedestal's breakdown of communications, since *Eskimo* and *Somali* had established hours earlier that the cruiser was lost. After it was all over, it emerged that incompetent wireless-operators may have muddled *Ledbury*'s orders by mistaking ships' call-signs: Burrough probably intended the destroyer to search not for *Manchester* but for *Brisbane Star*, which was believed to be limping down the North African coast. Hill's exasperated first lieutenant deplored the fact that the ship's telegraphists 'got most of the signals tied up'.

Confusion persists about some messages that Burrough dispatched at this time. One version holds that he ordered *Ledbury* to return to Gibraltar with her large contingent of survivors – presumably assuming that she could do no more for the convoy – countermanded only after a suggestive signal from the senior officer Malta, Vice-Admiral Sir Ralph Leatham, urging that the destroyer should continue to support Pedestal. The other version holds that only faulty decryption by *Ledbury*'s telegraphists ever canvassed a suggestion that the destroyer was intended to quit the operation. In any event, around 1115 Roger Hill set course for the Gulf of Hammamet. *Ledbury* still had a prominent role to play in determining the fate of *Ohio*, but was parted from her for the ensuing twenty-odd hours.

At 1345, the destroyer once more passed the scene of *Waimarama*'s destruction. Her captain exploited a lull in the action to visit his desperately tired engine-room staff, together with badly burned survivors in the sick bay. Then he collapsed on to his bunk. Hollings, the first lieutenant, said: 'the captain turned in and told me to fight anything which came along and not to wake him unless we saw *Manchester* or the coast'. Yet Hill had been asleep for only twenty minutes when at 1545 he was roused by a familiar voice pipe: 'Captain, sir, captain!' He dashed up on to his bridge to see gunners preparing to loose everything they had at two low-flying aircraft approaching from the starboard bow.

Half-asleep, he shouted: 'Don't shoot, they're Beaufighters!' First Lieutenant Tony Hollings disagreed: 'I'd never seen anything less like Beaufighters than those torpedo-bombers. Luckily the guns all thought the same as I did and opened fire almost at once.' One Italian torpedo missed by a few feet, as the ship swung to port. Main armament, together with the stammering pom-pom and Oerlikons, blazed at the three-engined Savoias – and shot down both, a remarkable achievement. Hollings said: 'Everyone cheered like mad at this magnificent piece of good luck.' The *Ledburys* saw a dinghy putting out from one floating plane, but the threat from submarines seemed too great to risk stopping to retrieve the airmen. The exultant Hill ordered a rum issue to every man aboard, crew and survivors alike, except their lone German prisoner from the previous day – 'stand fast the Hun'. He told his yeoman of signals to hoist the flags for 'splice the mainbrace', saying: 'Let us look as cheerful as we feel.'

At 1630 *Ledbury* sighted land, the tip of North Africa. She turned south, holding a course four miles offshore, Hollings suffering dreadfully from swollen ankles, having been so long on his feet. As it became clear that most of *Manchester*'s survivors were now ashore in the hands of the French, Hill and some of his officers, light-headed with exhaustion, discussed landing an armed party in their motorboat, towing the whaler, to rescue the *Manchesters*: 'Anyway they were to knock off any Frenchmen or Arabs who tried to interfere … In this gay and rather piratical mood we continued down the coast.'

Ledbury hoisted a large French ensign in hopes of deceiving watchers ashore, which caused their doctor, emerging from below, to demand in bewilderment: 'Why are we surrendering?' Shaking his head, the medical man returned to his patients, muttering about 'a lot of bomb-happy bloody lunatics'. The captain himself afterwards admitted that some of this talk and conduct lurched between adolescence and madness, but who could be surprised when every hour of every day and night seemed to thrust upon them some new horror or fresh challenge?

The chief engineer climbed on to the bridge, his invariable corncob pipe clamped between his teeth, to warn Hill that their fuel situation was becoming precarious. By 1900, they had found no trace of *Manchester* in the Bay. *Ledbury* abandoned her futile mission, precipitated by the general failure of communications, and turned back east at her twenty-two-knot best speed, to close *Ohio*. Even in darkness, plentiful evidence remained of the tragedies of the day: the sea was still ablaze where *Waimarama* had blown up fourteen hours earlier. A German E-boat commander, Wuppermann of *S56*, likewise remarked in his report on the stench of oil that pervaded great tracts of the watery battlefield for days after the passing of Pedestal. For most of that night of the 13th Roger Hill dozed in his high chair on the bridge, in expectation that would be fulfilled of an eventful morrow.

* * *

At 1530 next day, Friday 14 August, Malta celebrated a new miracle: the freighter *Brisbane Star*, which had sailed a lone course for two hundred miles since taking a torpedo hit in the bow at 2058 two nights before, steamed into Grand Harbour, to berth alongside the hulk of a wrecked freighter. The ship's experiences during the intervening forty-odd hours were remarkable even by the standards of Pedestal. Her naval liaison officer, Lt. George Symes, took up the story from the previous night in his report, writing: 'It was decided that as we could make no more than 10 knots, we had best leave the convoy to whom we would be only a lame duck, make our way down the Tunisian channel … then strike across to Malta during the night, and hope that the enemy would be too busy to notice us.' It deserves emphasis that there was no unanimity among those on board about the narrative of events that follows. We can offer here what is merely the most plausible version of what took place.

As the ship moved inshore, her crew saw the remainder of the convoy disappear on its south-easterly heading, and resisted the chivvying of a destroyer captain to join them. 'By the time we reached Kelibia Point the E-Boats had left, and we were not troubled by them. We sighted a darkened vessel, either submarine or darkened patrol vessel, lying inshore of us but it took no action.' They also passed the stricken *Glenorchy*, with a destroyer standing by, and caught a distant glimpse of that ship's explosive ending next morning.

Throughout the night, as *Brisbane Star* crept along the shoreline, her crew saw occasional starshell to the north-east and heard explosions. Early morning found the ship alone, a mile off the Vichy French Tunisian shore. An Italian Caproni bomber appeared, and circled. The ship had lowered her colours, 'but our identity must have been obvious'. The plane eventually disappeared without taking hostile action beyond alarming the crew by making low-level passes overhead. The ship's captain, forty-six-year-old Irishman Fred Riley, observed sardonically that the pilot was obviously a gentleman who observed the rules of war, or rather those of neutral waters. An exchange of courtesies with the Vichy French authorities took place by lamp signal off Hammamet, where a shore station demanded the ship's signal letters.

There ensued one of the few comic interludes in the Pedestal saga. *Brisbane Star* professed bewilderment:

BS: PLEASE EXCUSE ME

Hammamet: YOU SHOULD ANCHOR

BS: MY ANCHORS ARE FOULED, I CANNOT ANCHOR

Hammamet: YOU APPEAR TO BE DRAGGING YOUR BOW
     AND STERN ANCHORS!

BS: I HAVE NO STERN ANCHOR

Hammamet: YOU SHOULD ANCHOR IMMEDIATELY

BS: I CANNOT ANCHOR, MY ANCHORS ARE FOULED

Hammamet: DO YOU REQUIRE SALVAGE OR RESCUE?

BS: NO

Hammamet: IT IS NOT SAFE TO GO TOO FAST

Without experiencing further interference, the ship proceeded slowly south towards Monastir Bay. A signal to Burrough, repeated to Malta, announcing her intentions, received no reply, though indeed there was little useful that any admiral could say. *Brisbane Star*'s fate, her survival through the two-hundred-mile passage to Malta, now depended on the determination of her captain and crew, together with an immense dose of luck. Riley determined to hold his south-bound course, close inshore, until darkness fell on that day of the 13th, then swing eastward and race for Grand Harbour through the night hours. Every moment of every hour of that day, captain and crew were on tenterhooks about the prospect of meeting Axis aircraft, submarines or E-boats, any one of which was likely to be fatal to them. There were repeated sightings of periscopes, real or imagined.

As the light began to fail, tension slackened: the perils of the day seemed over. Off the port of Sousse, Riley turned on to a new course for Malta, making ten knots. Yet suddenly a French naval gunboat overtook them, ordering them to halt. When the ship maintained her way unchecked, the Vichy vessel fired a shot across her bows, then sent an armed boarding party up *Brisbane Star*'s side. A naval officer, accompanied by Sousse's harbourmaster, demanded that the ship turn and enter the port, to accept internment.

Riley declined, but invited the two men to join him in his cabin for a glass of whisky. There, in a notable exercise of diplomacy, he appealed to the harbourmaster as a fellow seaman to show kindness; to forget his passage; to let the ship go. There was a pause during which the Frenchman looked quizzically at his naval companion. Then he suddenly gave way, offered his *meilleurs sentiments* and extended his hand, saying: 'Goodbye captain, a safe voyage and good luck.' He noted the ship's intended course, then the visitors returned to their gunboat. Moreover, Riley persuaded them to take with them sixty-year-old greaser Edward Corfield, who had been badly wounded by shrapnel on the 12th. Corfield died shortly after reaching the shore, but here was one of the few recorded moments in the Pedestal story during which representatives of Vichy behaved in a fashion sympathetic to the allied cause.

Yet the ship's ordeal was far from over. Scarcely had the Frenchmen gone down the side when a delegation from the crew, including some greasers and stewards, climbed to the bridge to present an imperious demand that their damaged ship should abandon her passage to Malta and enter Sousse. The torpedo-damaged hull was quite unfit to brave a dash for Grand Harbour, they said. Dismayingly, the ship's naval liaison officer endorsed their view. Riley recorded later: 'Lt. Symes approached me and stated that my chances of getting to Malta were nil, for as soon as I left the coast the submarine reported as following us would put ... torpedoes into us and we would be blown sky-high ... The time had come for me to scuttle.' Lt. Eva, the sea transport officer, concurred with Symes. It was plain, he said, that the enemy knew their exact position.

Riley, in his report, observed that their permitted escape from Sousse was the only happy incident 'in a day made most unpleasant for me by members of the ship's personnel whose sense of duty and ... honour I am sorry to say sunk that day to a spring low-water mark'. Riley consulted his chief officer, Bob White, who sympathized with the Master but felt that the dissidents had a strong case. Riley wrote: 'The atmosphere ... was against me, and I don't think I am far wrong when I say that close to 100% were wanting the ship scuttled, and to get ashore.' Yet suddenly Riley received a signal from Malta which gave new life to his own dogged determination: from first light next day, the 14th, *Brisbane Star* was promised fighter cover.

Here was a vivid demonstration of the contribution that contact with a higher power, so often absent that week, could make to morale and operational decisions. A sense of isolation, loneliness, prompted several captains and crews to abandon hope, when support from their commanders might have persuaded them to persevere. Riley's officers, perhaps hesitantly but nonetheless importantly, agreed that the Malta signal altered the odds in favour of the ship. The prospective mutineers – for that is what they were, men who embraced a vision of escaping from the war into neutral internment – were ordered off the bridge.

At 1815 the ship embarked on the last phase of her voyage. Symes now accepted that the captain was right, as did the senior telegraphist, who had also been a faintheart. Doubts persist about exactly what took place on *Brisbane Star* that day – there was a suggestion that army personnel manning the ship's guns threatened to turn them on the mutineers. This sounds extravagant, but there is no doubt that the merchant seamen and engine-room personnel were serious about wanting to make for the shore. They had already been torpedoed once, a terrifying experience, and were desperate to escape a repetition. Only the formidable personality of Fred Riley denied them this option and drove the ship on towards Malta.

All through the night they zigzagged, the wireless-operators listening intently to the known U-boat wavelengths. There was a brief alarm when a submarine was heard signalling at deafening volume, on a bearing of 207 degrees, obviously very close, but the scare came to nothing. At 0630 RAF Beaufighters appeared overhead, and thereafter for the last hundred miles the ship was continuously covered by one or more aircraft from Malta. This did not, however, prevent an early-morning Luftwaffe Ju88 from overflying the ship at deck height, meeting a frenzy of Bofors and Oerlikon fire which appeared to knock out the plane's port engine and prompted the pilot almost to crash into *Brisbane Star*'s starboard side. Two bombs fell into the sea abreast the funnel, only a few yards from the ship, before the German departed, pursued by a Beaufighter. Thirty minutes later, a Caproni torpedo-bomber approached, obviously from North Africa, and was preparing to attack when hit from behind by fire from another Beaufighter, which had apparently been unnoticed by the Italian pilot.

The last seven hours of the ship's approach to Malta were uneventful, though never for a second did any man aboard dare to relax – remember the fate of *Dorset*, fatally crippled almost within sight of the island. *Brisbane Star* had experienced so many dramas, so many hair's-breadth escapes over the previous three days that it would have tempted fate to presume upon anything as miraculous as reaching safety. The NLO wrote in his report: 'There were no brilliant feats of courage or devotion to duty; everyone from the Captain down to the deck hand *did their duty* [emphasis in original] in the face of the enemy as was expected of them.'

He added that it was thus difficult to recommend decorations for any member of the crew over his fellows, save obviously Fred Riley. Unsurprisingly, Lt. Symes made no mention in the report of his own alleged collapse of resolution, nor of the revolt at Sousse by some – by no means all – of the hands. Given the difficulties Riley overcame with his own crew, as well as in defying the enemy, those who knew his story thought him deserving of the highest honours. A somewhat crestfallen seventeen-year-old quartermaster on *Brisbane Star* noted of their unspectacular arrival in Grand Harbour: 'There was no band to greet us, like the early arrivals. We were late.'

There was a postscript to the story of *Dorset*, on the evening of 13 August still drifting, abandoned and on fire, sixty miles off Malta. At 1648 Helmut Rosenbaum of *U-73*, the nemesis of *Eagle* who had been trailing the convoy for fifty-four hours, came upon the hulk and fired a single torpedo. Seven minutes later, the ship vanished into the Mediterranean, to the chagrin of more than a few naval and Merchant Navy personnel, who believed that her loss was avoidable. The four vessels that anchored on 14 August delivered to Malta 32,000 tons of general cargo. A further 52,000 tons reposed at the bottom of the Mediterranean.

12

Ohio

During another battle of the Second World War, in France in 1944, Gen. George Patton famously pleaded with Gen. Dwight Eisenhower to send him fuel: 'My men can eat their belts, but my tanks gotta have gas!' It would have been unthinkable for the taciturn Lord Gort, governor of Malta, to resort to such histrionics, but the imperatives were the same. Even if the island's garrison and people could survive on starvation rations for a few months more, the warships and aircraft – 186 of the latter on 14 August – indispensable for Malta's defence, and for offensive operations against Rommel's supply line, must have oil and high-octane spirit. Small quantities of aviation fuel had been shipped to Malta by fast minelayer and submarine – *Otus* made such a run while Pedestal was at sea. However, only a tanker could assuage the thirst of the island's vehicles, cookers, generators and vital defensive equipment, as well as the planes and warships.

The one such vessel with the remotest prospect of reaching Malta in time to avert its surrender was the torn, battered, blackened, half-drowned *Ohio*, with the wrecks of two enemy aircraft protruding from her deck piping and derricks; an expanse of her side open to the sea; bullet holes pockmarking the superstructure. The British and Axis high commands shared the expectation that the ship must soon disappear beneath the limpid sea. Her tardiness in doing so owed little to British admirals or air marshals, instead almost everything to a few hundred officers and men of the Royal Navy and Merchant Navy. Congregated around the hull, they became determined to the point of obsession that they would drag the ship the last hundred miles to Malta. Dragged she must be: there was no realistic prospect of restarting the tanker's engines. This huge deadweight, with thousands of tons of seawater added to those of her cargo, was now reliant for mobility upon the energies of the little destroyers and mine-sweepers attending what appeared likely to prove her death throes.

As described above, the air attack at 1000 on the morning of Thursday the 13th disabled circuit-breakers to the tanker's electric fuel pumps, and thus stopped the engines. These were restarted fifty minutes later, using the standby steam fuel pump, but engineers could not achieve a sustainable vacuum, so that the propeller shaft speed fell to twenty revolutions a minute. This enabled *Ohio* briefly to creep eastward at four knots, falling to two. Then at 1130 there was a severe explosion inside the furnace of the port boiler: escaping steam then extinguished its fires.

Twenty minutes later, the same fate befell its starboard counterpart. The engines stopped, this time for good, and all twelve hands in the engine-room left their posts and clambered topside. The funnel continued to gush black and white smoke. For a time, sufficient steam pressure lingered to operate the emergency steering gear. Even this collapsed, however, just as *Penn* was connecting a ten-inch manilla towing hawser. When the destroyer took the strain, the two ships made little progress, because *Ohio* veered stubbornly to port. Then the tow parted.

Under renewed air attacks through the late morning of the 13th, *Penn*'s despairing James Swain decided that his own ship could aspire only to screen the tanker against submarine attack until more vessels arrived. At 1350 he signalled Burrough to this effect, to receive in return a stern message from Admiral Leatham on Malta: 'you must make every endeavour to tow'. Swain was thirty-seven and had already seen much action in the Mediterranean: in June 1941, the destroyer *Waterhen* was sunk under him by air attack off Tobruk. A sober figure who was ordained into the Church of Ireland priesthood after the war, he now bore responsibility for the fate of *Ohio*. Dudley Mason suggested that the only realistic prospect was to drag his ship from alongside, with additional cables attached ahead and astern, to keep her on some sort of course. This would require four vessels.

Pending the arrival of additional help, Mason proposed that his own exhausted crew should be taken off the tanker. Unlike some other evacuations in those days, this was a wise as well as humane decision. Until means were available to move *Ohio*, the men aboard her were merely prospective victims of another calamity, whenever the next deluge of Axis bombs descended upon the ship. NLO Denys Barton wrote: 'we all thought she was going to sink, then'.

Penn's captain agreed. By 1400, *Ohio*'s entire crew, including Mason, were crowded aboard the already overburdened destroyer. For a time thereafter, *Penn* merely reprised a circular course around the tanker, awaiting the next air attacks. Shortly before 1800 the minesweeper *Rye* and two big motor launches arrived from Malta, to assist in towing *Ohio*, an operation that naval headquarters in Valletta codenamed Statue. Ten minutes later, when some unenthusiastic enemy torpedo-bombers had been driven off, Mason and his crew returned to their ship, accompanied by a naval party. The worst of the debris was cleared from the foredeck, pushed into the sea. A towing cable was linked to *Penn*, in hopes that the 1,825-ton destroyer could move the 30,000-ton steel, oil and seawater deadweight of the tanker, which now drew thirty-seven feet: she was slowly, but inexorably, sinking. The little 656-ton *Rye* sought to steady the cable between the two ships by joining her sweep wires to both. Mason's men disconnected *Ohio*'s rudders from the steering engines and substituted a chain linkage, with the intention of correcting her swings to port.

During the ensuing ninety minutes, this unholy tangle of ships and cables made a little progress. Then at 1830 a new wave of Axis aircraft appeared overhead. They were engaged by Spitfires from Malta, prompting some fierce dogfights, but several enemy aircraft broke through. A Ju88 dropped a bomb which passed through the tanker's boat deck, then the cabin accommodation, before exploding on the boiler tops. Clouds of white powder from the asbestos lagging swirled around the engine-room, choking and temporarily blinding those who descended to examine the damage. Flying fragments mortally wounded a Bofors gunner. At 2024, yet another enemy aircraft bracketed *Rye*, near-missing her with twelve bombs, one of which drenched men on the upper deck with water and splattered her hull with shrapnel, though inflicting no substantial damage. *Penn*'s captain decided that the risk to his own ship from the continuing bombing was too great to sustain the tow, and cast off. He shouted through a loudhailer from the bridge: 'Volunteers on the *Ohio*, stand by! We'll come back after the attack!'

Penn had aboard two hundred-odd survivors from *Santa Elisa* and *Empire Hope*. Some burns patients, liberally smeared with gentian violet, had to find sitting space on the upper deck because there was none below. There were long queues for plates of porridge, tea and biscuits, produced in continuous relays by cooks sweating in the galley. *Santa Elisa*'s purser Jack Follansbee reflected ruefully on the $2,000 in cash he had left in a briefcase at the bottom of the lifeboat in which he had escaped from the ship. Worse still, hours before the sinking he had secured Captain Thompson's permission to secrete a bottle of whisky in each boat. Many times during the twenty-four hours after they were taken aboard the destroyer, Follansbee felt stabs of regret he had also forgotten to salvage the liquor: 'I sure could have used a shot.'

During air attacks, the survivors panicked unapologetically. When bombers approached from *Penn*'s port side, they rushed to starboard and vice versa. Men cowered and crouched, covering their ears, closing their eyes. Follansbee wrote: 'our nerves were almost completely shot, now. Nothing to do but lie in the corner of a torpedo-room and hold your ears to try to eliminate the whistle of the bombs and the deafening detonation of the guns.' A British survivor said likewise: 'we used to get into a big rugby scrum. I don't know why we done that. It was self-preservation, I suppose. I remember [*Penn*'s] cook standing there and saying soothingly: "It's alright, they won't get you. The old man knows what he's doing."' An admiring American passenger, Ensign Suppiger, described the warship's crew as 'at all times calm and uncomplaining'.

Twenty-eight minutes later, yet another bombing attack was beaten off, having achieved only the severance of *Rye*'s sweep wires. Dudley Mason's opinion was reinforced, that it was intolerable to ask his crew to remain aboard *Ohio* while she could not move. The seamen were once more taken off, dividing their number between *Penn* and the two motor launches. Swain called across to Mason from the bridge of *Penn*: 'You might as well come aboard, we've got hot cups of tea!' *Ohio*'s captain had somehow burned both hands, which were heavily bandaged. Thus encumbered, he quit his ship, lay down on a settee in the destroyer's wardroom and lapsed into a sleep of the dead. Meanwhile Eddie Baines' destroyer *Bramham* arrived on the scene, and between 2030 and 0100 plied a circular course around the little group of ships to maintain an asdic anti-submarine watch.

A significant mystery persists: why Air Vice-Marshal Park, on Malta, failed to provide continuous air cover over *Ohio* through that apparently interminable day of the 13th, which should have been within his force's range and capabilities. Once again, there was a failure of command and control: nobody seems to have highlighted to the senior airman the supreme significance of the tanker, the overarching priority of her cargo, above all for the continued operation of the RAF's squadrons. As it was, the Germans were able to mount four successive attacks as the ship lay unmoving, with only inneffectual interference from British aircraft. Many, if not most, of the travails of Pedestal were inescapable, given the precarious condition of British arms in the Mediterranean. It seems blameworthy, however, that *Ohio* and her naval nursemaids were for so long so ill-protected in the air.

Admiral Burrough, in *Ashanti*, was already leading what was left of Force X – *Kenya*, *Charybdis* and their escorts – on the first leg of its perilous passage back to Gibraltar. He gave a wide berth to the tanker and her consorts, to avoid their airborne persecutors, but signalled *Ohio*: 'Proud to have met you'. It seems possible, albeit unrecorded, that this message inspired less gratitude than cynicism. The departure of Burrough's ships, while the fate of *Ohio* – and indeed, at that time, also of *Dorset* – remained unresolved, was by no means uncontroversial.

Once again, a delicate balance of advantage had to be determined – the plausible loss of two more cruisers next day against preservation of the tanker, which was probably doomed. No course of action was definitively right or wrong, but those aboard *Ohio* and her attendant destroyers could be forgiven for thinking themselves abandoned to their fate.

The coming of darkness temporarily secured the cluster of ships from enemy aircraft. In lieu of Mason's exhausted crewmen, Swain sent aboard the tanker his first lieutenant, George Marten, with a party of seamen, to reconnect the tow. At 2220 she once more began to move, and an hour later was making four knots, with *Rye* towing forward and *Penn* behind her stern. Less than three hours later, however, at 0107 on 14 August, *Ohio* sheered violently, parting both hawsers. Yet another attempt was made, but, Eddie Baines noted succinctly, 'little progress was made'. At 0230 the tows parted once more.

The captains were acutely sensitive to the approach of daylight, now less than three hours away, and with it renewed enemy attacks. It is hard to overstate the difficulties endured by the desperately tired officers and ratings hauling lines, marrying cables and manoeuvring their vessels in full darkness, not daring to show lights, because the submarine menace never waned. Baines of *Bramham* concluded that his ship could do better service helping to move *Ohio* than by standing off to screen her. He persuaded his fellow skippers that *Penn* should secure to her port side, *Bramham* to starboard. Yet when they set about this, the much bigger *Penn* – seventy feet longer than *Bramham* – proved unable to do her part, hampered by the huge jagged steel plate, displaced by *Axum*'s torpedo, protruding from the tanker's hull.

At some point during the night, Dudley Mason appears to have returned on board. *Deucalion*'s Master, Ramsey Brown, had clambered across the decks and rails from *Bramham*. He was restlessly roaming *Ohio*'s deck when challenged by a shadowy figure, who demanded to know his identity. He answered, and counter-challenged, to receive the stern response: 'I am the captain of this ship.' Brown realized that Mason was less than pleased to find a fellow skipper on his deck, and returned to the destroyer, 'leaving the *Ohio* in the hands of her own Captain'.

Swain concluded that no more could be done until daylight came: the distorted steel, in effect a huge blade, might pierce his own ship's thin sideplates, with disastrous consequences. The ships held their positions around the motionless tanker, while crews snatched an hour or two's desperately needed rest. Few men forgot that the submarine threat persisted through every hour; that at any moment a devastating torpedo explosion might shatter the unnatural quiet around those unnaturally motionless ships. Exhaustion, however, prevailed over almost all other sensations.

The most significant activity of those early-morning hours was that naval ratings and some of the four hundred survivors aboard the destroyers boarded the tanker and looted her rich reserves of provisions and alcohol. Petty officer Reg Coaker of *Bramham* said: 'by jingo, we had hit the jackpot, [she] was well equipped with all kinds of stores like grapefruit, peaches, things we hadn't seen for ages. It was very much a matter of "well, she's in a very bad state, we don't know if we shall reach Malta," and so the lads were going aboard and we lived quite well.' Cigarettes, gin, chocolate, biscuits were passed back to the destroyers' mess decks, in quantities such as the Pedestal crews had not seen for weeks. Some men carried away trophies, metal fragments wrested from the two bomber carcasses on her deck.

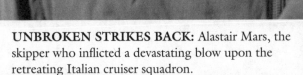

UNBROKEN STRIKES BACK: Alastair Mars, the skipper who inflicted a devastating blow upon the retreating Italian cruiser squadron.

Vincenzo Costantino, Second Trumpeter on board the *Bolzano*.

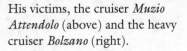

His victims, the cruiser *Muzio Attendolo* (above) and the heavy cruiser *Bolzano* (right).

A typical Pedestal bridge group, in this case, that of *Furious*.

The most terrifying station aboard every ship in the fleet and the most vulnerable: a warship's engine-room.

A bomb near-misses cargo-liner *Glenorchy*, seen from a warship's pom-pom mounting.

A merchant vessel's wheelhouse, with its almost nineteenth-century technology.

SOME PERSONALITIES: (above) Freddie Treves, Fred Jewett, and Anthony Kimmins broadcasting. (Left) Richard Onslow inspects some *Ashantis*. (Below) Syfret on his flag bridge; Charlie Walker; and Giorgio Manuti, who torpedoed *Manchester*.

(Above) *Nigeria*, listing and badly down by the head, after being hit and (below) a typical scene during the air battles. (Right) Renato Ferrini of *Axum*, who delivered one of the deadliest submarine attacks of the war.

Kenya being near-missed by bombs. This dramatic image was shot on the ship's westward passage to Gibraltar, but it shows the sort of scene that every ship faced almost hourly during the preceding days.

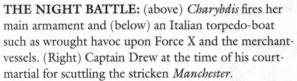

THE NIGHT BATTLE: (above) *Charybdis* fires her main armament and (below) an Italian torpedo-boat such as wrought havoc upon Force X and the merchant-vessels. (Right) Captain Drew at the time of his court-martial for scuttling the stricken *Manchester*.

(Above) An Italian Sparrowhawk plunges towards the sea after being hit and (below) *Dorset* during the ordeal by air attack that finally proved fatal.

AMERICANS: (top) Minda and Fred Larsen on their wedding day and (above) Larsen with Lonnie Dales after receiving their US decoration for Pedestal. (Left) Ensign Suppiger, the unhappy commander of the US Navy armed guard aboard *Santa Elisa*.

SOME MERCHANT NAVY CAPTAINS: (top left) Riley of *Brisbane Star*; (top right) MacFarlane of *Melbourne Star*; (bottom left) Mason of *Ohio*; (bottom right) Wren of *Rochester Castle*.

FORCE X: (above) *Ledbury* throws up a fine bow wave; (left) Harold Burrough, heavily stubbled after the Pedestal ordeal; (bottom left to right) Roger Hill of *Ledbury*, dressed for the Arctic; Alf Russell of *Kenya*; Eddie Baines and David Milford Haven of *Bramham*.

Ohio is hit – again.

Baines's *Bramham* stands by *Deucalion* during that ship's first misfortune. One of the boats in which some crew sought prematurely to escape is visible to her right.

Ohio completes her last, triumphant voyage.

A band plays *Melbourne Star* into Grand Harbour amidst rejoicing crowds.

Rescued survivors of many lost ships are assisted to disembark from *Ledbury*.

At 0420, an hour before dawn, *Rye* made a new attempt to tow *Ohio* from ahead with a ten-inch manilla rope secured to her sweep winch, while *Penn* attached herself to the tanker's undamaged starboard side. At 0503, however, yet again the rope parted. All tows were cast off, in readiness for enemy air attacks with the coming of daylight. One of the motor launches' twin diesels had been damaged during the previous evening's bombing, and this vessel now used its serviceable engine to creep towards refuge at Malta, carrying thirty-three survivors, who arrived safely. These included Denys Barton, the tanker's NLO, 'since I did not consider I could be of any further assistance in *Ohio*', a pragmatic but unheroic judgement. At 0745 *Rye* once more began to tow from ahead, with six hundred yards of cable attached to each of her two sweep winches, every man of the little flotilla conscious of the renewed presence of Axis 'snooper' shadowing aircraft.

Fifteen minutes later, *Ledbury* rejoined, following her night passage from the Gulf of Hammamet and triumph in shooting down two Italian torpedo-bombers. Roger Hill attached a cable from his own midships bollard to the tanker's stern in an effort to correct her chronic veer to port. He wrote: 'the chaos of wires, ropes and cables hanging down into the sea had to be seen to be believed'. The new arrangement, however, imposed too great a strain on *Rye*'s wires, which promptly parted. It is easy to sympathize with the frayed tempers and nerves on all the bridges as exhausted officers shouted between ships. Throughout the towing *Bramham*'s first lieutenant, David Milford Haven, worked tirelessly with a young rating named George Vernon, replacing parted springs – cables. Roger Hill was convinced that he had better ideas than did *Penn*'s captain about how to manage *Ohio*, but was obliged to defer to Swain, whom he had never before met, as the more senior captain.

The latter may have been unimpressed by the lavishly bearded Hill's urchin appearance, clad in trousers rolled up to the knee.

Whosever fault it was, if fault there could be, the tanker was once more at a standstill when, at 0900 on the 14th, the first air attackers of the day descended. Amid a noisy barrage from the warships, a bomb near-missed *Ohio*'s stern, carrying away her rudder and flooding the engine-room. The tanker began to settle by the stern. Spirits aboard all the ships sank to their lowest ebb, indeed the crews came close to despair. Behind the main stage of the drama, there were lesser cameos. Each time *Ohio* bumped the side of *Ledbury*, in the words of the destroyer's first lieutenant, 'two badly burned survivors in the sickbay, who were now suffering secondary shock, would start screaming and shouting that the ship was being blown up'. Survivor Jack Follansbee wrote: 'Malta suddenly seemed very far away.'

A sailor heard one Master, a merchant-ship survivor aboard the destroyer whom he did not identify, say despondently: 'Why doesn't the old scow sink, so we can all get ashore.' There were several stories, unconfirmed and unmentioned by Swain in his report, of merchant-vessel officers clambering up to his bridge and urging that he should abandon a futile endeavour. Two thousand miles away, the First Sea Lord might not have endorsed such counsel, but shared his subordinates' pessimism. From the Admiralty on 14 August, Sir Dudley Pound signalled to Churchill reporting the arrival at Malta of three merchant vessels, adding: 'there is a possibility that the oiler [*Ohio*] and one other ship [*Brisbane Star*] may get in, but chances are not good ... All forces appear to have played their part well, but the odds which concentrated against them were too heavy.'

Swain, on the bridge of *Penn* in the oil-heavy sea, brought out his portable gramophone, which began to broadcast by

tannoy through the decks the sunniest records he could find
– Glenn Miller's 'Chattanooga Choo-Choo', and its flip side
'Elmer's Tune', which were replayed many times before the day
was out. Jack Follansbee of *Santa Elisa* was disbelieving, hear-
ing the big-band sound between intermittent bomb blasts:
'Music? I must be getting crazy. Crazy as a bedbug. I listened.
It was music, all right.'

At 0915 the minesweeper *Speedy* arrived from Malta. Since
her captain outranked Swain, Hill and Baines, Lt. Cmdr.
Edward Doran thus assumed the direction of operations. A
column was formed, with *Rye* followed by *Ledbury* towing the
tanker from ahead, while *Penn* was secured to the starboard
side. An eternity, or rather a fortnight, earlier at Gourock,
Ledbury's officers had been crestfallen to hear that they were
bound for Malta. Tony Hollings, the destroyer's first lieutenant,
said: 'secretly I was a tiny bit disappointed – I had visions of our
opening a second front at Dakar … [Then] we drank a lot of
wine and ended by saying how grand it would be if we could
end by towing a ship into Malta – a strangely prophetic thought.'

Now this ramshackle procession began to creep painfully
eastwards. At 1030 Dudley Mason reboarded his ship to assess
her condition. He was accompanied by just six volunteers,
having discovered to his chagrin that the thirty-three men who
departed for Malta early that morning, aboard the damaged
launch, comprised the bulk of his crew. He explored *Ohio*, then
reported back to the naval officers. The sea was still pouring
into her damaged port side, he said, forcing up kerosene which
swilled and swirled in a treacherous film across her main deck:
gaseous fumes almost overwhelmed the boarding party. The
tanker was down by the stern, drawing forty feet, showing that
she had subsided substantially deeper in the water since the
previous evening. Hoses coupled to pumps aboard *Penn* were

battling against the flooding in the engine-room, but they were losing the fight at the rate of six inches an hour.

Mason nonetheless reached an important positive conclusion, which surely strengthened the resolve of the Royal Navy's officers on the scene. His ship was descending towards the sea bottom, but so slowly that, unless she broke in half, the hull should survive for at least a further twelve hours. Moreover, even if the stern section was lost, most of the vessel and 75 per cent of her priceless fuel ought to retain buoyancy. Indeed, losing the most damaged, distorted portion of the ship's hull could make the remainder easier to tow. The American designers, who had created the most densely compartmentalized vessel afloat, served Pedestal's cause brilliantly well. *Ohio* might yet make it to Malta.

A mixed party of naval ratings and merchant seamen from sunken ships now boarded the tanker, both to manage her tows and to reman her guns. Roger Hill called for volunteers aboard his own ship, and all the unwounded survivors agreed to go: 'I thought this was just about the bravest act I had ever known. If *Ohio* was hit she would go up even higher than the merchant-ships they had been on.' *Ledbury*'s gunner Charlie Musham led a party manhandling the link to the destroyer and organized the remanning of her Bofors. His men also joined the looting, appropriating for their own ship two Oerlikons to replace worn-out weapons, a megaphone, typewriters, sound-powered telephone instruments and all the food they could carry. Musham possessed himself of Dudley Mason's uniform cap. One of the tanker's officers later complained that, beyond the theft of his personal kit, he found even his bedclothes gone.

A handful of *Santa Elisa*'s American ship's company, including Fred Larsen and Lonnie Dales, were among the new gun crews, as was thirty-five-year-old bosun John Cook, who had been dragged from the burning sea only twenty-four hours earlier, after his ship *Waimarama* blew up. Also among the volunteers was *Santa Elisa*'s coxswain, who had behaved unselfishly in the water, mustering survivors and improvising a raft from driftwood tied with his belt: once aboard *Penn*, he had immediately offered his help in shifting ammunition. Meanwhile, carpenter Norm Owen was a survivor from *Deucalion*.

It was immensely to the credit of those who had already suffered such traumas that they returned to the fray on that Friday morning of 14 August, embracing yet again the risk – some would say, overwhelming likelihood – that the ship which they now crewed would explode into an inferno around them. But Lonnie Dales explained: 'My biggest reason ... for volunteering with Mr. Larsen ... was I felt much more comfortable having something to do than just sitting on the deck of *Penn* and watching the bombs fall.' Others assuredly felt the same.

That morning, the Malta RAF at last made serious efforts to establish an air umbrella, dispatching successive flights of fighters, each able to patrol overhead for twenty minutes, before being relieved to refuel. When the next air attack came, at 1044, sixteen Spitfires and several Beaufighters were circling. One of the former was flown by Geoff Wellum, who had been among the pilots who had ferried fighters to the island from *Furious*. Viewing the convoy's survivors from the air, he observed: 'It's becoming painfully obvious that they have had one hell of a battle and taken an awful pasting.' Far below one of the men on the ships, Fred Jewett, said: 'when we saw the Spitfires coming out we thought, God, we'd made it'.

Not quite. Although the fighters immediately engaged the nine black-painted Italian Stukas and their twenty-three-strong fighter escort, the disparity of numbers meant that many Axis planes were left free to do their worst. Several Ju87s broke through, one of which dropped a thousand-pound bomb just behind *Ohio*'s stern, twisting her propeller and tearing yet another hole in the hull. Hill, on *Ledbury*, said: 'it was horrible to be secured each end, moving at two knots and quite unable to dodge'. He shouted to his own men on the bridge and at the guns 'Lie down!' as a fourth attacker's load seemed destined to explode on their deck. Instead two oil bombs – essentially, large incendiary devices, perhaps a variant on the earlier German *Flammenbomben* – splashed into the sea beside the bow, drenching Hill's men with water and oil when they exploded. Charlie Musham emptied six 20mm Oerlikon drums at the attackers, while fifty other guns on the tanker and the warships also put up a barrage. One man in the crew of *Ohio*'s Bofors, now laid and trained by Larsen and Dales, was employed to dip buckets in the sea, then repeatedly to douse the red-hot barrel with water.

Two Spitfires caught a Stuka as it dived. 'I overtook it rapidly,' wrote one of the pilots, Sqn Ldr Tony Lovell, 'opened fire at 300 yards and broke away at 30 yards. I saw strikes all over the engine and fuselage. White smoke poured from both sides. He lost height, smoke stopped, and he did a steep turn to port and flew west, losing height.' As Lovell turned back towards *Ohio*, he saw the plane crash into the sea, sharing credit for its destruction with his New Zealand wingman. The Italian pilot, Captain Antonio Cumbat, was fortunate enough to survive the ditching and clamber into his dinghy with his wounded gunner, Michele Cavallo, who lapsed into semi-consciousness. For four hours they remained alone on the sea under the usual blazing sun, until at 1430 Beaufighters appeared overhead. Their British pilots believed that they had spotted one of their own crews, shot down during the previous day. Three planes went home, leaving one circling the castaways … which then found itself attacked by Spitfires. The fighters mistook the 'Beau' for a Ju88, causing it to dive hastily away. More aircraft then arrived, to orbit Cumbat's dinghy, pending the arrival of rescuers.

This represented an outbreak of humanitarianism, which would not have taken place had the RAF known that the survivors belonged to the enemy. Towards last light, a Dornier flying-boat arrived. Escorted by six fighters, it landed on the sea unimpeded by the RAF aircraft, which received radioed orders from Malta to leave the rescuers alone. Cumbat and Cavallo were picked up, having survived their twelve-hour ordeal.

On 14 August, flying a new aircraft after wrecking the engine of his own Reggiane in the previous day's battles, fighter pilot Giacomo Metellini was sent from Sicily to Pantelleria to escort a flight of three Ju88 torpedo-bombers, tasked with attacking *Ohio* and her escorts. Metellini failed to find the Ju88s, made a sweep towards Malta, then returned to base without encountering friend or foe. All three Reggianes which did participate in the attack were shot down by Malta-based aircraft, and only one pilot – Lt. Pocek – was rescued from the sea by a German floatplane.

The lost leader of the flight, Major Luigi Scarpetta, was described by Metellini as more notable for courage than judgement: 'he lacked experience of the cruel and challenging battlefield that was Malta. Before taking off he liked to enthuse us by saying "Come on boys, let's go whack the English!" We who were more experienced, responded "Major, let's go carefully. It's the English who are likely to whack us."' Metellini described his squadron as 'decimated' in the Pedestal battles, so that it had to be withdrawn from operations to replace the lost crews and repair damaged aircraft. He characterized his weeks of operations against Malta and the August convoy as 'the most testing of my entire career as a fighter pilot'.

The Italian airmen had delivered what proved to be the final Axis air attack on ships of Pedestal, though the British were denied this comforting foreknowledge. The blast of the morning's last Stuka bomb had thrust *Ohio* forward in the water, yet again severing her tows. Now *Bramham* secured to the tanker's port side while *Penn* remained on the starboard, with *Rye* towing ahead and *Ledbury* screening. The minesweeper's cable soon broke, but the other ships discovered that they could make good headway without her. The two destroyers assumed responsibility for hauling *Ohio* into Malta, knowing that they were engaged in a race between their eastward progress and the tanker's descent towards the bottom.

She was taking in more water by the hour, even with every accessible warship's pumps working furiously to keep her afloat, hoses snaking from both sides across her deck and down her companionways to the flooded machinery spaces. Roger Hill observed that the afternoon seemed interminable. On the warships it was deemed too hot to drink anything save lime juice. Aboard *Ohio*, however, in the absence of higher authority some men attacked her rum: one rating was found to be too drunk to stand. Charlie Musham, implausibly, donned a paper party hat that he found in a locker. Gunner Ron Linton, lately of *Dorset*, quit his station at the port bridge Oerlikon and found a bunk, on which he slumbered unmoving through the ensuing six hours. Larsen and Dales, still at their Bofors, burst into song to alleviate weariness and boredom.

Ledbury's doctor suggested an issue of Benzedrine, to keep key men awake, but Hill demurred, wary of the drug's after-effects. Hill wrote: 'we longed and longed for darkness when the enemy air attacks must cease'. It was a boundless relief that the light began to fail without any further sign of Axis planes: two Luftwaffe Ju88s spent the evening searching in vain for *Ohio*, before landing back at Grottaglie in southern Italy in full darkness at 2230. The warships' crews were told that all save telephonists, asdic and radar-operators, lookouts and suchlike could sleep at their action stations. One of Hill's sub-lieutenants lapsed into unconsciousness while standing upright at the chart table. A Malta-based minesweeper that was then nowhere near the scene signalled almost hysterically to naval headquarters in Valletta: 'situation deteriorating, *Ohio* sinking slowly'. Hill, seeing an intercept of this message, dictated an enraged contradiction: 'situation has never been better since *Ohio* was first torpedoed', but then wisely told his signallers not to send it.

By 2030, the awkward little cluster of ships had crept to a position south of Malta. They needed only to round its coastline to the eastern side, where lay Grand Harbour. *Ledbury*'s crew members returned from their brief secondments to *Ohio*. Slumbering sailors sprawled in duffel coats, using mates' legs or backs as pillows. Their captain wrote: 'Through the growth of beard and sunburn, their faces looked so young and peaceful. I felt a great surge of affection and pride for what they had achieved ... Empty brass cartridge cases were all over the place and the survivors had spilled out from the messdecks and wardroom onto the deck around the funnel.'

Progress was once more impeded, however, by a black comic incident. The garishly painted old paddle-driven tug *Robust* arrived on the scene and was ordered by *Speedy* to pass a line to *Ohio*'s bow and assist towing. Just ten minutes later, in darkness the tanker overran the tow and veered to port, dragging

Robust around with her at the extreme end of a cable too taut to be slipped. The tug smashed into *Penn*'s side, driving in the side of the wardroom where Thompson, Mason, Follansbee, Suppiger and other merchant-vessel officers were eating vegetable soup, roast beef and baked potatoes. *Penn*'s affronted captain ordered the unhelpful little vessel to retire forthwith to Malta.

The British were fortunate to be spared further attention from enemy torpedo-boats through many hours when *Ohio* and her accompanying warships were vulnerable. The night of the 13th had passed without further E-boat or MAS boat attacks, because following the boats' exertions twenty-four hours earlier only two still had torpedoes and were available for further operations: the engines of both these craft faltered before they could engage any British ships. A further attack was ordered against *Ohio* on the night of the 14th, but the German 3rd Flotilla's log recorded laconically and enigmatically: 'no enemy contact was made'.

All through the hours of darkness, small dramas interrupted the dogged progress of the flotilla. A petty officer on *Bramham*, Reg Coaker, claimed to have heard Dudley Mason shout across to the destroyer's bridge around 0300, abusing her exhausted captain, Ed Baines, saying: 'Hey over there! You! You've taken my ship from me! I want my command, and I want my ship back!' Coaker commented: 'To me it appeared as if he'd hit the bottle and quite frankly you can't blame him, can you? His behaviour led me to believe he was absolutely fed up. Mason's function was done. He'd done a good job getting the tanker to where it was.' The naval officer made no response, and Mason again disappeared below. Although there is no other witness support for this alleged incident, it seems believable. If indeed it took place, Mason deserves sympathy, given the terrors and stresses he and those on the accompanying ships had endured during the past days and nights. *Ohio*'s captain, together with chief engineer James Wyld and mate Doug Gray, now remained aboard the tanker for the last phase of her extraordinary voyage.

Ohio's stern continued repeatedly to swing to port, with a violence that threatened to impale the ship on the mines of Malta's defensive field. *Ledbury* assumed responsibility for pushing the tanker back on to her course, and every twenty minutes dropped a depth-charge as a 'scarer' to deter enemy submarines. Then, as they rounded the last promontory before Valletta, there was a final spasm of drama: shore batteries mistook them for hostile warships, switched on searchlights, opened a barrage which prompted a stream of enraged expletives on the warships' bridges. Neither firing flares in the colours of the day nor illuminating recognition lights stemmed the shooting until Roger Hill lamp-signalled the gunners with understandable intemperance: 'For Christ's sake, stop firing at us!'

Just after dawn Valletta's assistant King's harbourmaster – a wonderfully imperialistic title – came alongside *Penn* in a steam picket boat before climbing to her bridge with a senior pilot. The ship's officers were still on their feet only thanks to the adrenalin rush of their achievement, amid an exhaustion that afflicted every man aboard the five ships of the little convoy. A tug passed a wire to the stern of *Ohio* to assist in straightening her passage through the narrow harbour entrance. Operations were yet again briefly halted while another tug connected a cable to her fo'c'sle, enabling her to pass through the narrow gate in the defensive boom at Bighi Bay. Then, in stately procession, they crept onwards into Grand Harbour. Roger Hill of *Ledbury* wrote: 'It was the most wonderful moment of my life. The battlements of Malta were black with people ... It was the most amazing sight, to see all these people, who had suffered so much, cheering us.'

Sir Dudley Pound signalled to Churchill from the Admiralty about the successful towing of *Ohio*: 'Considering position and extreme value of cargo, weakness of enemy's air attack and complete absence of surface or submarine attack is remarkable. Seems probable Axis air forces had suffered such heavy casualties unable to stage or unprepared to risk full-scale attack.' In truth, failures of command and control by Axis air chiefs, matching those of the British, seem the most plausible explanation for their failure to coordinate strikes against the tanker. The Italians messaged their submarine *Asteria*, pinpointing the presence of 'a tanker under tow', then thirty-five miles from Gozo lighthouse, which *Asteria* was instructed to 'attack with utmost resolution'. The boat was spotted, however, bombed and damaged by British aircraft, and never caught sight of *Ohio*. It would have been unbearable for the ship's saviours to lose her at this last gasp.

John Jackson, the wireless-operator rescued from the sea by Freddie Treves after the sinking of *Waimarama*, and still aboard *Ledbury*, recorded: 'After the hell of the past three days, it was unspeakable joy to see the entrance ... The Fourth Officer of *Melbourne Star* was sitting opposite me ... I saw his face as he caught sight of [his own ship], and he burst into tears ... It was as if he had seen a ghost. A band was playing and crowds were lining the quayside, cheering. It was real *Boy's Own* stuff. You actually felt you had done something.'

'All ship's officers', wrote NLO Denys Barton in his report, 'were unanimous in their opinions that no ordinary ship could have withstood the amount of damage sustained, and remained afloat.' One of *Ohio*'s men wrote of their arrival with a wry understatement that was perhaps unintended: 'It was a real thrill, and almost made the trip feel worthwhile.' At 0945 on Saturday 15 August 1942, the tanker berthed in Grand Harbour and within minutes began to discharge her priceless oil. It was the Feast of Santa Marija, celebration of the Assumption of Our Lady into Heaven, one of the foremost events in Malta's religious calendar, because she was the island's Patroness. Operation Pedestal was completed.

13

Grand Harbour

1 BERTHING

Soon after *Ohio* tied up, uncoupled from the pumps of her supporting destroyers, the broken-backed tanker's bottom slumped on to the harbour bed, a few feet below. This no longer mattered. The ship had fulfilled her destiny, passed into maritime legend. Her hull's strength was a manifestation of the technological and industrial genius of the United States, her builders, which would become a dominant force in allied victory in the Second World War. Only around 15 per cent of the contents of the ship's tanks had been lost – 464 tons of kerosene, together with 1,705 tons of diesel.

The ship disgorged on to grateful Malta 1,430 tons of kerosene, 8,695 tons of aviation fuel, 902 tons of bunker fuel, 2,000 gallons of lubricating oil. On the surrounding shore, vile-smelling chemical smoke dischargers began to lay a screen above the newly arrived ships, in anticipation of Axis air attacks. Aboard *Penn*, seaman William Wilkinson gazed upon the shambles of Grand Harbour. Rather than feeling safe-havened, he was seized by impatience to sail away: 'it was just one mass of sunken ships. The floating dock was sunk, there were ships all over the place sunk. We wondered how long we was going to be there and if we was going to be sunk.'

Yet the Luftwaffe, to the amazement of the British, did not come. Ever more of Kesselring's shrinking squadrons were required to protect Axis supply convoys to Rommel in North Africa via Crete. Even during August, month of Pedestal, II Fliegerkorps committed 565 bomber and reconnaissance sorties to the protection of its own shipping movements to North Africa, as against 272 missions against British vessels. X Fliegerkorps in North Africa launched only twenty-five anti-shipping strikes that month, compared with 640 sorties flown in defence of its own sea routes.

Malta had become better armed to defend itself, especially when a new contingent of twenty-nine Spitfires, once again flown off *Furious*, reached the island on 17 August. After the losses of the Pedestal battles and inexorable operational attrition, II Fliegerkorps mustered just fifty-seven serviceable Ju88s, sixty-three twin-engined fighters, of which only forty-one were serviceable, and forty-five single-engined fighters. These numbers continued to decline thereafter. After a last big 26/27 August night attack, few more followed. While in July Malta had been the objective of 180 Axis sorties, in September, there were just sixty. There was one further spasm of serious bombing in October, but Kesselring recognized that he had lost the strategic battle to neutralize the island.

The offload of the Pedestal cargoes thus continued uninterrupted. Some men of the warship crews overcame weariness to venture ashore, where they discovered that a bottle of beer that once cost fourpence now sold for ten shillings, though the convoy delivered sufficient beer to issue each man of the garrison with two litres. A queue of ravenous soldiers formed at ships' gangways to eat the unwanted suppers of sailors who had gone ashore. In the blitzed streets, John Waters of *Dorset* marvelled at the sight of exuberant Maltese children dancing around him and his mates, shouting 'convoy, convoy'. Women kissed sailors' hands, though the increase in their rations made possible by the arriving ships was tiny, and remained so for a further three months. *Empire Hope* steward Jim Parry said of the local people: 'they were so pleased, you can't get over the feeling when you see the ruins and everybody wants to shake your hand … The whole place was charged with emotion.'

Dudley Mason wrote into *Ohio*'s log a formal pardon for the seaman who had been 'crimed' before leaving the Clyde for throwing dishes overboard, 'owing to his subsequent good behaviour'. Next day Fred Larsen received a cable from his sister via the Grace Line, reporting the safe arrival in the United States, from Norway via Portugal, of his little family: 'Minda and son are in Brooklyn with us they are well.' David MacFarlane was astonished to receive a visitor on *Melbourne Star* who greeted him by saying 'Hello, Uncle Mac'. The Master had no idea that a nephew, Johnnie Mejor, was completing a tour as an RAF Spitfire pilot on Malta, indeed had overflown the convoy a day or so before.

The pay of many British merchant seamen was stopped from the moment their ships sank, and this harsh rule was applied to some Pedestal survivors. Jim Parry of *Empire Hope* was awarded a mere £14 'and a few shillings' in compensation for the loss of all his clothes and personal possessions, 'and for that you lose your socks, your slippers, overcoat, raincoat, suits, ties, underpants, handkerchiefs. People don't know just how much you have with you.'

Following the return to Gibraltar of Syfret's Force Z, men granted shore leave flocked to the Royal and the UV, the Rock's favoured ratings' pubs, or settled down to play housey-housey

– bingo – at the Almeida Gardens. There were three cinema shows a day, all packed out. A few bold spirits donned civilian clothes and slipped into Spain on shopping trips, though court-martial awaited any man detected on such an outing. Garrulous yarn-spinners held forth about their experiences of the previous week, and many even told the truth. Chris Gould of the destroyer *Lightning* wrote to his fiancée: 'Now that we are all together, some of the tales are incredible.' Seaman Reg Gunn of *Nelson* told his family at home: 'I've come to the conclusion that a boat on the Serpentine is more in my line than being in the front line with the Navy. Still, <u>strictly in retrospect</u>, I'm glad I was there.'

George Blundell, also of *Nelson*, wrote in his diary on the 16th: 'A big day. All the commanding officers in the Fleet came on board for a conference. Most of us feel depressed by the party. Operation "M" for Murder, we call it. "The Navy thrives on impossibilities," said the BBC. Yes, but how long can it go on doing so?' Admiral Syfret's report to the Admiralty, delivered on 25 August, concluded: 'we are disappointed at not doing better but we should like to try again'.

A flag officer of the Royal Navy, emerging from a desperate battle, could scarcely fail to express such sentiments, nudging bravado, to emphasize his unshaken resolution. Yet privately Syfret, like most of the men who had fought ships through to Malta, knew that the Royal Navy could not readily repeat Pedestal. From September, Arctic convoys to Russia would once again assume priority, and for weeks thereafter Operation Torch, the allied invasion of North Africa scheduled for November, would absorb much of Britain's available warship strength. Syfret's losses strengthened the conviction of America's principal warlord, Gen. George Marshall, that it would be mistaken to include in the Torch plan an ambitious landing-place as far east as Bône in Tunisia, where Sicily-based Axis air power could be deployed against the invaders.

In the Pedestal post-mortems at Gibraltar, senior officers recognized that the destroyer anti-submarine screen had experienced mixed fortunes. The surprise achieved by *U-73*'s attack on *Eagle* was blamed on the complex hydrological conditions prevailing in the western Mediterranean, which confused asdic. Most of the subsequent losses were attributable to the hazards of the passage: it was almost impossible to escape scot-free from wolfpack attacks, such as the fleet had faced. Syfret and Burrough could point out that while *Axum* inflicted devastation on the evening of 12 August, earlier that day five other Italian submarines had been sunk, damaged or driven off. *Axum*'s captain enjoyed extraordinary luck, almost unique in the course of the war, in hitting three important ships with four torpedoes. It seems less easy to explain the sinking of *Waimarama* and *Dorset* by Axis aircraft, together with almost fatal damage to *Ohio*, within a hundred miles of Malta, where it should have been possible to provide effective fighter cover. The failure to do so was partly due to lack of voice radio fighter-direction from the warships, but this is an inadequate explanation. The RAF's commanders and controllers ashore did not distinguish themselves, and failed to make best use of the fighter resources they had.

In that formal era, the warship captains' official narratives, submitted to the admiral commanding, opened with the words 'Sir, I have the honour to report as follows …' and concluded 'I have the honour to be, sir, your obedient servant'. *Victorious*'s captain asserted that he did not believe the loss of fourteen (in reality sixteen) Hurricanes aboard *Eagle* 'materially altered the result … The conclusion is that carriers should not be brought within range of heavy attack by shore-based aircraft, particularly if escorted by fighters.' He said that it was no solution to cram flight-decks with additional aircraft because that merely slowed the rate of take-offs and landings.

Alf Russell of *Kenya* wrote: 'It cannot be too strongly emphasised that any ship in an escort for an operation of this nature must be fully fitted out with all the latest equipment. For a ship without RDF [radar] Type 283 for the four-inch armament, without RDF Type 282 for the pom-poms and six-inch close-range barrages to be exposed to dusk or night attacks by aircraft, makes the feeling of helplessness and ineffectiveness so great as to be a serious menace to morale for the Commanding Officer downwards ... apart from causing unnecessary expenditure of valuable ammunition.'

In short, warship anti-aircraft fire was seldom effective, and never became so even when the US Navy deployed its mighty 1945 armadas against the Japanese. Carrier-borne fighters remained the decisive weapons for defence of a fleet. Only in the Pacific in the last weeks of the Eastern war did the Royal Navy deploy a strong carrier force, and this performed indifferently, alongside the US Navy's task groups.

Lt. Cmdr. Tony Baines, who had been aboard *Santa Elisa*, was one of several Pedestal naval officers who urged that future convoys should be smaller – eight or ten ships. He also proposed that there should be a strong close escort of Hunt-class destroyers, while the heavier units – carriers, cruisers and fleet destroyers – should keep a distance 'with full freedom of speed and manoeuvre'. Such dispositions 'might prove less costly to the escort, and no more costly to the convoy'. This view was strongly influenced by the fact that four British cruisers and two carriers had been hit, while most of them – *Eagle* was at full speed, to fly off Hurricanes – were steaming relatively slowly in the

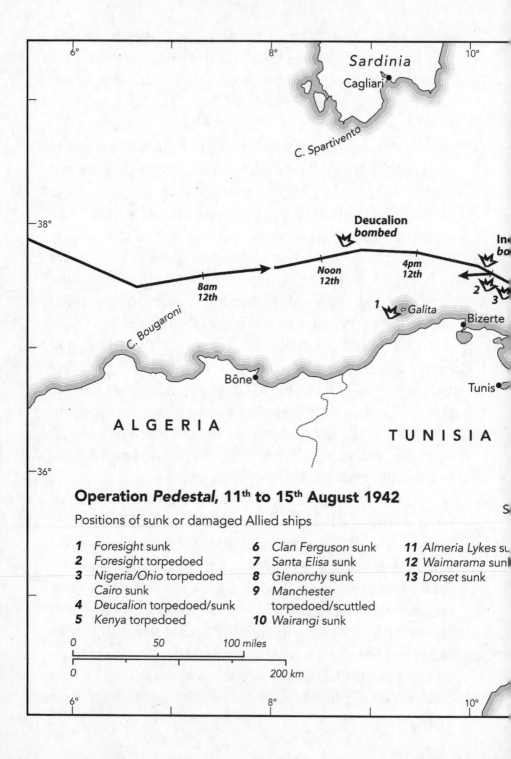

Sardinia

Cagliari

C. Spartivento

Deucalion
bombed

Noon
12th

4pm
12th

In●
bo●

8am
12th

2

3

1

Galita

Bizerte

C. Bougaroni

Bône

Tunis

ALGERIA

TUNISIA

Operation *Pedestal*, 11ᵗʰ to 15ᵗʰ August 1942

Positions of sunk or damaged Allied ships

1	*Foresight* sunk	**6** *Clan Ferguson* sunk	**11** *Almeria Lykes* su●
2	*Foresight* torpedoed	**7** *Santa Elisa* sunk	**12** *Waimarama* sun●
3	*Nigeria/Ohio* torpedoed	**8** *Glenorchy* sunk	**13** *Dorset* sunk
	Cairo sunk	**9** *Manchester*	
4	*Deucalion* torpedoed/sunk	torpedoed/scuttled	
5	*Kenya* torpedoed	**10** *Wairangi* sunk	

0   50   100 miles

0   200 km

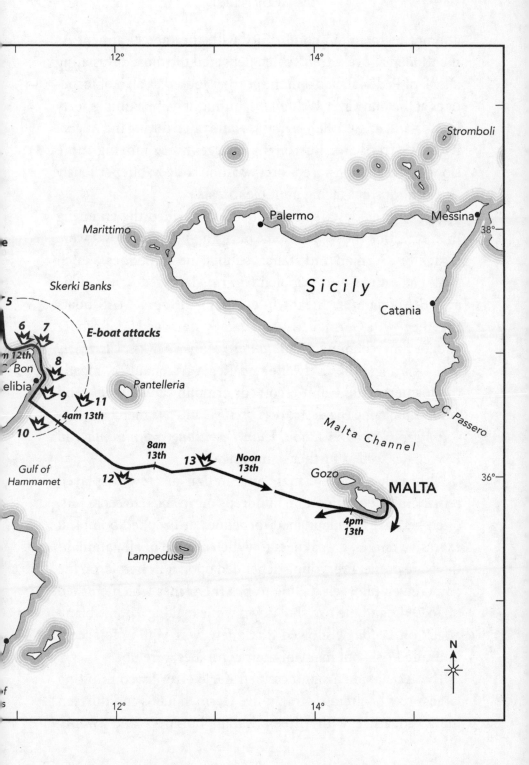

Stromboli

Palermo

Messina

Marittimo

38°

Skerki Banks

S i c i l y

5

6 7

E-boat attacks

Catania

m 12th

C. Bon

8

elibia

9   11

Pantelleria

C. Passero

4am 13th

Malta Channel

10

8am
13th

13

Noon
13th

36°

Gulf of
Hammamet

12

Gozo

MALTA

4pm
13th

Lampedusa

N

attempt to maintain conformity with the merchantmen. As mentioned above *Axum*, which delivered the most devastating attack of Pedestal, had a month earlier loosed a salvo of torpedoes at the minelayer *Welshman*, shifting at her maximum forty knots – and missed. Fifteen knots – and often during the August battles Syfret's and Burrough's charges were moving more slowly – was a high speed for a wartime convoy, but perilously sluggish for ships facing multiple threats.

Although Pedestal's air–sea battles provided the enduring memories for those who took part, the heaviest blows were struck by German and Italian submarine torpedoes, which sank *Eagle* and *Cairo*, crippled *Nigeria*, damaged *Kenya*, *Ohio* and *Brisbane Star*. Meanwhile Italian MS and MAS boats together with German E-boats fatally damaged *Manchester* and four merchantmen – *Santa Elisa*, *Almeria Lykes*, *Glenorchy* and *Wairangi*. The Luftwaffe and Regia Aeronautica, mostly the former, could claim credit for crippling *Indomitable* and fatally damaging the destroyer *Foresight* and five merchantmen – *Empire Hope*, *Deucalion*, *Clan Ferguson*, *Waimarama* and *Dorset* – as well as further damaging *Ohio*.

Although some German and Italian aircrew displayed remarkable determination, hundreds more failed to score hits, often because they launched torpedoes or dropped bombs at excessive ranges, a weakness to which airmen of all nationalities were prone. Everything depended upon how heroic, on the day, a given pilot felt willing to be. The planes which crashed on to *Ohio*, and the two shot down while making a point-blank attack on *Ledbury*, showed that a few were willing to accept unlimited risk, but many of their comrades were not.

The Axis made a significant strategic error by committing large forces of aircraft – thirty-five German Ju88s and thirteen Ju87s, together with fifteen Italian bombers and twenty torpedo-bombers – to attack Force X during its return passage to Gibraltar, without achieving a single hit. These aircraft could have been much more usefully employed against *Ohio* and *Brisbane Star*, which they should have been able to sink. Beyond the command misjudgement, it seems fair to guess that, by this late stage of the battle, many Axis aircrew were weary, and unwilling to take suicidal risks.

In the accountancy of total war, allied loss of life on Pedestal was small, given the ferocity of the action and number of ships that were sunk – just 457 men, more than a quarter of these victims of *Eagle*'s foundering. The modest 'butcher's bill' reflected the merciful character of the Mediterranean in summer, which spared thousands of distressed sailors, some of whom survived for hours in the water after their ships were hit. On similar Arctic or even Atlantic operations, men in life-jackets froze to death within minutes; in the Pacific they would probably have succumbed to sharks, or to the difficulty of recovering survivors across vast expanses of ocean. Moreover, had an angry sea been running, as can happen even in the August Mediterranean, the British would probably have found it impossible to save their crippled ships, above all *Ohio*.

Axis squadrons lost sixty-two aircraft during Pedestal, forty-two of these Italian and the remainder German. Forty-two were shot down by fighters – some of them Malta-based RAF aircraft – or ships' gunfire, the balance written off by ground strafing or accidents. Forty-five Axis aircrew were killed, together with forty-eight Italian submariners. The Fleet Air Arm lost thirteen planes in action and sixteen when *Eagle* sank, while the RAF lost one Beaufighter, five Malta-based Spitfires and a Sunderland.

The Axis trumpeted its triumph in what Mussolini's sailors and airmen dubbed *la battaglia di mezzo agosto*, the battle of mid-August: German and Italian newspapers feasted on the good news. The account in *Königsberger Allgemeine Zeitung* was headlined 'Voyage into the Sea of Death', and quoted a Wehrmacht source's assertion that 'all the tankers participating in the convoy have been sunk'. By lunchtime on 13 August, reported a Rome correspondent, more than two hundred British castaways had been rescued from the sea off the Tunisian coast.

On the 16th, said the journalist, the carrier *Illustrious* – a mistake for *Indomitable* – had limped into Gibraltar badly damaged. Axis papers reported that the USS *Wasp*, 'which cost $20 million to build and was one of the strongest and most heavily armed enemy ships', had also participated, and been crippled. German papers quoted London reports which 'stated that the English are facing an overwhelmingly superior enemy in the Mediterranean', and that further admissions of undisclosed shipping losses were expected from the Admiralty.

Supermarina's Admiral Arturo Riccardi wrote an effusive letter to the German high command on 15 August, paying tribute to the combined efforts of the Italian and German torpedo-boats. He asserted that 'the convoy is almost completely destroyed, and these important August days will go down in history as earnest of our unity of will and deed, to bring about a new order'. His extravagant words must have been intended in part to mask embarrassment about the Italian surface fleet's failure to play the decisive, conclusive role that it had been offered. As for the Italian submarines, Giorgio Giorgerini, author of their official wartime history, describes the massed concentration of boats into what the Germans called wolfpacks as 'representing a real revolution in [the Italian] use of submarines ... which brought about their first real successes in the Mediterranean ... and astonished the enemy'.

The German press embarked on an orgy of triumphalism about Britain's grievous Pedestal losses, though some of their reports were no more accurate than those of the British. The Germans mistakenly supposed that the USS *Wasp* was part of Pedestal, and failed correctly to identify the crippled *Indomitable*. The *Völkischer Beobachter* headline of 14 August reads:

German-Italian Naval and Air Forces on the Offensive

Huge Convoy Destroyed in the Mediterranean

46 ships stopped on their way to Egypt – apart from 'Eagle' 90,000 tons of shipping already sunk – the aircraft-carriers 'Furious' and 'Wasp' are badly damaged – further successes are in sight …

Shock in London – 'Extremely concerning'

The German after-action report revealed lingering doubts about the purposes of the British operation. Its author recognized that the absence of amphibious forces and special equipment ruled out the possibility that an opposed landing might have been intended, but continued: 'The size of the convoy and specific intelligence reports regarding vessels' cargoes indicate that it was unlikely to be a mere supply run to Malta.' While this might have been one objective, said the German document, given the imminence of Rommel's Egyptian offensive, the Admiralty might also have hoped to open a direct route to reinforce Eighth Army through the Mediterranean, rather than around the Cape: 'The British considered it worth committing significant naval elements to achieve this, and accepted the risk of large sacrifices.'

The Axis estimate of Pedestal ship losses added, beyond the reality, a further three destroyers allegedly sunk: 'despite the heaviest anti-aircraft fire and fighter defence, the [German and Italian air] attacks were successful'. The report added that, if the convoy's air umbrella had been strengthened by fighters flying off *Eagle* and *Furious*, the outcome might have been less favourable to the Axis. Though all these assessments caused Rome and Berlin to assert Pedestal's outcome as a victory, they also grudgingly conceded that sufficient stores had reached Grand Harbour to enable Malta to hold out for additional weeks, possibly months: 'the island can fulfil its role in disrupting German–Italian reinforcements to Africa at a possibly decisive phase of the North African campaign'. This was an important admission, restricted to the eyes of the German and Italian high commands, unavowed, of course, to Hitler's and Mussolini's peoples.

* * *

The Royal Navy was shaken by the losses of Pedestal, and asked itself, privately at least, whether the outcome justified them. An obvious conclusion was that even the deployment of the largest carrier force available did not suffice to provide adequate air cover. An Admiralty committee on the future composition of the Royal Navy reported on 18 November 1942 that the ideal British sea service of the future would be composed of twenty-two fleet carriers, each carrying forty-eight aircraft, twenty-four escort carriers, carrying twenty-four planes apiece, together with eighty-three auxiliary carriers; nine battleships; fifty cruisers; 191 destroyers and three hundred submarines. Winston Churchill could have told that committee: dream on. Only the United States possessed wealth to aspire to the creation of any such host. More realistic was the report's admission: 'The war has shown the extreme vulnerability of both warships and merchant-ships to air attack … The main role of the battleship becomes that of "aircraft-carrier support ship".

That assessment was drafted even as Montgomery's British Eighth Army advanced triumphantly, albeit falteringly, westwards, having defeated Rommel at El Alamein, and following Torch, the Anglo-American landings in French North Africa that began on 8 November 1942. Perversely, Pedestal exercised a marginal influence upon the Axis intelligence response, after the Torch armada had been sighted at sea. Back in August, German and Italian commanders had found it hard to believe that a fleet as large as that of Syfret was entering the Straits of Gibraltar with the sole purpose of relieving Malta: they suspected that a possible invasion of North Africa might be imminent. In a mirror reflection, when the Bodden Line monitors in Spain on 7 November reported the passage of the allied Torch fleet, for some important hours the Abwehr remained uncertain whether this presaged an amphibious landing or instead merely another desperate lunge to relieve Malta.

Posterity, and in particular students of the Second World War, remain divided in judgements upon Pedestal. American historian Vincent O'Hara has described the battle as 'the largest aero-naval victory won by Axis forces during the Mediterranean war'. He highlights the fact that while German air and naval units made an important contribution, the Italians inflicted most damage upon Syfret's fleet. British writer Simon Ball characterizes the convoy as a desperate throw which failed, by a Royal Navy at its last gasp in the Mediterranean. The operation, he writes, provided Malta with a bare sufficiency of supplies to subsist until November, when its salvation was secured not by British sea power but instead by Torch and El Alamein.

Martin Van Creveld asserts that the importance attached by many historians to the Mediterranean convoy battles is 'grossly exaggerated. At no time, except perhaps November–December 1941, did the aero-naval struggle in the central Mediterranean play a decisive part in events in North Africa.' Rommel's difficulties, argues Creveld, derived from his 'impossibly long' supply lines inside Africa. Likewise Correlli Barnett brands Malta a 'strategic burden and moral obligation glorified into a heroic myth'.

In assessing any campaign, it is tempting for a nation and its historians to view the difficulties from one side, and to suppose that all the misjudgements were made by allied commanders. Yet if Churchill attached excessive importance to Malta, the Axis also committed large, scarce resources – especially combat aircraft – to the Mediterranean air–sea struggle. By the autumn of 1942, the dominance achieved by Tedder's Desert Air Force over Egypt and Libya was a significant factor in

Montgomery's victory at Alamein. If the Luftwaffe had not felt obliged to commit, and to lose, so many aircraft in the summer's battles over both the British and Axis Mediterranean convoy routes, the balance of air strength above the desert battlefield might have tilted sharply in Rommel's favour.

In contrast to the sceptical view adopted above by O'Hara, Ball, Crefeld and Barnett, Dr Milan Vego of the US Naval War College wrote in a 2010 study: 'Operation Pedestal was in retrospect a clear operational success for the Allies ... Despite great odds, the Allied airmen and sailors displayed a superb fighting spirit. This was especially true of the merchant mariners ... The Axis forces did not accomplish their stated operational objective.' Karl Gundelach wrote in his authoritative history of the Luftwaffe in the Mediterranean: 'this was an outcome the British could not have dreamt of achieving in the spring ... An attack by the Italian fleet ... could have destroyed the rest of the convoy ... The balance in the central Mediterranean began to shift in favour of the British, who had plainly regained air superiority over Malta.' The Luftwaffe recognized that the accession of RAF strength on the island made it necessary to shift II Fliegerkorps to a defensive role.

The view adopted by such participants as George Blundell, Commander of *Nelson*, that Pedestal was a failure because so many ships had been lost seems mistaken. What was remarkable about the four-day battle was that the Axis was offered perhaps its best chance of the war to sink British capital ships, of which six were deployed including *Furious*, and accounted for only one, during battles in which the odds and the geography were strongly in the attackers' favour. It was indeed painful that nine ships of the Merchant Navy and US Merchant Marine vanished to the bottom. But five, including the vital *Ohio*, reached Malta, when more effective German and Italian direction of the sea and air campaigns could and should have completed the convoy's annihilation. Albert Kesselring makes no mention of Pedestal in his memoirs, and this seems significant. Had he seen anything to celebrate in its outcome, to embellish his own laurels, he would have said so. The Axis failure to prevent a narrowly sufficient portion of the convoy from reaching Malta was as significant in charting the shifting balance of advantage in the Second World War as was the British success in achieving this.

The operation may be viewed almost as a last hurrah, inappropriate as that word would have seemed to those who endured it, of Royal Navy battlefleets, as for centuries the world had known such fighting units. Only the United States Navy now had resources to build task forces big and powerful enough – though its April 1945 experience off Okinawa suggested, only just so – to defend themselves against shore-based enemy air forces. British losses in August 1942 chiefly reflected the equilibrium of strength and technology which existed between the warring nations at that point of the war, before allied superiority became explicit and eventually overwhelming.

Much that went wrong reflected the limitations of wireless links, a chronic weakness in all naval operations. Forty years later, many of the difficulties that beset the British campaign in the 1982 South Atlantic war were caused by communication failures. A senior naval officer told the author when it was over: 'I have seen commanders who understand each other perfectly face to face get into terrible tangles and rows when separated by ten miles, and communicating by signal.' So it was during Pedestal, only more so.

Sir Arthur Tedder, who after leaving the Middle East served as Eisenhower's deputy in the campaign for North-West

Europe, suggested in 1948 that warriors educating themselves for future conflicts should study not the victorious later campaigns of past wars but instead the early phases: 'There are no big battalions or blank cheques then,' he wrote, with the rueful wisdom of experience. Pedestal reflected this reality, of a battle fought at a time when the Axis still held some strong cards. This made its outcome all the more important and impressive, at a time when faith in Britain's will and power to fight was at its lowest ebb, especially in the United States.

Winston Churchill understood better than did most of his commanders that the moral issues at stake in the conduct of a war are quite as great as the material ones. No battle can be justly assessed by a mere profit-and-loss account of casualties, of tanks, aircraft, ships destroyed. Perception is also critical and often decisive. The British display of will, fighting the surviving vessels of Pedestal through to Malta despite repeated onslaughts and punitive losses, gave victory to the allied cause.

Churchill was still in Moscow on 14 August when he wrote to Stalin reporting this outcome. Assuredly, he took the trouble to correspond personally because he was still smarting from the Soviet warlord's insult a few days earlier about PQ17: 'your navy runs away'. Thanks to the arrival in Grand Harbour of three ships out of fourteen – as the prime minister then knew the story – 'the fortress, which is of great consequence to the whole Mediterranean position, can hold out until after both the impending battle in the Western Desert of Egypt and TORCH have taken place. For this we have paid a heavy price.' The prime minister rehearsed the list of warships sunk or crippled, piling on the agony in his determination to emphasize Britain's willingness to pour forth blood, to sacrifice tonnage, mentioning the 'nine or perhaps eleven fast merchant-ships, of

which we have not many left'. He concluded: 'I am of the opinion that the price was worth paying. The dangers of warships
operating amid all this shore-based enemy aircraft have had
another painful illustration … The Italian cruisers and battleships did not venture to attack the remnants of the convoy …
The enemy will no doubt proclaim this as a great victory at sea,
and so it would be, but for the strategic significance of Malta
in view of future plans.' Beneath these typed paragraphs, he
added in his own hand: 'Yours Sincerely Winston S. Churchill'.
After enduring so much at the hands of his Kremlin host,
however, he withheld any of the warm protestations of personal
friendship that sometimes accompanied his messages to Stalin.

Next day, 15 August, he returned to Cairo to be greeted by
Lord Gort, who had flown from Malta to report personally on
the battle and the wondrous late arrivals of *Brisbane Star* and
Ohio. Charles Wilson, Churchill's doctor, who was with the
party, wrote in his diary: 'The P.M.'s relief is a joyful sight. The
plight of the island has been distracting him. [Gort] is hardly
recognizable – stones lighter … years older, with sunken
cheeks and tired eyes'. The news the governor conveyed,
however, and his tale of the experience of Pedestal caused the
prime minister to shed tears such as often beset him in the face
of emotional tidings. He dispatched a stream of euphoric
signals, congratulating the Royal Navy on what he called its
'magnificent smash-through to Malta'. He messaged the First
Sea Lord: 'Prolongation of life of MALTA was worth the heavy
cost.' He took the trouble to send a personal 'WELL DONE'
signal to *Penn*, *Bramham* and *Ledbury* for their decisive contributions to saving *Ohio*.

The prime minister understood better than did some of his
admirals that warships existed to fight, and if necessary to
sink, in pursuit of national purposes. It was British acceptance

of this imperative, contrasted with Italian rejection of it, that made Mussolini's fleet an object of scorn despite the personal courage of many of its seamen. On 10 November, at the Mansion House in London, Churchill delivered his great speech applauding the victory of Montgomery's Eighth Army at El Alamein, and asserting memorably: 'Now, this is not the end. It is not even the beginning of the end. But it is, perhaps, the end of the beginning.'

British forces thereafter drove west, to liberate sufficient of the North African shore to make it much easier to provide the air cover necessary to succour Malta. While the supplies delivered by Pedestal in August had sufficed to avert the island's surrender, it was a November convoy, profiting from powerful North Africa-based air support, which delivered the Maltese people from privation and finally lifted the siege. Though the North African campaign continued until final Axis surrender in Tunisia the following May of 1943, the tide of the Mediterranean war had plainly turned by the end of the previous year. Moreover, the Soviet victory at Stalingrad, where Paulus' German Sixth Army surrendered in January 1943, removed the last realistic possibility of Hitler winning the Second World War.

By the time of Italy's surrender in September 1943, the war had cost its merchant fleet 1,324 vessels lost, totalling just over two million tons. Of these, British submarines accounted for 810,093 tons, air attack in ports 378,058 tons, air attack at sea 396,181 tons, surface action 122,791 tons; other causes (mines etc.) 399,398 tons. This compared with British total wartime merchant shipping losses of twelve million tons.

Operation Pedestal did not demonstrate the ability of a fleet of the Royal Navy to defy powerful and skilful enemy air and submarine forces in the confines of the inland sea. Contrarily, indeed, it highlighted its vulnerability, especially that of cruisers, together with the limitations of available British technology for shooting down aircraft and repelling submarines and even mosquito craft. It is also true that if the land campaign in North Africa had not developed so rapidly to the advantage of the allies in the last months of 1942 Malta would once more have been threatened with starvation, and even Churchill at his most bellicose would have been unlikely to demand a repeat of Pedestal. But this is a counterfactual such as appeals to the sort of historians who assert crossly that the Royal Navy's huge, bloody August 1942 fleet effort did not deserve to work. It did.

John Pudney was a poet whose sentimentality commands little respect among twenty-first-century literary pundits, but who enjoyed immense popularity during the war years for his evocation of realities that were intimately familiar, profoundly sympathetic, to his readers. They were exemplified by such lines as these in 'Convoy Job':

Convoy the dead:
Those humble men who drown,
Dreaming of narrow streets, of alleys snug
In lamplight, love in a furrowed bed,
Pints in a *Rose and Crown*.

There is a pathos in the names, ages and descriptions that fill the casualty lists of Pedestal, for instance this random selection from among the 162 men who died with *Eagle*: Baldwin, Edward George (26), Shipwright 3rd class; Goulding, Thomas James (35), Engine-Room Artificer 2nd class; Venvell, Thomas Edwards (44), Chief Painter; Wead, James Richard (21), Leading Stoker; Smith, Thomas Hill (26), Shipwright 3rd class;

Lett, Ronald Frederick (22), Electrical Artificer 4th class; Judd, Noel Arthur James (19), Stoker 2nd class; Vaggers, Denys John (18), NAAFI Canteen Assistant.

Likewise, a few of the eighty-three merchant seamen who perished aboard *Waimarama*: Atkins, Charles Frederick (21), Third Refrigeration Engineer Officer; MacLeod, Allan (52), Boatswain; Meaghan, George Thomas (47), Greaser; Phillips, Stanley (15), Deck Boy; Roberts, John (52), Lamptrimmer; Suleman Dawood (51), Donkeyman; Willett, John Raymond (20), Sailor; Pearce, Robert Strasenburgh (54), Master. These were typical in their years, roles and tragedies of the tens of thousands of seamen who paid the 'price of Admiralty'.

A time was close at hand when the role of Britain as the principal antagonist of fascism would be eclipsed by the two superpowers, as the United States and Soviet Union would be dubbed, which would play the dominant roles in bringing the Second World War to a tolerably successful conclusion. Pedestal was almost a last earnest of Winston Churchill's unflinching fortitude in adversity, his refusal to yield in the face of setbacks and indeed disasters, before the coming of better times. The operation did not demonstrate the Royal Navy's command of the Mediterranean, but became instead an epic of warrior virtues, displayed by a few thousand men, from the prime minister at the apex to those who sailed aboard *Ohio*, *Penn*, *Bramham*, *Ledbury* and their kin at the base. Loyston Wright, captain of the destroyer *Derwent*, wrote proudly about his own crew: 'in all respects they conducted themselves in a way that reflects well upon the Royal Navy's health and happiness'. Such men redeemed from the brink of disaster one of the most hazardous naval operations of the Second World War.

2 HONOURS AND OBSEQUIES

After the battle was done, it is likely that Syfret and Burrough had plenty to say privately about their ordeal, including strictures upon certain officers and ships, both of their own service and of the Merchant Navy. But a decision was made, in keeping with the Royal Navy's code of conduct, that little dirty washing was to be laundered outside the Admiralty. With the important exception of the loss of *Manchester*, about which more below, the written reports on the operation omitted its discreditable episodes. In the list of decorations awarded, Cdr. Arthur Venables, the convoy commodore who attempted to flee, was awarded the DSO, alongside warship and merchant-ship captains deemed to have conducted themselves well, such as the wholly admirable Fred Riley. After Pedestal, the truth was again demonstrated of the old saw that 'the only man who knows what a decoration is really worth is he who receives it'.

Stern measures were adopted to impose secrecy about the events on *Brisbane Star* off Sousse – what was, essentially, an attempted mutiny. Whereas several naval liaison officers whose ships reached Malta were decorated, and an exceptional five of *Brisbane Star*'s officers received DSCs, her NLO got nothing, which lends plausibility to the reports that this officer failed to match the resolution shown by the Master. A few men on other ships displayed a comical lack of self-knowledge about their own omission from the awards list. A Royal Navy signalman, conspicuously absent from his post on *Rochester Castle*'s bridge at moments of crisis, complained bitterly to shipmate Jack Harvey, who told him contemptuously that he was lucky not to have been court-martialled: 'the yellow streak was deeply ingrained'.

The only two Americans to receive Distinguished Service Medals were Fred Larsen and Lonnie Dales, survivors from *Santa Elisa* who afterwards manned guns on *Ohio*. Three British gunners, obviously blameless for *Almeria Lykes*' scuttling, received Mentions in Dispatches. On 5 September, the latter ship's junior engineer Henry Brown, who had behaved better than his crewmates following the torpedo strike, jumped from a New York hotel window, killing himself. It is impossible to guess what stresses might have driven him to such an act following the collective disgrace of the ship's company.

It seems a mistake that the Admiralty allowed, far less required, ill-motivated American merchant seamen, still new to the war and unfamiliar with the British, to risk their lives to save Malta, to participate in such a desperate venture as Pedestal, especially after US merchant shipping losses had been so heavy on PQ17. The Anglo-American alliance became a remarkable working partnership, but in August 1942 mutual trust was lacking at the summits of command. How, then, could it exist in the bowels of American merchant vessels, crewed by civilians? It is striking that the only seaman on either *Santa Elisa* or *Almeria Lykes* who displayed animosity towards the Axis was the Norwegian-American Larsen.

The Second World War was the first conflict in which British Merchant Navy personnel became eligible for armed-forces decorations. Although they constituted only a small proportion of those who sailed to Malta in August 1942, almost one-third of all awards made for courage or distinguished service on Pedestal – 151 out of 507 – were presented to them, and justly so. Most of what befell the warship crews was in the familiar line of duty. The men of the cargo vessels, however, endured one of the most gruelling experiences of their service's war. Riley, MacFarlane and the other successful captains were among the first-ever recipients of DSOs, and few could have been more worthy of them.

As for the Royal Navy, as always a disproportionate share of awards went to officers, and especially captains – the men who made the decisions, directed the fighting of their ships. This practice was rational, but bore hardly on those who kept the warships afloat, manned guns, sustained machinery and technology. It is hard to imagine a more grim and terrifying role in which to serve the King than within the engine-room of one of his ships. Yet relatively few such men received more than a Mention in Dispatches: it seems dismaying that, for instance, the engineers on *Indomitable* who laboured so hard to restore power and order after the carrier had been crippled went almost unrecognized.

Neville Syfret was knighted for his command of Pedestal, and continued to command Force H until in 1943 he became vice-chief of the naval staff, first under Pound as First Sea Lord then, on the latter's death, under Cunningham; he died in 1972.

In September 1942, St John Tyrwhitt, commanding *Tartar*, faced a court-martial for sinking *Foresight* unnecessarily, having allegedly made insufficient effort to save her. He was honourably acquitted, which was the only plausible verdict. The crippled destroyer might have been towed to Gibraltar in the absence of the enemy, but not under renewed air attack.

The case of the lost *Manchester* was far more contentious. In February 1943, following the liberation of those of her crew interned by the Vichy French, a court-martial hearing at Portsmouth addressed the conduct of Harold Drew and his officers after the cruiser was torpedoed off Cap Bon in the early hours of 13 August. Its findings were brutal. The captain was severely reprimanded for dishonourable failure, 'in that he negligently performed the duty imposed upon him ... whereby he gave orders on 13 August 1942 for His Majesty's Ship *Manchester* to be abandoned and scuttled when, having regard to the conditions prevailing at the time, it was his duty to stand by the ship and do his utmost to bring her into harbour'.

Several of Drew's subordinates were convicted, essentially, of having succumbed to panic. Lt. Cmdr. Daniel Duff, the gunnery officer, was found guilty of ordering the abandonment of gun positions and of the transmitting stations without due cause, and also that 'he failed to take proper steps to restore the fighting efficiency of the gun armament'. Lt. Allan Daniels RNVR was found to have quit his action station, directing the ship's high-angle guns, without due cause; Sub-Lt. John Tabor RNVR and Ordnance Warrant Officer Albert Reddy were found guilty likewise. Another petty officer, who had been serving as chief quartermaster at *Manchester*'s lower steering position, was found to have abandoned his post after the torpedo hit, leaving it unmanned and without reporting to the Navigating Officer.

Harold Drew was inevitably appalled by these verdicts. He wrote a long and bitter letter to C-in-C Portsmouth about its procedures and findings, including one passage: 'It was distressing in court to see officers and men, some of whom I had noticed that night [of 12/13 August 1942] almost tottering with fatigue, and yet doing their duty without fear or hesitation, failing to understand why, under the searching light of an enquiry seven months later, their own conduct appeared less good and less worthy than in fact it was.' Yet the court-martial could not have reached the conclusions that it did had not some of the convicted men's shipmates and fellow officers testified against them.

There is no doubt that Drew and some of his subordinates were treated with a harshness that would be unthinkable in peacetime, when passing judgement on a disaster at sea that befell exhausted men who possessed some cause to consider that the entire Pedestal venture had been shown to be doomed. In February 1943, however, Britain and the Royal Navy were still in the midst of a life-and-death struggle. Since the late eighteenth century, courts-martial had been held to examine the circumstances of the loss of any King's ship. It was the purpose of the navy to ensure that every officer who wore its uniform understood his duty: to strive to the limits of human courage and endurance to fight, and to preserve, a warship damaged in the face of the enemy. Drew never received another sea-going command. The navy placed him in the pillory, for the same reason that Admiral Byng was shot on his own quarterdeck in 1759 after failing to relieve Minorca: *pour encourager les autres*. Some senior figures, however, came to believe that this verdict was overly cruel: in 1948 Drew was made a naval aide-de-camp to the King, and promoted to the rank of commodore before his retirement.

Roger Hill, following his fine deeds in the Mediterranean, was bitterly disappointed that he never received promotion to commander and had to be content with a DSO. Yet this becomes less surprising in the light of his incorrigibly cheeky conduct. On returning to Portsmouth, he was enraged to be informed that his ship, her crew dominated by 'Pompey' men, was to be dispatched to refit at Hull. He marched into the office of the commander-in-chief to deliver an intemperate protest, an act of lèse-majesté that earned him a stinging reprimand. Hill also conceived the idea that he and the other Royal Navy captains who had brought in *Ohio* should submit a formal claim for salvage money, such as often enriched the captains of Nelson's navy. James Swain, his counterpart on *Penn*, refused to have any part in such an undignified proceeding, but Eddie

Baines of *Bramham* joined Hill in submitting an application. Eighteen months later, both officers received cheques for the peculiar sum of £19 18s 4d, while the ratings on their ships, together with those of *Penn*, got a pound or two apiece.

After Hill left *Ledbury*, he belatedly understood how close he had come to a nervous breakdown, conceding with disarming honesty: 'I did not realise at the time what a bad state I was in and how unpleasant I had become to serve with.' He commanded destroyers for the rest of the war, but was obliged to quit the navy in 1946 following a bad car crash. Captain Jackie Broome, a Pedestal destroyer-flotilla leader, said of him: 'in another age, he would have made an excellent – if humane – pirate'. Hill emigrated to New Zealand, and wrote a lively memoir of his time as a destroyer captain. On his death in 2001, his ashes were scattered in the Mediterranean off Malta.

Some other Pedestal participants rose to high rank. Tom Troubridge became a vice-admiral before dying young, aged only fifty-four, in 1948, perhaps a victim of his extravagant weight. Peter Gretton of *Wolverine* also retired as a vice-admiral; he died in 1992, aged eighty. Richard Onslow of *Ashanti* became Admiral Sir Richard Onslow, C-in-C Plymouth, dying in 1975, aged seventy-one. Terence Lewin, a lieutenant on *Ashanti*, rose to admiral-of-the-fleet, and served as chief of defence staff during the 1982 Falklands War. He died in 1999, aged seventy-eight. Eddie Baines, in his Pedestal report, paid effusive tribute to his twenty-three-year-old first lieutenant, David Milford Haven, who received a DSC: 'This officer was outstanding in his leadership of the men.' The Marquis of Milford Haven, a great-great-grandson of Queen Victoria, in 1947 acted as best man to his former Dartmouth classmate Prince Philip, Duke of Edinburgh, at the latter's marriage to Princess Elizabeth. He died in 1970, aged fifty-one.

More than a few other men who came through Pedestal failed to survive the subsequent three years of war. Fleet Air Arm ace Dickie Cork was killed in a 1944 flying accident. Arthur Thorpe, the war correspondent who defied probability by surviving the sinkings of both *Ark Royal* and *Eagle*, succumbed to the odds two years later, when he went down with yet another lost ship which he accompanied in action. Captain David MacFarlane, who brought *Melbourne Star* into Grand Harbour in the finest style, was fortunate enough to be on leave in Britain when his ship was torpedoed in the West Indies on her next voyage to Australia, and all save four of her crew then perished. He himself remained a Master until his retirement in 1961, and died in 1984, aged eighty-nine.

Jack Tuckett, who commanded *Dorset* on her final voyage, took another merchant vessel to Australia a few months later, only to die in his sleep in December 1942, while berthed in Sydney Harbour. It seems reasonable to assume that Tuckett, who was aged fifty-nine, suffered from a weak heart, which cannot have been strengthened by his experiences in the Mediterranean.

Kapitänleutnant Helmut Rosenbaum of *U-73*, which sank *Eagle*, was royally feted on his return to Germany, especially by Döbeln, his birthplace, where the town turned out to garland its local hero and to present him with a porcelain representation of a seagull with outstretched wings. 'Your work is not finished,' declared the beaming mayor. 'Indeed, it has hardly begun.' The triumphant submariner signed the town's Golden Book of distinguished citizens, writing beneath his signature: 'Be faithful. Fight bravely. Die smiling. May we be victorious and come home happily'. Rosenbaum spent two years in shore postings before being killed in a 1944 plane crash. He received a belated promotion to lieutenant-commander, unlikely to have given him much satisfaction since it was posthumous.

Captain Alf Russell, the much loved captain of *Kenya*, died of cancer in August 1945, just days after the war ended. It seems likely that the stress of his experiences contributed to this early end.

Captain Giorgio Manuti, who with Franco Mezzadra achieved the greatest wartime triumph of the Italian Navy's mosquito craft by fatally crippling *Manchester*, was killed in a 1945 car crash in Rome. Giovanni Cafiero, who claimed to have sunk *Santa Elisa*, served with allied forces following the Italian Armistice in September 1943, commanding MAS boats that dropped British commandos and agents on the Yugoslav coast.

After Germany's 1945 surrender Field-Marshal 'Smiling Albert' Kesselring, all teeth and domed forehead, profited from the ill-judged chivalry of Field-Marshal Sir Harold Alexander, who solicited clemency for a skilful battlefield adversary in the 1943–5 Italian campaign. Yet Kesselring's ruthlessness was manifested in innumerable acts of authorized brutality towards Italian civilians by his forces, as well as in the above-mentioned devastation of Maltese homes. By the criteria that determined the fate of most of those indicted at Nuremberg, Kesselring should have hanged. As it was, his death sentence for the massacre of 355 Roman civilians in the Ardeatine Caves was commuted, and he was released from imprisonment in 1952. He died in 1960, aged seventy-four.

When the war was over Frederick Treves, the seventeen-year-old cadet who was decorated for his part in saving men from the stricken *Waimarama*, abandoned the sea and embarked on a successful career as a character actor. One of his last roles was to play Lt. Col. John Layton, father of Sarah, in the 1984 classic Granada TV series *The Jewel in the Crown*. He died in 2012, aged eighty-six. Charlie Walker, *Ledbury*'s heroic cook, died in 2011, aged ninety-seven, and Alan Bennett, the teenager whom he rescued from the burning sea, also lived to a ripe age. Lt. John Manners, gunnery officer of *Eskimo*, remained a naval officer but resumed his career as a first-class cricketer, scoring 121 for Hampshire against Kent in 1947, and 123 against New Zealand in 1949. He died in 2020, aged 105. Commander Anthony Kimmins, who served with Burrough's staff as a war correspondent, returned to writing light comedies, such as *The Amorous Prawn*, a 1959 West End hit, later filmed with Kimmins directing. He died in 1964, aged sixty-two.

Alastair Mars, the young captain of *Unbroken* who made a notable contribution towards redeeming Pedestal's losses by wrecking two Italian cruisers, became one of the most successful British submarine commanders of the war, emerging with a DSO and two DSCs. His post-war career was less happy. He fell out spectacularly with the Royal Navy, which he accused of paying him so miserably in overseas postings that he was unable to support his wife and children. He was dismissed the service after a sensational 1952 trial by court-martial for insubordination and absence without leave. Thereafter, he wrote two best-selling memoirs of his wartime experiences, and served as a junior officer in the ocean weather service. He died in 1985, aged seventy. As with so many men who achieved remarkable things in the extreme circumstances of war, nothing in Mars' subsequent life yielded the same fulfilment.

Dudley Mason GC was flown back to Britain from Malta with his chief engineer, James Wyld, but nobody from the Admiralty, or indeed any other official body, troubled to meet and greet them: they spent their first night back in a grubby boarding-house. Nothing much happened to Mason for the ensuing four decades of his life. He captained various ships until retirement in 1967, married again and had one son. He died in Hampshire in 1987, aged eighty-five. Fortune had singled him out for a few days of glory on the bridge of *Ohio*, but thereafter he retired into that uneasy twilight existence occupied by many war heroes. The hulk of Mason's tanker was used as an accommodation base for naval personnel through the ensuing two years. When peace came, *Ohio* proved an embarrassment to Grand Harbour; unromantically, she was towed out to sea and scuttled. Her ship's wheel, however, continues to be proudly displayed aboard *Wellington*, headquarters ship of the Company of Master Mariners, moored beside London's Temple Stairs. *Rochester Castle*, *Port Chalmers* and *Brisbane Star* were the only three Pedestal merchant vessels to survive the war and rejoin the peacetime carrying trade, proudly repainted in company liveries.

Fred Larsen, of *Santa Elisa* and thereafter *Ohio*, returned to New York on the *Queen Mary*, and hastened to his sister's house in Brooklyn. He learned that Minda was out at the movies, went to the theatre where he spotted her in the darkness, crept up behind and put his hands over her eyes before whispering the song line 'You are my sunshine'. Back at sea, by the war's end he had participated in sixty-five convoys, from 1944 as Master. He remained a ship's captain for almost forty years, dying in 1995, aged eighty.

Commander Giuseppe Roselli Lorenzini, the staff officer who pleaded with his chiefs at Supermarina to commit the Italian Navy's cruiser force to destroy the remains of Pedestal on the morning of 13 August 1942, became a post-war admiral, and was chief of staff of the Italian Navy in the years 1970–3. He died in 1985.

The drastic depletion of Britain's merchant naval strength in the Second World War was never afterwards made good.

Alfred Holt & Co. lost half its hundred-strong 1939 fleet. The Blue Star Line started the war with thirty-eight ships, of which twenty-nine were lost, including all its passenger liners. Aboard those latter vessels 646 personnel died, including eleven captains, forty-seven navigating officers and eighty-eight engineer officers. These were among a total of 30,248 Merchant Navy personnel who perished in the entire fleet. The US Mercantile Marine meanwhile lost 8,251 men, likewise a higher rate of loss than that of the US Navy.

In the twenty-first century, anyone who wishes to catch the mood and to feel the fabric of such British warships as fought the Pedestal battles need only travel to London Bridge and board the Edinburgh-class light cruiser *Belfast*, a slightly larger version of Harold Drew's *Manchester*, scuttled off Tunisia. Although *Belfast*, since 1971 a museum ship, was modernized and refitted in the late 1940s, the visitor gains a vivid impression of her scale and apparent might. Those who descend to the machinery spaces below the waterline can conceive the terrors that beset many men who served in such places in action, a mere three inches of steel away from the ocean, and conjure a vision of the fate that so often befell them if their ship was hit: those sweaty, clamorous, ill-accessible working spaces were within a millisecond laid open to the inrushing sea.

In chronicling such extraordinary tales as that of Pedestal, I have often reflected that, whatever troubles oppress us in our own times, they are less terrible than those which encompassed the men and women who participated in the Second World War or fell victim to it. Only those who know no history can today be foolish enough to express nostalgia for its experiences, glorious or no. And few could forbear to pay homage to the men of the Royal Navy and Merchant Navy, who fought such battles as this one, and ultimately prevailed.

Appendix
Losses during Operation Pedestal

Fatal casualties aboard British warships, British and American merchant vessels:

HMS *Eagle* 162
HMS *Manchester* 13
HMS *Nigeria* 53
HMS *Indomitable* 52 (inc. aircrew lost in combat)
HMS *Victorious* 8 (all aircrew)
HMS *Cairo* 24
HMS *Foresight* 5
HMS *Kenya* 3
HMS *Ledbury* 1
MV *Waimarama* 91
MV *Clan Ferguson* 12
MV *Glenorchy* 8
MV *Melbourne Star* 14
MV *Brisbane Star* 1
MV *Santa Elisa* 3
MV *Deucalion* 1
MV *Ohio* 1

Syfret's command lost thirteen vessels: one carrier (*Eagle*), two cruisers (*Manchester* and *Cairo*), one destroyer (*Foresight*) and nine merchantmen. *Indomitable*, *Nigeria* and *Kenya* were also badly damaged.

The Fleet Air Arm lost thirteen aircraft on operations and a further sixteen Hurricanes with the sunken *Eagle*. The RAF lost a Beaufighter and five Spitfires out of Malta, together with the Sunderland flying-boat shot down by the Italian submarine *Giada*.

Axis losses:
The Italian cruisers *Bolzano* and *Murio Attendolo* were crippled by HMS *Unbroken*; the Italian submarines *Dagabur* and *Cobalto* were sunk by Force Y; the Italian submarine *Giada* and German E-boat *S58* were damaged.

The Luftwaffe's II Fliegerkorps mounted 650 fighter and bomber sorties against British warships in the Mediterranean between 11 and 14 August, claiming to have destroyed twelve British aircraft in the air for the loss of nineteen of its own.

The Regia Aeronautica lost a further forty-two aircraft, including those destroyed on the ground by RAF airfield strafing.

During and immediately following Pedestal, the following Axis transport ships were sunk by the British in the eastern Mediterranean: *Ogaden* off Derna on 12 August, *Lerici* on the 15th – both by the submarine HMS *Porpoise*; *Pilo* by RAF air attack on the 17th, likewise the tanker *Pozarico* on the 21st.

Acknowledgements

Some wit observed that to borrow from one source is plagiarism, while to borrow from many is research. Every honest history or biography should start by acknowledging a debt to those writers who have addressed the same theme in the past. Brian Crabb's exhaustive 2014 *Operation Pedestal* is especially valuable for its appendices, detailing a mass of factual data. He has also provided some excellent images, collected by himself and his father Percival 'Buster' Crabb, who was a stoker petty officer aboard *Kenya* throughout the battle. Brian corrected a draft of my text. Peter Smith's earlier 1970 *Pedestal* and Sam Moses' 2008 *At All Costs* profited from their authors' interviews and correspondence with survivors of the battle who were then alive and are now long gone. James Holland's excellent 2004 *Fortress Malta* deals with Pedestal only incidentally, but provides a vivid general account of the island's 1940–3 experience. James also introduced me to Laura Bailey, who has done some splendid research for this work, especially in the audio archive of the Imperial War Museum. He also found for me the important detail of *Ohio*'s loading manifest, and provided some images given to him by participants. Francesco Roselli Lorenzini shared some memories of his father, who served in Italy's submarine fleet and was thereafter a staff officer at Supermarina in Rome during Pedestal.

Beyond these, I owe much to Richard Woodman. For many years I have admired his books about the sea, especially the 2000 *Malta Convoys*. While I was researching Pedestal, he transcended the generosity that authors often display to each other, presenting me with the files he accumulated for his own book, which would otherwise have taken months of labour to trace in archives and replicate. He has also read and commented upon a draft of this work. Here I pay tribute to Richard's kindness as well as to his exceptional grasp of the issues, and commitment to sustaining the legacy and heritage of the Merchant Navy.

Many other works that have been invaluable, especially those addressing the Italian and German experiences of the battle, are listed below in the Bibliography. Serena Sissons has translated much Italian material, some of it available online, and Silke Montague has done likewise with German accounts and records, culled from the Freiburg Archives by Axel Wittenberg. It was fascinating to meet Pauline Lee, ninety-six-year-old daughter of Admiral Sir Harold Burrough, and to share her vivid memories of her father. I am indebted to my old friend and colleague Nicholas Shakespeare for putting me in touch with Mrs Lee. She, in turn, introduced me to Hugh Ellerton, who enabled me to read the papers of his father John, who served for fifteen years as Burrough's naval secretary, an exceptionally happy and devoted professional partnership.

The great Professor Nicholas Rodger of All Souls College, Oxford, who possesses an encyclopaedic knowledge of the history of the Royal Navy and ranks among its finest historians, shared his thoughts on Pedestal, as did that other giant of this subject, Professor Paul Kennedy of Yale. Ralph Erskine, whose mastery of wartime intelligence matters immensely assisted me in writing *The Secret War*, again provided generous insights, especially about the Abwehr's Bodden Line. Robert Woods, a dear friend and former CEO of P&O, is a mine of information about the Merchant Navy and its history, some of which he generously shared with me, together with important books and his magnificent builder's model of HMS *Ledbury*, which I envy him. William Spencer is now, alas, retired following his long career at the National Archive with specialist responsibility for military documents, but he continues to counsel researchers with wonderful generosity and insight. Andrew Webster of the Imperial War Museum guided me through their superb collection of Pedestal images, taken by service photographers aboard the warships. It must only be a matter for regret that no photographers sailed on merchant vessels, to preserve a close-up visual record of their extraordinary experiences.

Arabella Pike and Jonathan Jao, of HarperCollins respectively in London and New York, have been their accustomed wonderful selves, as has my agent Andrew Wylie, and Rachel Lawrence, my secretary with only a brief interruption since 1987. My wife Penny has played an exceptionally precious role, as the custodian of my sanity, at some risk to her own, through lockdown.

References and Notes

For this book, as for my previous works, in pursuit of accessibility I have made free translations and adaptations of German and Italian material, including quotations, into idiomatic English. In several places in the text, clearly indicated, where no witness account survives from Pedestal, for instance about the Italian submariners' experience of being depth-charged, I have used quotations from similar episodes, comparable witnesses, at the same period of the war.

TNA indicates a file held at the British National Archive, Kew, and IWM refers to a document in the archive of the Imperial War Museum. Ellerton Papers refers to the personal archive of John Ellerton, wartime secretary to Admiral Sir Harold Burrough, which I explored through the kindness of his son Hugh. Copies of the papers were also deposited at the IWM. LHA indicates the Liddell Hart Archive at King's College London. AI denotes an author interview. As in all my books, I have not referenced familiar quotations, such as Churchill's 'end of the beginning'. Almost all the quotations given from senior officers' reports derive from the voluminous Pedestal files in the National Archive, cited below, and where brief mentions are not detailed in the references, can be assumed to come from this source. My intention to check all the numbers by a post-edit visit to the National Archive was frustrated by CV-19, in consequence of which this great institution remained – deplorably – almost entirely inaccessible to historians from early 2020 into 2021.

Introduction

xvi 'the high-water mark' Roskill, Captain S. W. *The War at Sea* vol. II *The Period of Balance* HMSO 1956 p. 192

xvii 'The Germans have most things' TNA WO208/1777

xvii 'The prospects [in the Mediterranean theatre]' Ciano, Count Galeazzo *Diary 1937–1943* Phoenix 2002 p. 510

xix 'If Rommel's army' Kennedy MS diary, LHA 18.7.42

xx 'We must have a victory!' Harvey, Oliver *The War Diaries of Oliver Harvey 1941–1945* ed. John Harvey Collins 1978 22.6.42

xxi 'There was a strong touch' Hill, Roger *Destroyer Captain* Periscope 2004 p. 46

1: 'It Would Be a Disaster of the First Magnitude'

1 MALTA

1 'the Mediterranean bored' Porch, Douglas *Hitler's Mediterranean Gamble: The North African and the Mediterranean Campaigns in World War II* Weidenfeld & Nicolson 2004 p. 69

1 'driven out of' SKL *Seekriegsleitung* 14.11.40 quoted 'Fuehrer Conferences on Naval Affairs, 1939–1945', in *Brassey's Naval Annual 1948* New York: Macmillan 1948 p. 155

4 'The novelist and traveller' Theroux, Paul *The Pillars of Hercules: A Grand Tour of the Mediterranean* Hamish Hamilton 1995 p. 310

4 'I could not imagine' Lloyd, Air Marshal Sir Hugh *Briefed to Attack: Malta's Part in African Victory* Hodder & Stoughton 1949 p. 14

4 'To Italian seamen' Cocchia, Admiral Aldo *Submarines Attacking: Adventures of Italian Naval Forces* Kimber 1956 p. 186

4 'Malta was a lonely place' Lloyd p. 32

5 'The war … was not' Kesselring, Albert *The Memoirs of Field-Marshal Kesselring* History Press 2015 p. 106

7 'though we could not' ibid. p. 114

7 'the *abracci* – embraces' ibid. p. 121

7 'Keep your shirt on' ibid. p. 109

8 'it is to the credit' ibid. p. 122

8 'The destruction is inconceivable' Alanbrooke, Field Marshal Lord *War Diaries 1939–1945* ed. Alex Danchev and Dan Todman Weidenfeld & Nicolson 2001 2.8.42 p. 288

11 'deplorably weak' Tedder, Arthur [Lord] *With Prejudice* Cassell 1966 p. 269

11 'Too many old' ibid. p. 263

11 'Only those who have suffered' IWM 8/114/1 Sprot

12 'if you can call' ibid.

12 'Oh, what a glorious' IWM 06/27/1 Nicholls

13 'as each dawn broke' Lloyd p. 173

13 'Among the graveyards' Ciano 21.2.42 p. 495

13 'One can hardly' Boog, Horst and others *Germany and the Second World War* vol. VI *The Global War* Clarendon Press 2002 p. 658

14 'every effort must' TNA PREM3/266

15 'Such an attitude' Tedder *With Prejudice* p. 345

15 'Personal intrigues' ibid. p. 344

16 'If we cannot get' TNA PREM3/266/2

17 'Roll out the *Nelson*' Tedder *With Prejudice* p. 162

18 'Thanks to [the blitz's]' Kesselring p. 122

18 'It would have been easy' ibid.

18 'we just had to have' ibid. p. 129

18 'proved too weak' ibid.

18 'one of the most disastrous' Howard, Michael *Grand Strategy* vol. IV *August 1942–September 1943* HMSO 1970 p. 63

19 'The answer, in my opinion' Boog vol. VI p. 720

19 'I should be very glad' TNA PREM3/266/2

19 'morale remains good' TNA PREM3/266/5

2 CHURCHILL'S COMMITMENT

22 'what do you think' Colville, John *The Fringes of Power: Downing Street Diaries 1939–1955* Hodder & Stoughton 1985 20.5.41 p. 388

22 'a disaster of the first magnitude' Playfair, I. S. O. and others *The Mediterranean and the Middle East* vol. III *British Fortunes Reach their Lowest Ebb* HMSO 1960 p. 203

23 'the island served less' Barnett, Correlli *Engage the Enemy More Closely: The Royal Navy in the Second World War* Hodder & Stoughton 1991 p. 491

23 'on the success' Playfair and others p. 316

23 'The Mediterranean was not' Porch pp. xii & 284

23 'the dominant factor' Playfair and others p. 200

23 'we are determined' ibid. p. 203

24 'I was on watch' IWM 02/56/1 C. A. Gould

24 'a very widespread' TNA PREM3/119/6 folio 8

25 'Lord Gort must' Playfair and others p. 316

26 'the Naval Staff' TNA COS (42) 52nd meeting 10.30pm 15.6.42

28 'when we sent you' IWM 77/122/1 Nicholl

2: Men and Ships

1 THE FLEET

32 'We are now approaching' TNA MEP02/1663 quoted Woodman, Richard *Malta Convoys 1940–1943* John Murray 2000 p. 373

33 'very quiet and perhaps' TNA ADM196/92/67

33 'more wine, food' IWM 90/38/1 Blundell

33 'a tower of strength' quoted Moses, Sam *At All Costs: How a Crippled Ship and Two American Merchant Mariners Turned the Tide of World War II* Random House 2006 p. 115

33 'a clear thinker' IWM 90/38/1 Blundell

35 'the unattractive policy' TNA PREM3/1714

36 'general unsettling effect' TNA ADM199/1242

37 'War experience creates' TNA ADM1/1950 Boyd 19.2.42

38 'in September 1941' MO File Reports 886–7, University of Sussex

38 'publicity is anathema' Cunningham of Hyndhope, Viscount *Sailor's Odyssey* Hutchinson 1951 p. 410

39 'we came together' IWM 96/2/1 Shackleton

40 'made them unintelligible' Gretton, Sir Peter *Convoy Escort Commander* Cassell 1964 p. 4

40 'a poor crowd' IWM 87/15/1 Osborne pp. 13–14

40 'He means stagnant' IWM 90/38/1 Blundell

40 'What excitement you can' IWM 06/6/1 Somers p. 24

41 'We both felt' Crosley, R. 'Mike' *They Gave Me a Seafire* Airlife 2001 p. 79

41 'what a super wedding' Mackenzie, Hector *Observations* Pentland 1997 p. 81

42 'The chief drawback' IWM 04/31/1 Walker 3.9.44

42 'whenever I hear' IWM 84/36/1 F. Calvert

42 'People didn't talk' IWM 91/17/1 Goodbrand pp. 26–7

43 'we lived rather narrowly' IWM 08/117/1 Cmdr. H. Hounsell

43 'The good regulars' Gretton p. 100

43 'Right through the Navy' IWM 93/1/1 Hughes

44 'a heroic figure' Popham, Hugh *Sea Flight: The Wartime Memoirs of a Fleet Air Arm Pilot* Seaforth 2010 p. 123

44 'once faced a strike' IWM audio 22128 Wilkinson R4 3:42

44 'I have no faith' IWM 79/13/1 Thomas 15.3.41

45 'What I've seen' IWM 1783 Hutchinson 18.8.40

45 'we had cabins' Gretton p. 74

45 'Our food has been lousy' IWM 02/56/1 Gould

45 'I didn't want any more' IWM audio 17117 Coaker

46 'The previous commanding officer' Mackenzie p. 108

46 'A chief of staff' ibid. Blundell October 1941

46 'Though acquitted of cowardice' IWM 10/1/1 Manners

46 'the whole ship's company' IWM 77/22/1 Nicoll

46 'Another rating was discharged' IWM 03/14/1 Keith Rail

47 'that silly little ship' IWM audio 22614 Jewett

47 'Onslow in his chair' Kimmins BBC broadcast 19.8.42

48 'dirty, unromantic and damp' Mallalieu, J. P. W. *Very Ordinary Seaman* Gollancz 1944 p. 277

48 'Whenever I visit' IWM 79/13/1 Thomas

2 THE CONVOY

50 'Get out!' IWM 88/42/1 Fyrth

50 'with the weather fine' ibid.

51 'Blimey' IWM audio 17293 Parry

52 'two-berth cabins' Morton quoted Pearson, Michael *The Ohio and Malta: The Legendary Tanker that Refused to Die* Pen & Sword 2011 p. 39

52 'It all seemed too good' ibid. p. 40

53 'dockers are not' TNA ADM199/1243 *Deucalion* report

58 'this Europe is certainly' Suppiger private narrative, courtesy of Richard Woodman

61 'the Surrey man' IWM 91/31/1 Isard

62 'This was wonderful news' Hill p. 5

63 'he looked more like' Kimmins, Anthony *Half-Time* Heinemann 1947 p. 131

63 'admirals are heavenly' oral testimony to the author in 1981

64 'for a moment' Moses p. 107

65 'most impressed with' Burrough (1999) p. 121

66 'The Admiral might have' Moses p. 107

66 'her courage is' Pearson p. 47

67 'We are bound to do' IWM 90/38/1 Blundell

67 'There is great speculation' IWM 82/13/1 Norsworthy

67 'Gawd, just to 'ave' Hill p. 31

67 'You are young' IWM audio 17117 Coaker

3: Sailing

69 'Maybe that escort' Moses p. 116

69 'The Malta run' Popham p. 122

70 'station-keeping at this stage' TNA ADM199/1243

70 'Many men on all' IWM 85/24/1 Dickens

70 'so we had an idea' Follansbee letter of 26.12.94 to Richard Woodman

72 'On that first day' Ellerton Papers

72 'portioned out like a holy oblation' IWM 02/56/1 Gould

73 'what I didn't cater' Wellum, Geoffrey *First Light* Viking 2002 p. 264

74 'I used to worry' AI Pauline Lee 11.10.19

75 'a little tired' Hill p. 56

75 'the importance of not' Binfield in TNA ADM199/1243

76 'Keep well clear' Hill p. 62

76 'When an aircraft' IWM 90/38/1 Blundell

77 'fighter direction is' TNA ADM199/1243

78 'his temper was volcanic' Popham p. 60

78 'an indestructible boyishness' ibid. p. 69

78 'an immaculate pilot' ibid.

79 'To achieve it' ibid. p. 85

79 'We were using batsmen' IWM audio 15533 Parker

80 'one attack by' Gretton p. 89

80 'as dusk drew in' Popham p. 124

80 'Meeting up with' Wellum p. 314

81 'this atmosphere of suppressed' IWM 82/13/1 Norsworthy

81 'The prologue is done' Hill p. 61

82 'everyone knew what' Popham p. 124

82 'a very simple service' Kimmins BBC broadcast 22.8.42

82 'I can't think' IWM 90/38/1 Blundell

83 'we are now' IWM 09/43/1 Penrose

83 'the tendency seemed' Binfield in TNA ADM199/1243

84 'The Ceuta station' see Erskine, Ralph 'Eavesdropping on Bodden: ISOS v. The Abwehr in the Straits of Gibraltar' *Intelligence and National Security* 12(3) 1997 pp. 110–29 for an authoritative and detailed exposition of this intelligence saga

86 'sailors are lonely' Hill p. 53

86 '[Hill] wasn't very popular' John Nixon to Sam Moses, quoted Moses p. 118

86 'Escorting another convoy' Hill p. 54

87 'You know what happened' ibid. p. 62

88 'almost to the blue bowl' *Daily Express* 24.8.42

90 'No time-outs' Metellini, Giacomo *Un pilota racconta* ed. Alessandro Metellini, Mursia 2011 p. 75

91 'This makes me tremble' Paravicini, Pier Paolo *Pilota da caccia 1942–1945* Mursia 2011 p. 28

4: First Blood

1 HUNTERS

92 'The very word Mediterranean' Theroux p. 8

92 'The Mediterranean is' Durrell, Lawrence *Balthazar* Faber & Faber 1958 p. ??

93 'The air has a dead' Popham p. 74

93 'Submerged beneath the surface' ibid. p. 124

93 'It cannot be emphasized' TNA ADM199/1242

94 'There was a great volume' Hinsley, F. H. and others *British Intelligence in the Second World War* vol. II *Its Influence on Strategy* HMSO 1981 p. 418

95 'it was hard' Gretton p. 109

96 'Serving in submarines' Mars, Alastair *Unbroken: The Story of a Submarine* Frederick Muller 1953 p. 9

96 'We passed the towering' ibid. p. 32

96 'The struggle possesses beauty' Roselli Lorenzini MS from Francesco Roselli Lorenzini

97 '*antipasto* of meats' Comandante Sergio Parodi, quoted Hood, Jean ed. *Submarine: An Anthology of First-Hand Accounts of the War under the Sea 1939–1945* Conway 2007 p. 93

98 'On patrol the most' ibid.

98 'It is scarcely' Villa, Livio *Fino alla fine: Diario di guerra del Regio Sommergibile 'Scirè'* pp. 38–9

98 'the dominant clatter' Mars p. 24

99 'A submarine obliged' IWM 91/38/1 Commander Peter Bartlett, CO of *Perseus*

101 'and all the time' Mars p. 29

101 'the boat was an oven' ibid. p. 105

102 'look very scruffy' IWM 09/70/1 Frank

103 'Frankly, I couldn't face' Barca, Luciano *Buscando per mare con la decima Mas* Riuniti 2001 p. 49

103 'Barca's batman had served' ibid. p. 34

103 'the depth-charge' Gretton p. 85

104 'the attack was badly' TNA ADM199/167

105 'as a submarine commander' Vause, Jordan *Wolf: U-Boat Commanders in World War II* Annapolis, Md: Naval Institute Press 1987 p. 83

105 'Commanding a submarine' Mars p. 29

2 THE NEMESIS OF *EAGLE*

107 'everyone had been acting' MccGwire, Michael Kane *A Midshipman's Tale: Operation Pedestal, Malta Convoy, August 1942* Leaping Boy Publications 2016 p. 13

108 'the old girl' Wellum p. 318

109 'was making a downwind' IWM audio 15533 Parker

110 'I am longing' Bundesarchiv Freiburg RM98/180

112 'that at least half' IWM audio 16788 Northover

113 'My soup plate' Crosley p. 9

114 'is she going?' *Exchange Telegraph* report quoted Winton, John ed. *The War at Sea 1939–45* vol. I *Freedom's Battle* Book Club Associates 1974 p. 233

114 'Then the ship started' IWM audio 16355 Amyes

117 'it was so sudden' Kimmins BBC broadcast 19.8.42

118 'For a few seconds' Popham pp. 147–9

119 'What a tragic failure' IWM 90/38/1 Blundell

119 'Now then lads' IWM 95/5/1 Bill Haworth

119 'I'm not very happy' IWM 02/56/1 Gould

120 'Poor old St. Croix' IWM 90/38/1 Blundell

120 'It was quite clear' IWM audio 17117 Coaker

120 'it will be a happy' TNA ADM199/1243

120 'like a brown leaf' Wellum p. 322

121 'I don't want word' ibid. p. 324

121 'I was so very pleased' Crosley p. 78

5: 'Stand By to Ram'

122 'the immense formation' IWM 83/13/1 MR Creasey

123 'when these large shells' IWM audio 15533 Parker

124 'but gave much satisfaction' IWM 11/5/1 D. M. Scott

124 'The noise was deafening' IWM 08/60/1 Stockwell

125 'cowering in a dark corner' *Laforey*'s gunnery officer, account to Richard Woodman

125 'The din was terrific' Kimmins BBC broadcast 19.8.42

125 'the attacks were not pressed' TNA ADM199/1243

125 'We were listening' Hill p. 65

126 'enclosed in a sparkling' Popham p. 128

126 'for Christ's sake' IWM 82/13/1 Norsworthy

126 'had expressly told' TNA N/T9/3727 Brown interview

127 'You bloody useless' Popham p. 134

128 'only achieved at the expense' TNA ADM199/1242 Henry Bovell report

128 'It is evident' TNA ADM199/1143

128 'no AA fire' Winton, John *Cunningham: The Greatest Admiral since Nelson* John Murray 1998 p. 211

130 'Oh Lord' Gretton p. 93

132 'one of the few aircraft-carriers' *Döbelner Anzeiger* 13.8.42

132 'expressed doubts in his diary' Ciano 12.8.42 p. 568

133 'the attack was especially' Santoro, Giuseppe
L'aeronautica italiana nella seconda guerra mondiale vol.
II Edizioni Esse 1957 p. 399

134 'Members of Special' Moses p. 195

134 'As far as we were' ibid.

135 'the soldiers surrendered' Santoro p. 400

135 'Churchill gave orders' TNA PREM3/266/2/234

136 'I've got my wallet' IWM 02/56/1 Gould

6: The Twelfth

1 'SHALL I BE KILLED TODAY?'

139 'What do you think' Hill p. 63

140 'The standby squadron' Crabb, Brian James *Operation
Pedestal* Shaun Tyas 2014 p. 69

141 'the white water' Popham p. 99

145 'The real war' Ciano p. 535

145 'The officers were' www.regiamarina.net Andrea
Piccinotti 2002 interview

145 'A young midshipman' Renato d'Ottaviano unpublished
memoir

146 'A senior naval officer' 1995 interview by Marc de
Angelis

146 'just as well' IWM 77/122/1 Rear-Admiral Angus Nicholl

147 'A pre-war Italian study' Cocchia p. 121

149 'Italian torpedo-bombers' Aichner, Martino *Il Gruppo
Buscaglia: Aerosiluranti nella seconda guerra mondiale*
Mursia 2013 p. 111

150 'the role of torpedo squadrons' Gibbs, Patrick *Torpedo
Leader* Grub Street 2002 p. 159

150 'In the first two hours' Action diary of flag officer Force
R, Ellerton Papers

151 'you cannot see' IWM Clifford Simkin, 'War From the Engine-Room'

2 DOGFIGHTING

154 'our eyes burned' Metellini p. 71
155 'like a fool' Rees memoir BBC People's War website
156 'which should suffice' quoted Greene, Jack and Massignani, Alessandro *The Naval War in the Mediterranean 1940–1943* Chatham 1998 p. 249
157 'We release all' Aichner p. 83
159 'Lt. Campbell in the director-tower' IWM 11/5/1 Scott
160 'some burnt-out shells' Aichner p. 84
161 'became chronically constipated' IWM 09/24/1 Harvey
161 'the difference between' IWM 88/42/1 Fyrth
161 'resembled the inside' Le Bailly, Louis *The Man around the Engine* Kenneth Mason 1990 p. 82
161 'Most of the time' *Times* obituary 16.3.20
162 'Danger is relative' Hill p. 63
162 'I would like to lash' IWM 82/13/1 Norsworthy
162 'I did not then' IWM 85/24/1 Brunskill
163 'all of whom remained' TNA N/T9/3727 Brown interview
164 'in some ways' *Daily Telegraph* obituary 1993

3 MINES, *BOMBAS*, TORPEDOES AND A CANARY

166 'completely carbonized' Raffaelli's narrative, provided by Col. Antonio Sasso
166 'The sky seemed' Moses p. 149
166 'The gap was no more' Rees memoir BBC People's War website
167 'gradually we were getting' IWM 85/24/1 Hollings
167 'almost rioted' IWM 84/36/1 Calvert

168 'I was depressed' Popham p. 133

168 'the pilots snatched meals' ibid. p. 135

169 'it became more and more' Mackenzie p. 131

169 'The chief mate hastily' Suppiger narrative to Woodman

169 'they could do something' IWM audio 22614 Jewett

169 'The noise was' MccGwire p. 13

170 'refusal laced with expletives' IWM 85/24/1 Brunskill

170 '3 destroyers are <u>not</u> required' Action diary of flag officer commanding Force F, Ellerton Papers

170 'He first sighted' Giorgerini, Giorgio *Uomini sul fondo: storia del sommergibilismo italiano dalle origini a oggi*, Mondadori 2002 pp. 334–5

173 'Hard astarboard' report in the *Listener* quoted Winton *War at Sea* p. 235

174 'I am received with military' Giorgerini *Uomini sul fondo* p. 335

175 'bombs fell all over' IWM 90/38/1 Blundell

7: Cruel Sea

1 'I BELIEVE THEY'VE BUGGERED US!'

177 'great courage and some recklessness' Metellini p. 76

178 'circling maybe five thousand' Moses p. 143

178 'every near-miss' ibid.

179 'down below it sounded' IWM 85/24/1 Hollings

179 'Enemy aircraft bearing' gunnery officer's narrative to Woodman

180 'The main problem' ibid.

181 'shells exploding in front' Aichner p. 113

181 'I felt very depressed' ibid. p. 117

181 'exhausted, our nerves shredded' ibid. p. 112

182 'The carrier looks as if' Hill p. 69

183 'pressed home their attack' Binfield in TNA
      ADM199/1243

183 'I thought that she would roll' IWM 82/13/1 Norsworthy

183 'It really was rather' IWM 86/11/1 Barton

183 'all eyes tended' Mackenzie p. 135

184 'She completely disappeared' IWM 90/38/1 Blundell

184 'suddenly the ship shuddered' quoted Smith, Peter C.
      Pedestal: The Convoy that Saved Malta Crécy 1999 p. 134

185 'smoke and steam' Popham p. 136

185 'It was not an easy time' Mackenzie p. 136

185 'everyone prayed for nightfall' IWM 92/27/1 Venn

186 'I ... saw part' Crabb *Operation Pedestal* p. 66

186 'My most vivid memory' quoted Smith p. 134

2 THE PARTING OF COURSES

189 'a brave concourse' IWM 90/38/1 Blundell

190 'It had ... much of' Woodman p. 454

190 'Air power had chased' Kennedy, Paul *The Rise and Fall
      of British Naval Mastery* Penguin 1976 p. 322

190 'we were shocked' IWM 90/38/1 Blundell

190 'it was a most impressive' Taffrail *Blue Star Line at War
      1939–1945* Blue Star 1948 p. 9

191 'I'm just about' IWM 02/56/1 Gould

191 'the ship herself' Popham p. 140

192 'I do not know' Mackenzie p. 138

192 'it was evident' Suppiger narrative to Woodman

193 'I am positive that "Uncle"' Rees memoir BBC People's
      War website

193 'It makes me sweat' IWM 90/38/1 Blundell

194 'In view of ammunition' TNA ADM199/1243

194 'I can still see the flames' Gregson obituary *Bay of Plenty
      Times* 2017

194 'a really gallant' TNA N/T9/3727 Brown interview

195 'The great expanse' Metellini p. 74

195 'very sterile and impersonal' ibid. p. 71

8: Force X

198 'Surely this is hardly' IWM 90/38/1 Blundell

200 'he was born' AI Pauline Lee 16.11.19

200 'In view of the' TNA ADM199/1242

202 'The Asdic pings' Pracchi quoted Hood p. 44

202 'so many thoughts' Villa p. 41

204 'After sixty-three seconds' Giorgerini *Uomini sul fondo* p. 337

204 'silent routine' Rapalino, Patrizio and Schivardi, Giuseppe *Odissea di un sommergibilista: dal mar rosso al mediterraneo 1940–1943* Mursia 2008 p. 171

205 'a flash, heard' Kimmins BBC broadcast 19.8.42

205 'do we go over?' IWM 08/60/1 Stockwell

206 'Most merciful God' Burrough p. 73

206 'I had a hell' letter of July 1969 quoted Smith p. 148

207 'I hate leaving you' Kimmins BBC broadcast 19.8.42

207 'should have been heard' Rees memoir BBC People's War website

207 'You blokes' Jack Follansbee letter to Woodman

207 'Anybody going my way?' Rees memoir BBC People's War website

208 'from then on' TNA ADM199/1243

208 '*Nigeria* would obviously' Burrough p. 87

208 'in the wardroom' IWM 85/24/1 C. G. Crill

209 'We felt a tremor' IWM audio 23813 Bartholomew

210 'we all thought' Moses p. 165

211 'I've got to go on' Kimmins BBC broadcast 19.8.42

211 'no RDF signals' Appendix III to *Ashanti* gunnery officer's report 'Details of Air Attacks', Ellerton Papers

212 'the attackers swept in' Bundesarchiv Freiburg RL8/11 War Diary of the Luftwaffe's KG54

212 'two rather vague' Appendix III to *Ashanti* gunnery officer's report, Ellerton Papers

213 'A really intense' TNA ADM199/1243 Bernard report

213 'Thank you chief' Onslow letter of July 1969 quoted Smith p. 155

214 'The strange part' Peck letter of 2.7.95 to Woodman

214 'There's no bloody water!' IWM audio 17293 Parry

215 'there were screams' Moses p. 180

215 'the crew were jumping' IWM 85/24/1 Brunskill

216 '*Clan Ferguson* was observed' TNA ADM199/1243 Binfield

217 'one of my chief stokers' IWM 85/24/1 Crill

217 'about the same as driving' quoted Hastings, Max *Nemesis: The Battle for Japan 1944–45* Harper Press 2007 p. 165

217 'there was a crash' quoted Crabb *Operation Pedestal* p. 78

218 'this, as it proved' TNA ADM199/1242

218 'the effect of this series' TNA ADM199/1242 Part III

219 'The state of affairs' TNA ADM199/1242 Frederick

220 'his firm handling' IWM 85/24/1 Crill

221 'The admiral is waiting' Hill p. 71

221 'we felt naked' ibid.

222 'the action ... in the Narrows' quoted Smith p. 158

222 'PS the "glorious Twelfth"' IWM 10/19/1 Hazledean

9: Scuttling Charges

224 'shortly after we left them' IWM 90/38/1 Blundell

224 'Gor blimey' IWM audio 22614 Jewett

225 'It is almost impossible' Burrough p. 90

225 'To have a realistic chance' Bundesarchiv Freiburg
RM55/56 K10-2/41 OKM handbook for joint
Luftwaffe–E-boat operations 30.6.42

229 'The six engines' Hichens, Antony *Gunboat Command:
The Life of 'Hitch', Lieutenant-Commander Robert
Hichens* Pen & Sword 2007 p. 218

230 'The first explosion' quoted Smith p. 171

230 'the screaming went on' Moses p. 201

232 'if one was to go' IWM 95/5/1 Rambaut

233 'mixed feelings' ibid.

233 'groped his way through' IWM 85/24/1 Terry

233 'we were all delighted' 'Chicko' Roberts quoted Smith p.
177

236 'You are a young man' Woodman p. 425

236 'had served on the Alfred Holt & Co.' IWM 85/24/1
Brunskill

236 'I made up my mind' Brunskill interview transcript from
Richard Woodman

237 '*Glenorchy*'s radio officer' Bundesarchiv Freiburg
RM7/235 C11-34 'Interrogation of British Prisoners',
interview on 19.8.42 – German translation of an Italian
document

237 'They retrieved another two swimmers' ibid. RM7/235
PG/45028

238 'Our engine-room telegraph' quoted Smith p. 175

239 'As far as the eye' Moses p. 198

240 'for we had been fighting' IWM 85/24/1 Hollings

240 'just didn't know' IWM audio 17117 Coaker

240 'what should have been' IWM 09/16/1 Waters

240 'these two fingers' IWM audio 22614 Jewett

241 'sorry to wake' Moses p. 212

241 'At one moment' Rees memoir BBC People's War website

242 'abandon ship!' IWM 09/24/1 Harvey

243 'They were a mixture' TNA ADM199/1242

243 'Instead she went off' Hill p. 72

245 'you could even hear' IWM 85/24/1 Waters

246 'the bullets made' Moses p. 214

247 'it was a hell' ibid. p. 215

247 'As the men rowed' ibid. p. 217

252 'It was an ugly' Kimmins BBC broadcast 19.8.42

10: Retribution

255 'A further significant factor' O'Hara, Vincent, *Struggle for the Middle Sea: The Great Navies at War in the Mediterranean Theater, 1940–1945* Annapolis, Md: Naval Institute Press 2009 p. 185, citing Fabio Tani unpublished MS

255 'In darkness, his mistake' Mattesini, Francesco 'Operazione Pedestal: La Battaglia di Mezzo Agosto 1942' published online by SISM 2014

257 'If at dawn' Rocca, Gianni *Fucilate gli ammiragli: La tragedia della Marina italiana nella seconda guerra mondiale* Mondadori 1987 p. 264

257 'I was always grateful' Burrough p. 161

257 'the map track' Gibbs p. 159

257 'we have two jobs' Mars p. 16

258 'I decided I could not' ibid. p. 95

260 'I glanced around' ibid. p. 108

260 'There is always a fascination' ibid. p. 90

261 'we were disturbed' ibid. p. 109

262 'the Admiralty was vividly' IWM 91/38/1 Commander Peter Bartlett, CO of *Perseus*

263 'They were working hard' Mars p. 103

263 'the situation was such' ibid.

264 'I could see no other' ibid. p. 115

265 'My heart thumped' ibid. p. 114

265 'As [the first lieutenant] relayed' ibid. p. 117

266 'This was a "now or never"' ibid. p. 118

267 'she was too close' ibid. p. 121

268 'as dense black smoke' www.lavocedelmarinaio.com/ 2017/06/22

270 'The scene in the control-room' Mars p. 123

271 'It appears that things' Ciano 13.8.42 p. 569

11: Blenheim Day

272 'we don't know' IWM audio 17293 Parry

272 'three cruisers as well' *Königsberger Allgemeine Zeitung* 14.8.42

273 'There came a turning point' IWM 09/24/1 Harvey

273 'who should wash up' ibid.

273 'a telegraphist read off' IWM audio 22614 Jewett

275 'It is not in' Villa p. 47

280 'none of them' Taffrail p. 93

280 'We were all feeling' Burrough p. 94

280 'Admiral Burrough's stubble' Kimmins *Half-Time* p. 141

281 'Most of us were bloody' IWM audio 22614 Jewett

283 'one of the grimmest' IWM 86/11/1 Barton

286 'a lovely old man' Moses p. 223

288 'for Christ's sake' Hill p. 73

288 'Oh, he was really' Moses p. 228

288 'the spirit of the men' IWM 85/24/1 Hollings

289 'one knew that' Holloway, Adrian letter to Ronald Spector 22.2.98

291 'deplored orders imposed' Deichmann *Spearhead to Blitzkrieg* p. 213

292 'we are told' Moses p. 237

292 'That day there' Aichner p. 116

293 'they dived at so steep' Appendix III *Ashanti* gunnery officer's report, Ellerton Papers

294 'if I could find the bastard' Moses p. 237

297 'The clouds are getting' Aichner p. 118

298 'Avoiding turns taken' TNA ADM199/1243

298 'an interested ship's company' Burrough p. 91

299 'a heartbreaking scrapyard' Mars p. 43

300 'what a glorious sight' IWM 8/114/1 Sprot

300 'As the three ships' Wellum p. 328

300 'Through the mercy' *Times of Malta* 14.8.42

301 'got most of the signals' IWM 85/24/1 Hollings

301 'he ordered *Ledbury*' Crabb *Operation Pedestal* p. 108

302 'the captain turned in' IWM 85/24/1 Hollings

302 'I'd never seen anything' ibid.

303 'Anyway they were to knock' Hill p. 77

303 'stench of oil' Bundesarchiv Freiburg RM59/25S56 report on 15.8.42

307 'in a day made most unpleasant' Riley report to Blue Star Line reprinted in *Maltese National War Museum Newsletter* no. 101, July/September 1997

309 'There was no band' Woodman p. 447

12: Ohio

313 'we all thought she was' IWM 86/11/1 Barton letter of 1.9.42

315 '$2000 in cash' letter of 27.1.95 to Woodman

315 'we used to get' IWM audio 17293 Parry

315 'at all times calm' Suppiger narrative to Woodman

318 'I am the captain' TNA N/T9/3727 Brown interview

318 'by jingo' IWM audio 17117 Coaker

319 'the chaos of wires' Hill p. 82

320 'the two badly burned' IWM 85/24/1 Hollings

320 'Why doesn't the old' Moses p. 243

321 'Music? I must' Moses p. 266

321 'secretly I was' IWM 85/24/1 Hollings

322 'I thought this was just' Hill p. 79

323 'My biggest reason' Moses p. 253

323 'It's becoming painfully obvious' Wellum p. 326

323 'when we saw the Spitfires' IWM audio 22614 Jewett

324 'it was horrible' Hill p. 82

325 'he lacked experience' Metellini p. 77

326 'the most testing' ibid. p. 80

327 'we longed and longed' Hill p. 83

327 'Through the growth' Hill p. 85

328 'Hey over there!' Moses interview p. 275

329 'It was the most wonderful' Hill p. 86

330 'After the hell' Crabb *Operation Pedestal* p. 131

331 'It was a real thrill' quoted Smith p. 235

13: Grand Harbour

1 BERTHING

332 'it was just one mass' IWM audio 22128 Wilkinson R4
    1:24

334 'convoy, convoy' IWM 09/16/1 Waters

334 'they were so pleased' IWM audio 17293 Parry

335 'Now that we are all' IWM 02/56/1 Gould

335 'I've come to the conclusion' IWM 05/59/1 Gunn letter
    of 31.8.42

335 'A big day' IWM 90/38/1 Blundell

336 'materially altered' TNA ADM199/1242

337 'with full freedom' ibid.

342 'Voyage into the Sea' *Königsberger Allgemeine Zeitung* 17.8.42

342 'the convoy is almost' Bundesarchiv Freiburg N316/V.38 log of the 3rd E-Boat Flotilla 15.8.42

342 'representing a real revolution' Giorgerini *Uomini sul fondo* p. 332

344 'The size of' Bundesarchiv Freiburg RM7/235

345 'The war has shown' TNA ADM116/5150

346 'the largest aero-naval' O'Hara p. 185

346 'desperate throw which failed' Ball, Simon *The Bitter Sea: The Brutal World War II Fight for the Mediterranean* Harper Press 2009 p. 150

346 'grossly exaggerated' Van Creveld, Martin *Supplying War: Logistics from Wallenstein to Patton* Cambridge University Press 1977 p. 199

346 'strategic burden' Barnett p. 526

347 'Operation Pedestal was in retrospect' Vego, Milan 'Major Convoy Operation to Malta 10–15 August 1942' *Naval War College Review* 63(1) Winter 2010

347 'this was an outcome' Gundelach, Karl *Die deutsche Luftwaffe im Mittelmeer 1940–1945* 2 vols, Frankfurt: Peter Lang 1985 vol. I p. 161

348 'I have seen' Captain Jeremy Larken to MH

349 'There are no big battalions' Tedder, Arthur [Lord] *Air Power in War* Hodder & Stoughton 1948 p. 41

350 'The P.M.'s relief' Moran, Lord *Winston Churchill: The Struggle for Survival* Constable 1966 p. 181

351 '1,324 vessels lost' Cocchia p. 120

353 'in all respects they conducted' Ellerton Papers

2 HONOURS AND OBSEQUIES

354 'an attempted mutiny' see Pearson p. 108
354 'the yellow streak' IWM 09/24/1 Harvey
356 'faced a court martial' TNA ADM156/225
356 'in that he negligently' TNA ADM156/209 & 210
357 'It was distressing' Harold Drew letter of 27.2.43
359 'I did not realise' Hill p. 96
359 'in another age' *Daily Telegraph* obituary 22.5.01

Bibliography

Articles and online material

http://www.alieuomini.it/catalogo/dettaglio2_catalogo/8/5
(details of Italian aircraft)

http://www.alieuomini.it/pagine/dettaglio/uomini,5/una_
sors_coniungit,303.html (everyday life in the Italian Air
Force)

https://www.barlettaviva.it/notizie/l-eroe-barlettano-del-
canale-di-sicilia/ (biography of Giorgio Manuti, MS
captain)

http://www.bbc.co.uk/history/ww2peopleswar/ (BBC
People's War website: assorted personal recollections

http://www.casinadeicapitani.net/giannettino.asp (obituary
of Giovanni Battista Cafiero, known as Giannettino, MAS
captain)

http://www.combattentiliberazione.it/movm-dal-1935-al-7-
sett-1943/mezzadra-franco (biography of Franco
Mezzadra, MS captain)

http://conlapelleappesaaunchiodo.blogspot.com/2017/11/
ms-16.html (further info about MS attacks/sinking of
Manchester)

https://en.wikipedia.org/wiki/BETASOM (information about
Italian submarines)

Erskine, Ralph 'Eavesdropping on Bodden: ISOS v. The Abwehr in the Straits of Gibraltar' *Intelligence and National Security* 12(3) 1997 pp. 110–29

Hague, Arnold 'The Supply of Malta 1940–42' naval-history. net

http://www.icsm.it/articoli/ri/assita.html (air aces)

http://www.icsm.it/regiamarina/carmezas.htm (info about MAS)

https://www.lavocedelmarinaio.com/2017/06/22-6-1944-affondamento-regia-nave-bolzano/ (the destruction of *Bolzano*)

https://www.lavocedelmarinaio.com/2018/05/14-5-1941-in-ricordo-di-giovanni-cafiero (biography of Giovanni Cafiero)

https://www.marinaiditalia.com/public/uploads/2011_05_24. pdf (Testimonianze – chief electrician on *Bolzano*)

Mattesini, Francesco 'Operazione Pedestal: La Battaglia di Mezzo Agosto 1942' http://www.societaitalianastoria militare.org/COLLANA%20SISM/MATTESINI%20 L'operazione%20Pedestal.%20Mezzo%20Agosto% 201942.pdf

Rees, L. D. W. 'HMS Penn in Operation Pedestal' WW2 People's War, online archive gathered by the BBC

www.regiamarina.net (several first-hand accounts)

Rosselli del Turco, Niccolò via email (info about MAS and their captains at Pedestal)

Smith, Thomas T. 'The Bodden Line: A Case-Study of Wartime Technology' *Intelligence and National Security* 6(2) 1991 pp. 447–57

Vego, Milan 'Major Convoy Operation to Malta 10–15 August 1942' *Naval War College Review* 63(1) Winter 2010

http://www.villacidro.net/zzz/storia/1942can.htm (*canarino* story)

Books

Admiralty, The *The Fleet Air Arm* HMSO 1943

Aichner, Martino *Il Gruppo Buscaglia: Aerosiluranti nella seconda guerra mondiale* Mursia 2013

Alanbrooke, Field-Marshal Lord *War Diaries 1939–1945* ed. Alex Danchev and Dan Todman Weidenfeld & Nicolson 2001

Ball, Simon *The Bitter Sea: The Brutal World War II Fight for the Mediterranean* Harper Press 2009

Barca, Luciano *Buscando per mare con la decima Mas* Riuniti 2001

Barnett, Correlli *Engage the Enemy More Closely: The Royal Navy in the Second World War* Hodder & Stoughton 1991

Battistelli, Pier Paolo *Albert Kesselring* Osprey 2012

Blair, Clay *Hitler's U-boat War* vol. I *The Hunters 1939–1942* Random House 1996

Boog, Horst and others *Germany and the Second World War* vol. VI *The Global War* Clarendon Press 2002

Bradford, Ernle *Siege: Malta 1940–1943* Pen & Sword 2011

Bragadin, Marc'Antonio *La marina italiana nella seconda guerra mondiale 1940–43* Lega Navale 1950

Ciano, Count Galeazzo *Diary 1937–1943* Phoenix 2002

Cocchia, Admiral Aldo *Submarines Attacking: Adventures of Italian Naval Forces* Kimber 1956

Colville, John *The Fringes of Power: Downing Street Diaries 1939–1955* Hodder & Stoughton 1985

Crabb, Brian James *In Harm's Way: The Story of HMS Kenya* Paul Watkins 1998

— *Operation Pedestal* Shaun Tyas 2014

Crosley, R. 'Mike' *They Gave Me a Seafire* Airlife 2001

Cunningham of Hyndhope, Viscount *Sailor's Odyssey* Hutchinson 1951

Durrell, Lawrence *Balthazar* Faber & Faber 1958

Edwards, Kenneth *Seven Sailors* Collins 1945

Ehlers, Robert *The Mediterranean Air War* University Press of Kansas 2015

'Fuehrer Conferences on Naval Affairs, 1939–1945', in *Brassey's Naval Annual 1948* New York: Macmillan 1948

Gibbs, Patrick *Torpedo Leader* Grub Street 2002

Giorgerini, Giorgio *La Guerra Italiana sul mare* Mondadori 2001

— *Uomini sul fondo: storia del sommergibilismo italiano dalle origini a oggi* Mondadori 2002

Greene, Jack and Massignani, Alessandro *The Naval War in the Mediterranean 1940–1943* Chatham 1998

Gretton, Sir Peter *Convoy Escort Commander* Cassell 1964

Gundelach, Karl *Die deutsche Luftwaffe im Mittelmeer 1940–1945* 2 vols, Frankfurt: Peter Lang 1985

Hartcup, Guy *Camouflage: A History of Concealment and Deception in War* David & Charles 1979

Harvey, Oliver *The War Diaries of Oliver Harvey 1941–1945* ed. John Harvey Collins 1978

Hastings, Max *All Hell Let Loose: The World at War 1939–1945* Harper Press 2011

— *Finest Years: Churchill as Warlord 1940–45* Harper Press 2009

— *Nemesis: The Battle for Japan 1944–45* Harper Press 2007

Hay, Doddy *War under the Red Ensign: The Merchant Navy 1939–45* Jane's 1982

Hichens, Antony *Gunboat Command: The Life of 'Hitch', Lieutenant-Commander Robert Hichens* Pen & Sword 2007

Hill, Roger *Destroyer Captain* Periscope 2004

Hirschfeld, Wolfgang *The Story of a U-boat NCO 1940–1946* as told to Geoffrey Brooks, Leo Cooper 1996

Holland, James *Fortress Malta: An Island under Siege 1940–1943* Weidenfeld & Nicolson 2004

Hood, Jean ed. *Submarine: An Anthology of First-Hand Accounts of the War under the Sea, 1939–1945* Conway 2007

Kaplan, Philip and Currie, Jack *Convoy: Merchant Sailors at War 1939–1945* Aurum 1998

Kennedy, Major-General Sir John *The Business of War: The War Narrative* Hutchinson 1957

Kennedy, Paul *The Rise and Fall of British Naval Mastery* Penguin 1976

Kesselring, Albert *The Memoirs of Field-Marshal Kesselring* History Press 2015

Kimmins, Anthony *Half-Time* Heinemann 1947

Le Bailly, Louis *The Man around the Engine* Kenneth Mason 1990

Lloyd, Air Marshal Sir Hugh *Briefed to Attack: Malta's Part in African Victory* Hodder & Stoughton 1949

Mackenzie, Hector *Observations* Pentland 1997

Mallalieu, J. P. W. *Very Ordinary Seaman* Gollancz 1944

Mansfield, Angus *I Wish I Had your Wings: A Spitfire Pilot and Operation Pedestal* History Press 2016

Mars, Alastair *Unbroken: The Story of a Submarine* Frederick Muller 1953

Massimello, Giovanni and Apostolo, Giorgio *Gli assi italiani della seconda guerra mondiale* Osprey 2000

MccGwire, Michael Kane *A Midshipman's Tale: Operation Pedestal, Malta Convoy, August 1942* Leaping Boy Publications 2016

Metellini, Giacomo *Un pilota racconta* ed. Alessandro Metellini, Mursia 2011

Mizzi, John A. *Operation Pedestal: The Story of the Santa Marija Convoy* Midsea Books 2012

Moran, Lord *Winston Churchill: The Struggle for Survival* Constable 1966

Moses, Sam *At All Costs: How a Crippled Ship and Two American Merchant Mariners Turned the Tide of World War II* Random House 2006

Murray, Williamson *Luftwaffe* Allen & Unwin 1985

Nassigh, Riccardo *Operazione Mezzo Agosto* Mursia 1976

New Zealand Shipping Co. *The Carriage of Cargo* NZS 1948

O'Hara, Vincent *Struggle for the Middle Sea: The Great Navies at War in the Mediterranean Theater, 1940–1945* Annapolis, Md: Naval Institute Press 2009

Paravicini, Pier Paolo *Pilota da caccia 1942–1945* Mursia 2011

Pearson, Michael *The Ohio and Malta: The Legendary Tanker that Refused to Die* Pen & Sword 2011

Popham, Hugh *Sea Flight: The Wartime Memoirs of a Fleet Air Arm Pilot* Seaforth 2010

Porch, Douglas *Hitler's Mediterranean Gamble: The North African and the Mediterranean Campaigns in World War II* Weidenfeld & Nicolson 2004

Prysor, Glyn *Citizen Sailors: The Royal Navy in the Second World War* Viking 2011

Rapalino, Patrizio and Schivardi, Giuseppe *Odissea di un sommergibilista: dal mar rosso al mediterraneo 1940–1943* Mursia 2008

Rocca, Gianni *Fucilate gli ammiragli: La tragedia della Marina italiana nella seconda guerra mondiale* Mondadori 1987

The Royal Navy and the Mediterranean Convoys: A Naval Staff Study Routledge 2007

Sadkovich, James J. *The Italian Navy in World War II* Greenwood Press 1994

Santoro, Giuseppe *L'aeronautica italiana nella seconda guerra mondiale* vol. II Edizioni Esse 1957

Smith, Peter C. *Pedestal: The Convoy that Saved Malta* Crécy 1999

Spector, Ronald *At War at Sea* Penguin 2001

Taffrail *Blue Star Line at War 1939–1945* Blue Star 1948

Tedder, Arthur [Lord] *Air Power in War* Hodder & Stoughton 1948

— *With Prejudice* Cassell 1966

Theroux, Paul *The Pillars of Hercules: A Grand Tour of the Mediterranean* Hamish Hamilton 1995

Twiston Davies, David ed. *The Daily Telegraph Book of Naval Obituaries* Grub Street 2004

Van Creveld, Martin *Supplying War: Logistics from Wallenstein to Patton* Cambridge University Press 1977

Vause, Jordan *Wolf: U-Boat Commanders in World War II* Annapolis, Md: Naval Institute Press 1987

Villa, Livio *Fino alla fine: Diario di guerra del Regio Sommergibile 'Scirè'* Marvia 2006

Wellum, Geoffrey *First Light* Viking 2002

Williamson, Gordon *German E-Boats 1939–45* Osprey 2002

Wilson, Alastair *A Biographical Dictionary of the Twentieth Century Royal Navy* vol. I *Admirals of the Fleet and Admirals* Pen & Sword 2013

Winton, John *Cunningham: The Greatest Admiral since Nelson* John Murray 1998

— ed. *The War at Sea 1939–45* vol. I *Freedom's Battle* Book Club Associates 1974

Woodman, Richard *Malta Convoys 1940–1943* John Murray 2000

Zara, Alberto da *Pelle d'ammiraglio* Mondadori 1948

Official histories

Hinsley, F. H. and others *British Intelligence in the Second World War* vol. II *Its Influence on Strategy* HMSO 1981

Howard, Michael *Grand Strategy* vol. IV *August 1942–September 1943* HMSO 1970

Jones, Ben *The Fleet Air Arm in the Second World War* vol. II *1942–1943* Navy Records Society 2018

Playfair, I. S. O. and others *The Mediterranean and the Middle East* vol. III *British Fortunes Reach their Lowest Ebb* HMSO 1960

Richards, Denis and St George Saunders, Hilary *The Royal Air Force 1939–1945* vol. II *The Fight Avails* HMSO 1993

Roskill, Captain S. W. *The War at Sea* vol. II *The Period of Balance* HMSO 1956

Index

About the Author

MAX HASTINGS is the author of twenty-eight books, most about conflict, and between 1986 and 2002 served as editor in chief of the *Daily Telegraph*, then as editor of the *Evening Standard*. He has won many prizes, for both his journalism and his books, the most recent of which are the bestsellers *Vietnam*, *The Secret War*, *Catastrophe*, and *All Hell Let Loose*. Knighted in 2002, Hastings is a fellow of the Royal Society of Literature, an honorary fellow of King's College London, and a Bloomberg Opinion columnist. He has two grown children, Charlotte and Harry, and lives with his wife, Penny, in West Berkshire, where they garden enthusiastically.